Textbook of Ion Channels Volume I

The *Textbook of Ion Channels* is a set of three volumes providing a wide-ranging reference source on ion channels for students, instructors and researchers. Ion channels are membrane proteins that control the electrical properties of neurons and cardiac cells; mediate the detection and response to sensory stimuli like light, sound, odor, and taste; and regulate the response to physical stimuli like temperature and pressure. In non-excitable tissues, ion channels are instrumental for the regulation of basic salt balance that is critical for homeostasis. Ion channels are located at the surface membrane of cells, giving them the unique ability to communicate with the environment, as well as the membrane of intracellular organelles, allowing them to regulate internal homeostasis. Ion channels are fundamentally important for human health and diseases, and are important targets for pharmaceuticals in mental illness, heart disease, anesthesia, pain and other clinical applications. The modern methods used in their study are powerful and diverse, ranging from single ion-channel measurement techniques to models of ion channel diseases in animals, and human clinical trials for ion channel drugs.

Volume I, Part 1 covers fundamental topics such as the basic principles of ion permeation and selectivity, voltage-dependent, ligand-dependent and mechano-dependent ion channel activation mechanisms, the mechanisms for ion channel desensitization and inactivation, and basic ion channel pharmacology and inhibition. Volume I, Part 2 offers a practical guide of cardinal methods for researching ion channels, including heterologous expression and voltage-clamp and patch-clamp electrophysiology; isolation of native currents using patch clamping; modeling ion channel gating, structures and its dynamics; crystallography and cryo- electron microscopy; fluorescence and paramagnetic resonance spectroscopy methods; and genetics approaches in model organisms.

All three volumes give the reader an introduction to fundamental concepts needed to understand the mechanism of ion channels; a guide to the technical aspects of ion channel research; a modern guide to the properties of major ion channel families; and includes coverage of key examples of regulatory, physiological and disease roles for ion channels.

Textbook of Ion Channels Volume I

Basics and Methods

Edited by
Jie Zheng
Matthew C. Trudeau

CRC Press
Taylor & Francis Group
Boca Raton London New York

CRC Press is an imprint of the
Taylor & Francis Group, an **informa** business

First edition published 2023
by CRC Press
6000 Broken Sound Parkway NW, Suite 300, Boca Raton, FL 33487-2742

and by CRC Press
4 Park Square, Milton Park, Abingdon, Oxon, OX14 4RN
CRC Press is an imprint of Taylor & Francis Group, LLC

Library of Congress Cataloging-in-Publication Data
Names: Zheng, Jie, 1966- editor. | Trudeau, Matthew C., editor.
Title: Textbook of ion channels / edited by Jie Zheng, Matthew C. Trudeau.
Description: First edition. | Boca Raton : CRC Press, 2023. | Includes bibliographical references and index. | Contents: volume 1. Basics and methods -- volume 2. Properties, function, and pharmacology of the superfamilies -- volume 3. Regulation, physiology, and diseases.
Identifiers: LCCN 2022043157 (print) | LCCN 2022043158 (ebook) | ISBN 9780367538156 (v. 1 ; hardback) | ISBN 9780367560492 (v. 1 ; paperback) | ISBN 9780367538163 (v. 2 ; hardback) | ISBN 9780367560812 (v. 2 ; paperback) | ISBN 9780367538194 (v. 3 ; hardback) | ISBN 9781032408040 (v. 3 ; paperback) | ISBN 9781003096214 (v. 1 ; ebook) | ISBN 9781003096276 (v. 2 ; ebook) | ISBN 9781003310310 (v. 3 ; ebook)
Subjects: LCSH: Ion channels--Textbooks.
Classification: LCC QH603.I54 T49 2023 (print) | LCC QH603.I54 (ebook) | DDC 571.6/4--dc23/eng/20230126
LC record available at https://lccn.loc.gov/2022043157
LC ebook record available at https://lccn.loc.gov/2022043158

Hardback ISBN: 9780367538156
Paperback ISBN: 9780367560492
Master eBook ISBN: 9781003096214

DOI: 10.1201/9781003096214

Typeset in Palatino
by Deanta Global Publishing Services, Chennai, India

eResources are available at: https://www.routledge.com/Textbook-of-Ion-Channels-Volume-I-Basics-Zheng-Trudeau/p/book/9780367538156

Contents

Preface

We view this *Textbook of Ion Channels* as a modern extension of the field's most authoritative textbook, *Ion Channels of Excitable Membranes* by Bertil Hille. It is an update of our previous *Handbook of Ion Channels* (CRC Press 2015). As Hille observed, "The field of ion channels has grown from tentative hypotheses to vigorous mainstream science" (*Progress in Biophysics and Molecular Biology*, 2022, vol. 169–170, p. 18). In order to offer an updated coverage of this vast and rapidly changing field, our approach was to invite leading world experts to write on selected topics that make up 58 different chapters.

The extensive coverage of this textbook makes it necessary to divide the contents into three volumes:

Volume I, Part 1 deals with the foundational concepts of permeation and gating mechanisms, with a balance of classic theories and latest developments. Volume I, Part 2 focuses on various ion channel techniques. This methods section covers both classic, well-developed techniques and newly developed powerful techniques spanning the basic principle of the method, application to channel research, and practical issues.

Volume II covers over 25 major ion channel types, with a combination of well-studied ion channels and newly identified channels, and includes in-depth discussion of physiological roles, permeation and ion selectivity, gating mechanisms, biophysics and pharmacology. The organization of this section follows the major superfamilies of ion channels.

Volume III offers examples of ion channel regulation, the in-depth role of ion channels in key physiological systems and the role of channels in major diseases, discussing genetics to mechanisms and developments in ion channel pharmaceuticals.

As a community-supported project, many colleagues have contributed to the textbook. In addition to the chapter authors, we are deeply indebted to the Editorial Advisory Board for its guidance. The Editorial Advisory Board for the *Textbook of Ion Channels* is Drs. Richard Aldrich, Henry Colecraft, Jianmin Cui, Lucie Delemotte, Teresa Giraldez, Merritt Maduke, Andrea Meredith, Andrew Plested, Gail A. Robertson and William N. Zagotta.

In addition, we thank special topic advisers Drs. Michael Pusch, Laszlo Csanady, Zack Sellers, Zhe Lu and Vivek Garg.

For their professional guidance, we thank our editors at CRC Press: Ms. Carolina Antunes and Ms. Emma Morley. The book project benefited from Ms. Betsy Byers's fantastic and patient assistance, for which we are deeply grateful.

Editors

Jie Zheng, PhD, is a professor at the University of California Davis School of Medicine, where he has served as a faculty member in the Department of Physiology and Membrane Biology since 2004. Zheng earned a bachelor's degree in physiology and biophysics (1988) and a master's degree in biophysics (1991) at Peking University. He earned a PhD in physiology (1998) at Yale University, where he studied with Dr. Fredrick J. Sigworth on patch-clamp recording, single-channel analysis, and voltage-dependent activation mechanisms. He received his postdoctoral training at the Howard Hughes Medical Institute (HHMI) and the University of Washington during 1999 to 2003, working with Dr. William N. Zagotta on the cyclic nucleotide-gated channels activation mechanism and novel fluorescence techniques for ion channel research. Currently, Zheng's research focuses on temperature-sensitive TRP channels.

Matthew C. Trudeau, PhD, is a professor in the Department of Physiology at the University of Maryland School of Medicine in Baltimore, Maryland. He earned a bachelor's degree in biochemistry and molecular biology in 1992 and a PhD in physiology in 1998 while working with Gail Robertson, PhD, at the University of Wisconsin-Madison. His thesis work was on the properties of voltage-gated potassium channels in the human ether-á-go-go related gene (hERG) family and the role of these channels in heart disease. Trudeau was a postdoctoral fellow with William Zagotta, PhD, at the University of Washington and the Howard Hughes Medical Institute (HHMI) in Seattle from 1998 to 2004, where he focused on the molecular physiology of cyclic nucleotide-gated ion channels, the mechanism of their modulation by calcium–calmodulin and their role in an inherited form of vision loss. Currently, Trudeau's work focuses on hERG potassium channels, their biophysical mechanisms, and their role in cardiac physiology and cardiac arrhythmias.

Contributors

Andriy Anishkin
Department of Biology
University of Maryland
College Park, MD

Rikard Blunck
Department of Physics
and
Department of Pharmacology and Physiology
and
Center for Interdisciplinary Research on Brain and Learning (CIRCA)
University of Montréal
Montréal, Canada

Baron Chanda
Department of Anesthesiology, CIMED
Washington University
St. Louis, MO

Yifan Cheng
Department of Biochemistry and Biophysics
and
Howard Hughes Medical Institute
University of California
San Francisco, CA

Sandipan Chowdhury
Department of Physiology and Molecular Biophysics
University of Iowa
Iowa City, IA

Lucie Delemotte
Science for Life Laboratory
Department of Applied Physics
KTH Royal Institute of Technology
Stockholm, Sweden

Eric G. B. Evans
Department of Chemistry
and
Department of Physiology and Biophysics
University of Washington
Seattle, WA

Chen Fan
Department of Anesthesiology
Weill Cornell Medical College
New York, NY

Moshe Giladi
Department of Physiology and Pharmacology
Sackler Faculty of Medicine
Tel Aviv University
Tel Aviv, Israel

Yoni Haitin
Department of Physiology and Pharmacology
Sackler Faculty of Medicine
Tel Aviv University
Tel Aviv, Israel

Frank T. Horrigan
Department of Molecular Physiology & Biophysics
Baylor College of Medicine
Houston, TX

Toshinori Hoshi
Department of Physiology
University of Pennsylvania
Philadelphia, PA

León D. Islas
Department of Physiology
Faculty of Medicine
National Autonomous University of Mexico
México City, México

Dorothy M. Kim
Department of Anesthesiology
Weill Cornell Medical College
New York, NY

Yun Lyna Luo
Department of Pharmaceutical Sciences
College of Pharmacy
Western University of Health Sciences
Pomona, CA

Matthew J. Marquis
Department of Physiology and Membrane Biology
School of Medicine
University of California, Davis
Davis, CA

Jason G. McCoy
Department of Anesthesiology
Weill Cornell Medical College
New York, NY

Andrea L. Meredith
Department of Physiology
University of Maryland School of Medicine
Baltimore, MD

Jeanne M. Nerbonne
Department of Medicine
and
Department of Developmental Biology
Washington University Medical School
St. Louis, MO

Phuong T. Nguyen
Department of Physiology and Membrane Biology
School of Medicine
University of California, Davis
Davis, CA

Crina M. Nimigean
Department of Anesthesiology
Weill Cornell Medical College
New York, NY

Victor De la Rosa
Department of Physiology
Faculty of Medicine
National Autonomous University of Mexico
México City, México

Jon T. Sack
Department of Physiology and Membrane Biology
School of Medicine
and
Department of Anesthesiology and Pain Medicine
University of California, Davis
Davis, CA

Alexander A. Simon
Department of Anesthesiology
Weill Cornell Medical College
New York, NY

Stefan Stoll
Department of Chemistry
and
Department of Physiology and Biophysics
University of Washington
Seattle, WA

Sergei Sukharev
Department of Biology and IPST
University of Maryland
College Park, MD

Vladimir Yarov-Yarovoy
Department of Physiology and Membrane Biology
and
Department of Anesthesiology and Pain Medicine
School of Medicine
University of California, Davis
Davis, CA

William N. Zagotta
Department of Physiology and Biophysics
University of Washington
Seattle, WA

Jianhua Zhao
Sanford Burnham Prebys Medical Discovery Institute
La Jolla, CA

Section 1

Fundamental Mechanisms

1

Ion Selectivity and Conductance

**Alexander A. Simon, Chen Fan, Dorothy M. Kim,
Jason G. McCoy, and Crina M. Nimigean**

CONTENTS

1.1 Introduction

Ion channels function to orchestrate an exquisite array of physiological processes, including nerve impulses, muscle contraction, regulation of cell volume, and cell signaling in all organisms. The electric current in a signaling event is generated by fast ion fluxes across the cell membrane controlled by the opening and closing of ion channels, including those permeable to potassium (K^+), sodium (Na^+), calcium (Ca^{2+}), and chloride (Cl^-) ions. The direction of the ion fluxes is determined by preset transmembrane ionic gradients established by passive transport through specific K^+ channels plus the Na^+/K^+ ATPase that pumps ions against their concentration gradients through hydrolysis of adenosine triphosphate (Figure 1.1). For certain signaling modes—such as the generation of the action potential—it is important that the ion channels involved are mostly permeable to a specific ion and have a high ion conducting rate.

As early as 1902, Julius Bernstein predicted that a change in membrane permeability resulted in an electrical excitation of the membrane. Bernstein hypothesized that cells at rest were only permeable to K^+ ions and permeability to other ions occurred during excitation. This provided the first suggestion of ion-selective components in the membrane (Bernstein 1902) (Figure 1.1). Breakthrough studies on the squid giant axon in the 1940s and 1950s identified the action potential in the axon to be the result of a composite of currents carried by different ions. Using the voltage-clamp technique, Alan Hodgkin and Andrew Huxley developed a model directly correlating Na^+ and K^+ fluxes with excitation

DOI: 10.1201/9781003096214-2

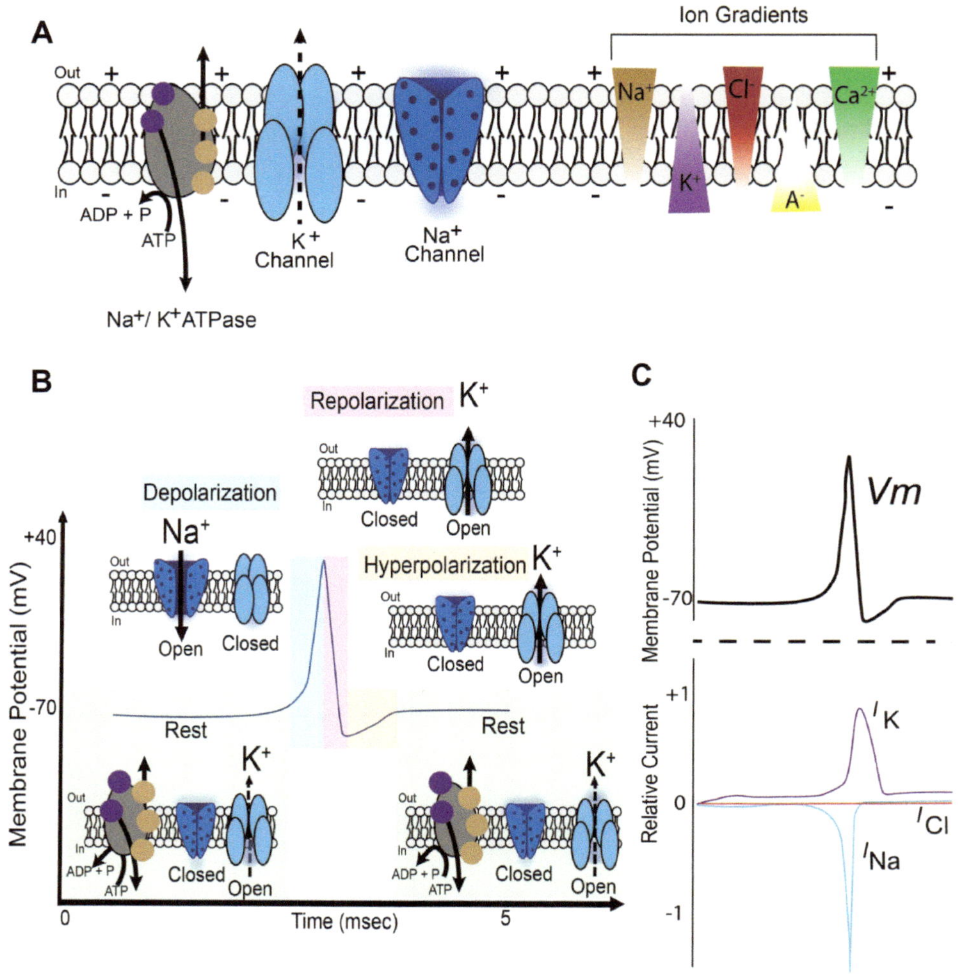

FIGURE 1.1
Ion flux choreography in a typical neuron at rest and during an action potential. (A) Na^+/K^+ ATPase and K^+ channels regulate the resting state of the membrane to establish standing ion gradients—particularly of Na^+ and K^+—across the membrane and maintain a charge separation resulting in a negatively charged interior. The Na^+/K^+ ATPase transports Na^+ (brown spheres) out of the cell and K^+ (purple spheres) into the cell with a 3:2 stoichiometry, respectively. The presence of large organic anions (A^-) in the intracellular space likewise contribute to a negative resting potential. Although gradients of other ions (Ca^{2+}, Cl^-) exist and are pivotal to other physiological processes in electrochemical signaling cascades, their low permeabilities across the membrane at rest minimize their contribution to the steady-state transmembrane potential according to the Goldman–Hodgkin–Katz equation. (B) General schematic of ion channel opening and closing during an action potential. In both (A) and (B), dashed lines through K^+ channels indicate a "leak" or background current. (C) Representation of the membrane potential changes (top) mirrored against corresponding ionic currents (bottom). Upward deflections represent outward currents, while downward deflections are inward currents by convention.

and electric conduction along the axon, earning them the Nobel Prize in 1963. They concluded that the resting membrane was indeed predominantly selective to K^+ ions, resulting in a negative transmembrane potential (according to the electrophysiological convention, the transmembrane potential is measured inside the cell relative to the outside). On the other hand, Na^+ ions were responsible for the inward current causing the cell membrane to become more positive than at rest, referred to as depolarization (Hodgkin and Huxley

1952; Hodgkin and Keynes 1955) (Figure 1.1B). These studies revealed a choreography of ion fluxes across the cell membrane underpins the electrical propagation and the fluxes would need to be associated with highly selective ion channels to maintain the observed specificity.

Shortly thereafter, in the 1970s, ion channels selective for Na^+ and Ca^{2+} were purified from membrane preparations. It was not until the 1980s that the first ion channel genes encoding a nicotinic acetylcholine receptor (nAChR) from *Torpedo californica* and a voltage-gated Na^+ (Na_v) channel from the electric eel *Electrophorus electricus* were cloned (Noda et al. 1984; Noda et al. 1982). The cation channel forming nAChR was also the first to be recorded with patch clamping at the single-channel level from native membranes, and then eventually purified, reconstituted, and recorded in a lipid bilayer (Montal et al. 1984; Neher and Sakmann 1976).

The first protein specifically involved in the conductance of K^+ was discovered through genetic analysis of *Drosophila* mutants exhibiting a neuromuscular defect manifested through uncontrollable shaking. The region of genomic DNA carrying the gene of interest became known as the *Shaker* locus and it encoded a member of the voltage-gated potassium channel (K_v) family. The encoded Shaker protein was also the first K^+ channel to be cloned (Papazian et al. 1987). Molecular cloning of these ion channels allowed for functional characterizations and was the foundation for deciphering structure–function mechanisms. Notably, heterologous expression of a K^+ channel for patch-clamp recordings revealed a clear selection preference for K^+, Rb^+, and NH_4^+ over Na^+ not observed in the other cation channel types (Heginbotham et al. 1994). The genetic and electrophysiological evidence provided by these studies on Na^+, Ca^{2+}, and K^+ channels enabled powerful taxonomic grouping of known channels into functional superfamilies as well as the predictions of novel channels by sequence analysis that became targets for subsequent cloning, providing a rich trove of amino acid sequence information. The rapid accumulation of information in the era of cloning led to a major achievement in the first crystal structure of an ion channel in 1998 (Doyle et al. 1998).

1.2 Structural Basis for Selectivity in K^+ Channels

The key for understanding the mechanism of selectivity in channels lies in the aqueous pore, a narrow canal comprising the permeation path for ions across a hydrophobic membrane (Figure 1.2A,B). The simplest illustration of the pore is that of a molecular sieve that can only pass ions not exceeding a certain radius (Figure 1.2C). Although pore size was initially found to be a major determinant of ion permeability, it was insufficient to explain the permeability sequences determined for some channels. The region within the channel pore that directly interacts with the conducting ions is called the selectivity filter, and both the dimensions and chemical properties of this region influence ion selectivity.

Prior to the elucidation of the atomic structures of ion channels, one early hypothesis asserted that for channels to discriminate between ions they would need to dehydrate upon entry into the pore (Bezanilla and Armstrong 1972). The energy required for this step would need to be compensated for by stabilizing interactions between the ion and the walls of the pore (Eisenman 1962). Binding sites contained in the pore and selectivity filter would only accommodate ions of certain size and valence, and consequently require ions to overcome various energy barriers to permeate the pore. Thus, the pore provides an

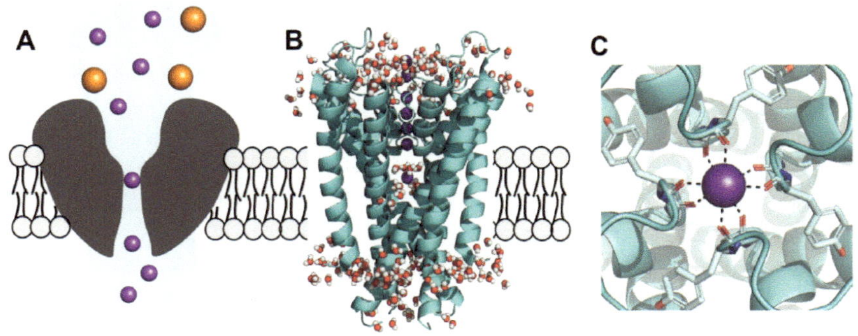

FIGURE 1.2
A model K^+ channel. (A) Cartoon of a simplified ion channel envisioned as the molecular sieve based on early functional electrophysiology studies. Small purple spheres represent permeant ions and large orange spheres represent non-permeant ions. (B) Side view of the K^+ selective ion channel, KcsA, determined by X-ray crystallography constituting an aqueous passageway spanning a lipid membrane. K^+ ions (purple spheres) within the selectivity filter form a single-file line. Water molecules are depicted with an oxygen atom shown as a red sphere and two hydrogen atoms as gray spheres. (C) Close-up view of K^+ ion within the selectivity filter at position S1 viewed from the extracellular side of the membrane.

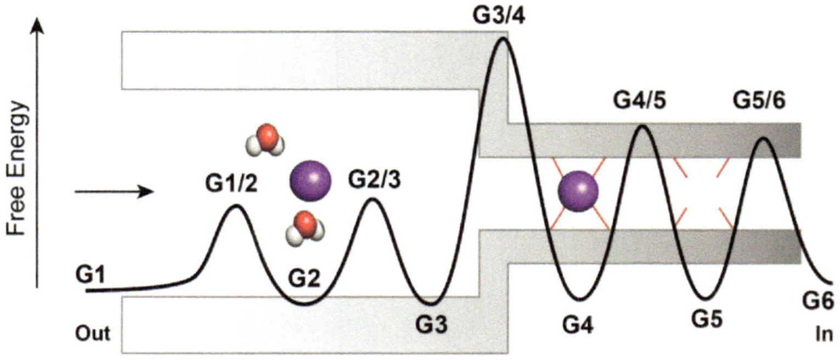

FIGURE 1.3
Energy landscape. Hypothetical energy landscape of an ion (purple sphere) traversing an envisioned form of a simplified ion channel pore with narrow selectivity filter. Ion-binding sites are shown as red sticks. A large energy barrier is indicated for ion dehydration and entry into selectivity filter.

energy landscape that favors permeation of certain ions. Within the energy landscape, ion-binding sites represent energetic minima separated by energetic barriers that can also be encountered upon entering and exiting the pore (Hille 2001) (Figure 1.3). Ions with lower permeability may face a higher energy barrier at external binding sites or alternatively may encounter very deep energy wells within the selectivity filter (Figure 1.3). The differential energy barriers experienced by various ions can also explain why channels often are permeable to several ions but have a preferential selectivity sequence.

The predicted binding sites "stabilizing" dehydrated K^+ ions would be composed of a "bracelet" of oxygen dipoles within the selectivity filter region of the K^+ channel pore (Bezanilla and Armstrong 1972). The selectivity filter would also be of a cylindrical shape with an inner diameter between 3.0 and 3.4 Å (shown in Figure 1.4A, in direct comparison with the predicted structure of a Na^+ channel selectivity filter in Figure 1.4D). Later, the selectivity filter was identified on the "P-loop" containing a highly conserved signature sequence of amino acids—TVGYG—that could contribute oxygen dipoles and found to instigate distinct changes in ion selectivity upon mutation (Heginbotham et al. 1994).

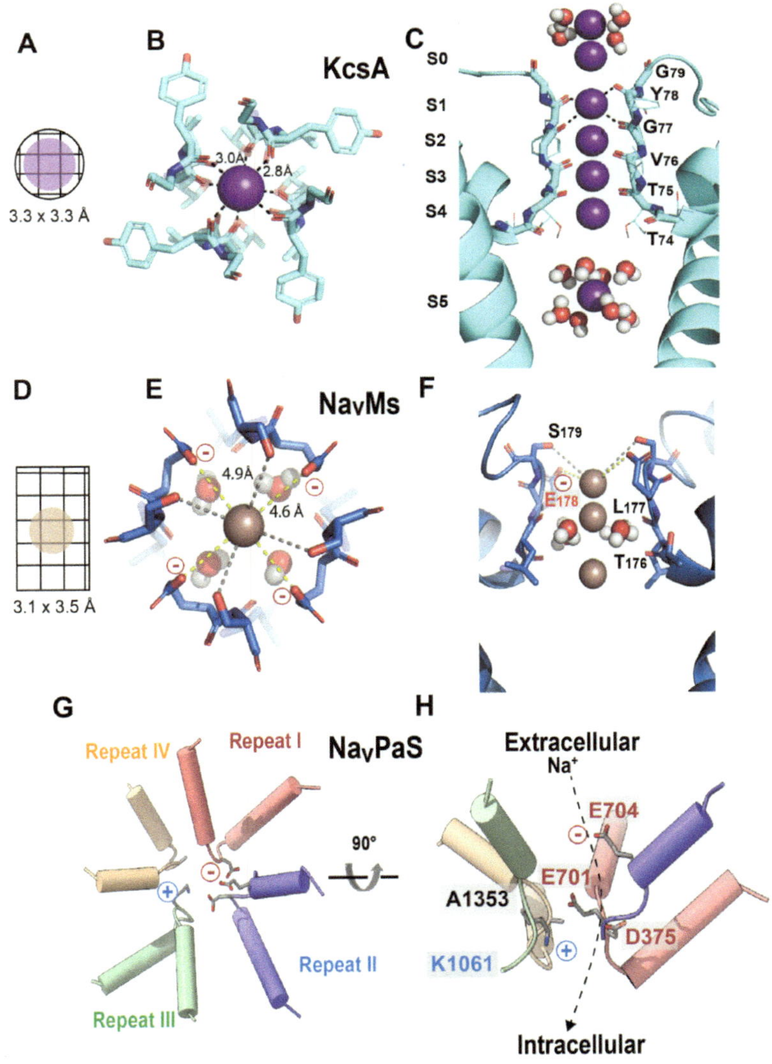

FIGURE 1.4

Pore dimensions and ion selectivity. (A and D) Predicted pore dimensions of a K^+ and Na^+ channel, respectively, based on Dwyer et al. (1980). (B and E) Structural determinations of a prokaryotic K^+ channel pore (KcsA, PDB: 1K4C) and prokaryotic Na^+ channel pore (Na_vMs, PDB: 5HVX) viewed for the extracellular side of the membrane. Both tetrameric structures are symmetric. (C) Close-up view of the KcsA selectivity filter comprised of TTVGYG sequence. Symmetrical subunits form a narrow constriction requiring the dehydration of ions and holds four K^+ ion-binding sites (S1, S2, S3, S4) coordinated by the carbonyl backbones of the signature sequence. (F) Close-up view of the Na_vMs selectivity filter designated by residues TLES from a side view showing a partially hydrated selectivity filter where multiple bound ions are largely coordinated by amino acid side chains, particularly the high field strength anion of glutamate. Yellow dashed line in (E, F) represents bond between ion and the negatively charged glutamate that comprises the characteristic EEEE ring of prokaryotic Na^+ channels. In both (C) and (F), only two opposing subunits are displayed for clarity. (G and H) Top and side views of the asymmetric selectivity filter of the eukaryotic Na_vPaS channel (PDB: 6A90). The four repeats are colored differently, critical residues are shown as sticks and labeled. The negative and positive charges are indicated. The proposed Na^+ conduction pathway is indicated by a dashed line.

These predictions—based entirely on functional data—turned out to be astonishingly accurate and were supported by the crystal structure of the KcsA potassium channel (Figure 1.4B, C).

The KcsA crystal structure revealed a selectivity filter with a radius of ~3 Å, formed by four symmetrical P-loops to harbor K^+ binding sites, which may not accommodate Na^+ ions or other large cations like Cs^+ or Ba^{2+} (Doyle et al. 1998; Zhou et al. 2001). The specific K^+ binding sites were formed by the backbone carbonyls of residues from the signature sequence that point directly into the center of the protein at its narrowest point (Doyle et al. 1998; Morais-Cabral, Zhou, and MacKinnon 2001). Oxygen dipoles coordinated the dehydrated K^+ ions in a cage-like structure by surrounding each K^+ ion with eight oxygen atoms (two from each subunit) (Doyle et al. 1998; Zhou et al. 2001). In addition to the four K^+ ions fully coordinated by the protein (at sites S1–S4), a fifth K^+ binding site (S0) at the extracellular surface of the channel was coordinated by the backbone carbonyl oxygens of a tyrosine and four water molecules (Figure 1.4B, C).

Structural data, in conjunction with the wealth of structure-based data that followed, revolutionized the ion channel field and inspired many new avenues of research. The structure of the KcsA potassium channel and the three-dimensional architecture of the selectivity filter yielded a detailed picture of the molecular mechanisms by which K^+ channels distinguish between different ions. These structures confirmed many of the predictions of early functional data including the physical properties of the pore such as the identities of functional groups that interact with the K^+ ions and the bond distances, the dehydration of ions within the selectivity filter, and the multi-ion nature of ion conduction. Moreover, they also serve as good starting models for molecular dynamic simulations, which can also provide a detailed understanding of selectivity.

1.3 Structural Basis for Selectivity in Na^+ Channels

Electrophysiological experiments determining permeability ratios indicated that Na_v channels can accommodate many different cations and have a large pore. These studies hypothesized that the Na^+ channel contains a rectangular opening of approximately 3 × 5 Å surrounded by oxygen atoms as well as a negatively charged carboxylate ion (Dwyer, Adams, and Hille 1980) (Figure 1.4D, E). The size of this opening further suggested that the Na^+ ions conduct in a partially hydrated state. K^+ channels, on the other hand, were observed to contain a narrower pore than Na^+ channels resulting in higher selectivity due to increased interaction and contact between the ion and the pore walls (Hille 2001) (Figure 1.4A–C). This prediction of relative pore size between Na^+ and K^+ channels was later confirmed by the crystal structures of Na_v channels from bacteria, which have a selectivity filter that is ~4.6 Å wide (Payandeh et al. 2011) (Figure 1.4E). More recent structures captured by cryo-electron microscopy (cryo-EM) confirm a water density around a Na^+ ion suggesting ions enter the pore in a partially hydrated state, collectively supporting hypotheses for mechanisms of ion selectivity based on pore diameter and ion hydration (Sula et al. 2017) (Figure 1.4E, F). Furthermore, since the selectivity sequence of Na^+ channels (Na^+~Li^+ > Tl^+ > K^+ >> Ca^{2+}) closely follows the electrostatic model of Eisenman, it was argued that selectivity was also dependent on the field strength of the binding site (Eisenman 1962). A high-field-strength anion, such as a glutamate side chain, would be required to increase the Na^+ selectivity relative to that of K^+.

Characteristic of several bacterial Na_v channels is a tetrameric architecture and a glutamate residue in the signature position within the selectivity filter, similar to eukaryotic Ca^{2+} channels. A ring of four glutamates (EEEE)—one glutamate from each of the four subunits—is positioned along the pore in the center of the channel near the extracellular opening (Figure 1.4E, F). The cavity between the four glutamate side chains is the most constricted region within the channel allowing just enough room for the conducting Na^+ to remain partially hydrated while directly interacting with one of the glutamates. Mutagenesis experiments illustrated the glutamate residue was responsible for Na^+ permeation relative to Ca^{2+} (Naylor et al. 2016). Interior to this glutamate, the backbone carbonyl oxygens of the two following amino acids (Leu and Thr) point into the pore (Figure 1.4F). Solvent molecules coordinated to these carbonyl oxygens presumably help rehydrate the Na^+ ion as it passes through the filter. The similarity of the EEEE sequence of prokaryotic Na_v channels that shows more homology to eukaryotic voltage-gated Ca^{2+} channels than eukaryotic Na_v channels invites questions regarding the evolutionary origins of Na^+ and Ca^{2+} channels. The similarities in the selectively filter further suggest that there are additional features that tune selectivity for Na^+ versus Ca^{2+}.

Cloning and sequencing of mammalian Na_v channels led to the identification of four key residues instrumental in maintaining Na^+ selectivity. These residues consist of an aspartate, a glutamate, a lysine, and an alanine that form the DEKA motif. Unlike K^+ channels and the bacterial Na^+ channels, eukaryotic Na^+ channels are monomeric proteins containing four similar repeated domains, thus lacking the fourfold symmetry (Figure 1.4G). Each residue of the DEKA motif is in a distinct domain starting with the aspartate in the N-terminal-most repeat but occupies the same location in the primary structure based on sequence alignments of each individual domain. As Hille predicted, the filter sequence contains two residues with carboxyl-containing side chains that evidently are conserved evolutionarily (Hille 2001). Mutations of the DEKA motif demonstrates its profound impacts on selectivity. Mutation of glutamate to alanine leads to an increase in K^+ permeability, while mutation of the aspartate to alanine has little effect. Glutamate therefore appears to play a larger role in selectivity in the Na^+ channel filter. Removal of the lysine side chain leads to an even greater loss in selectivity against K^+ as well as Ca^{2+} (Favre, Moczydlowski, and Schild 1996).

Recently, several Na_v channel structures from eukaryotes have been solved by cryo-EM (Shen et al. 2018; Pan et al. 2018). All these structures share similar asymmetric selectivity filter architectures lined by the side chains of the D,E,K,A signature residues found in repeats I to IV, respectively. In the Na_vPaS structure, the lysine (K_{1061}) side chain adopts a downward conformation, while an additional outer negatively charged glutamate residue (E_{704}) lies above the DEKA residues (Figure 1.4H). E_{704} guards the entrance to the selectivity filter vestibule by attracting cations and repelling anions allowing Na^+ coordination through a "DEE binding site" comprised of residues D_{375}, E_{701} and E_{704} (Shen et al. 2018) (Figure 1.4H). The DEE site is also predicted to comprise the energetic minimum along the ion conduction pathway (Zhang et al. 2018) (Figure 1.4H).

1.4 Mechanistic Models for Selectivity in Cation Channels

The aforedescribed structural data provides insights into selectivity mechanisms. There is now compelling evidence that key characteristics of the selectivity filter are observed in both prokaryotic and eukaryotic K^+ channels. However, it remains to be seen to what extent all cation channels share evolutionary origins of selectivity as both quaternary structures and molecular properties diverged. Future studies pinpointing molecular mechanisms of ancestral prokaryotic channels in tandem with those focusing on the eukaryotic descendants are critical for identifying deviations. These discoveries are paramount for understanding the evolution of selectivity in ion channels. Next, we describe several models that have been proposed as the basis of selectivity. Many of these models are not necessarily mutually exclusive.

1.4.1 The Close-Fit Model for Selectivity

The underlying assumption of the close-fit model is that the selectivity filter distinguishes between different ions based on ionic radius. In this model, the distances between the conducting ion and the selectivity filter backbone carbonyl oxygens are tuned specifically to optimize the coordination for K^+ ions relative to that of smaller or larger ions (Figure 1.5). This partially alleviates the energetic penalty of dehydrating the K^+ ion required for entry into the selectivity filter. Despite this high affinity, the ions are still rapidly processed through the selectivity filter presumably due to electrostatic repulsions between multiple K^+ ions in close proximity in the filter. In the published 2 Å-resolution KcsA crystal

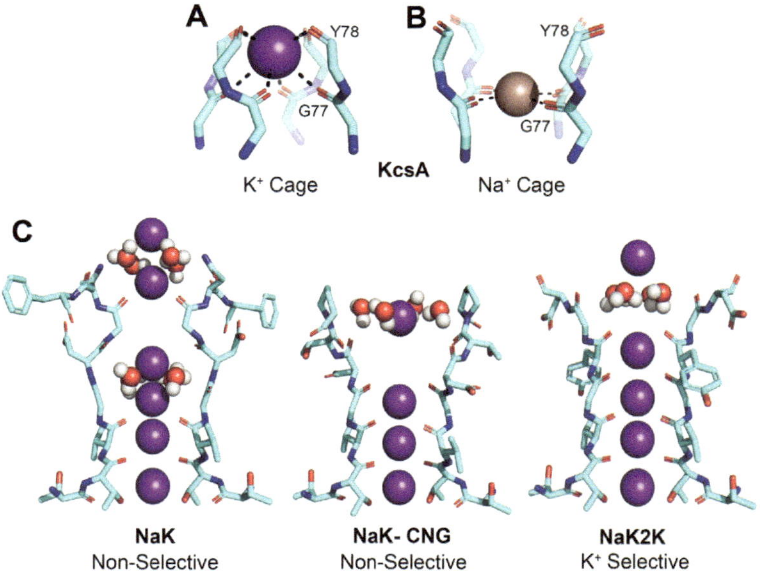

FIGURE 1.5
Selectivity models. (A) Ion coordination in the K^+ channel selectivity filter demonstrates close-fit, field-strength, coordination, and kinetic models. K^+ (purple sphere) binds with octahedral coordination in the center of a cage formed by eight carbonyl oxygens (left). In KcsA, the bond distance from G77 to the K^+ is 3.0 Å (PDB: 1K4C). (B) In contrast, Na^+ (tan sphere) binds with square planar coordination in the plane of the carbonyls (right). In the KcsA E71A variant, the bond distance from G77 to Na^+ is 2.5 Å (PDB: 3OGC). (C) Number of contiguous ion binding sites demonstrates the site number model of selectivity. The nonselective NaK channel (PDB: 3E8H) lacks the S1 and S2 binding sites (left panel). Restoration of the S2 site through mutagenesis of the selectivity filter to TVGDTPP (called NaK-CNG, PDB: 3K03), a sequence very similar to that of nonselective cyclic nucleotide-gated (CNG) channels fails to increase the K^+ selectivity of the channel (center). Restoration of site S1 through mutagenesis of the NaK selectivity filter to TVGYGDF (called NaK2K, PDB: 3OUFA), a sequence very similar to that of KcsA, results in a K^+ selective channel (right).

structure, the average coordination distance between the K^+ ions and the selectivity filter oxygen atoms range from 2.7 Å to 3.08 Å, nearly identical to that observed in the potassium-selective ionophores nonactin and valinomycin (Zhou et al. 2001). Na^+, with an ionic radius approximately 0.4 Å smaller than that of K^+, would require a positional shift of the selectivity filter in order to bind optimally. It should also be noted that crystal structures depict a positional average over time. Therefore, even though the structures show K^+ density in each of the selectivity filter binding sites, it is possible that the K^+ ions occupy only alternate binding sites as they pass through the channel (for example S4 and S2 or S3 and S1) (Morais-Cabral, Zhou, and MacKinnon 2001).

1.4.2 The Field-Strength Model for Selectivity

The field-strength model foregoes the idea of rigid cages (Figure 1.5A), and instead treats the selectivity filter as a flexible, "liquid-like" environment. The dipole moments of the selectivity filter backbone carbonyls generate electrostatic forces that repel each other and attract the cation leading to selectivity for K^+ (Noskov, Berneche, and Roux 2004). In this model, it is the physical properties of the ligand groups (i.e., the selectivity filter backbone carbonyls) that lead to selection for K^+ over other ions. In simplified simulations, a ligand dipole between 2.5 to 4.5 debye (similar to a carbonyl) selects for K^+, whereas decreasing or increasing the ligand dipole leads to optimal selection for larger or smaller cations, respectively (Noskov, Berneche, and Roux 2004).

1.4.3 The Coordination Model for Selectivity

The coordination model differs from the close-fit and field-strength models in that the number of groups that interact with the ligand also contributes to selectivity. In other words, a protein framework with eight liganding moieties, whether they belong to waters or carbonyls, as in the KcsA K^+ channel, will inherently select for K^+ over Na^+ (Bostick and Brooks 2007) (Figure 1.5A, B). In another version of the coordination model, the electric properties of the local environment of the binding sites create strong selectivity by over coordinating K^+ with eight carbonyl ligands instead of the usual six ligands coordinating K^+ in water (Varma, Sabo, and Rempe 2008).

1.4.4 Kinetic Model for Selectivity

In the aforedescribed selectivity mechanisms, the origins of selectivity are ultimately derived from a difference in the free energy of binding for different ions within the selectivity filter. An alternative kinetic-based model suggests that an additional layer of K^+ selectivity over other ions may be the result of a high-energy barrier preventing other ions from progressing through the selectivity filter. Molecular dynamics simulations of KcsA have predicted that the S4 site is slightly selective for both Na^+ and Li^+ over K^+; however, the energy minima for the Na^+ and Li^+ lies in the plane formed by the carbonyl oxygens dividing the S3 and S4 sites (Figure 1.5B) and supposed maxima at the preceding S1 or S2 (Thompson et al. 2009). Crystal structures of KcsA in the presence of Na^+ have also shown spherical densities between the carbonyl oxygens instead of centered in the carbonyl "cage" like K^+ (Cheng et al. 2011) (Figure 1.5A, B). The presence of K^+ may then occlude the preferred binding sites of smaller ions such as Na^+, generating a large Coulombic repulsion that prevents Na^+ from entering the selectivity filter (Thompson et al. 2009; Kim and Allen 2011).

1.4.5 The Site Number Model of Selectivity

Determination of the NaK channel crystal structure from *Bacillus cereus* led to the hypothesis of the site number model (Shi et al. 2006). Unlike KcsA, the NaK channel does not prefer K^+ to Na^+. This difference in selectivity appears to be largely due to a change in the sequence of the selectivity filter. The selectivity filter sequence in NaK is TVGDGNF, in contrast to the KcsA sequence of TVGYGDL. The result of this change is that sites S1 and S2 are disrupted, creating a large cavity in place of the two cage-like binding sites (Figure 1.5C). Modifying the NaK selectivity filter sequence to TVGDTPP—similar to what is observed in nonselective cyclic nucleotide-gated channels—reestablishes the S2 site but the channel still fails to demonstrate K^+ selectivity (Figure 1.5C). Modifying the NaK selectivity filter to TVGYGDF, similar to that of KcsA, reestablishes both the S1 and S2 sites, and results in heightened K^+ selectivity (Derebe et al. 2011) (Figure 1.5C). This suggests that the presence of four contiguous binding sites is a critical component for enacting K^+ selectivity.

1.4.6 Other K^+ Channel Selectivity Determinants

While it is clear that the selectivity filter is crucial to mediating channel selectivity, many aspects of selectivity cannot be explained entirely by the filter alone. Despite sharing the selectivity filter sequence TTVGYG, voltage-gated channel K_v2.1 conducts strong Na^+ currents in the absence of K^+, while K_v1.5, KcsA and Shaker do not (Korn and Ikeda 1995; Heginbotham and MacKinnon 1993; LeMasurier, Heginbotham, and Miller 2001). C-type inactivation, a process by which the selectivity filter changes conformation to halt K^+ conductance in response to prolonged opening of the channel, has also been shown to influence selectivity. Shaker channels, like many other members of the K_v family, can conduct Na^+ during C-type inactivation (Starkus et al. 1997). In contrast, Na^+ conduction can be increased in KcsA via a mutation behind the filter that rendered the channel non-inactivating (Cheng et al. 2011). Similarly, inward-rectifying K^+ channel K_{ir}3.2 was made permeable to Na^+ through mutation of a residue in the transmembrane helix behind the selectivity filter. The addition of a negatively charged amino acid into the K_{ir}3.2 channel cavity below the filter was able to restore K^+ selectivity (Bichet et al. 2006). These results demonstrate that residues beyond those forming the selectivity filter can influence ion selectivity and highlight the role that interactions between neighboring or distant amino acids have on the overall protein function.

1.5 Conductance

Ion conductance is a measure of the ease with which ions flow across the membrane. Ion conductance (g) typically given in units of picosiemens (pS) can be calculated from measurements of electrical current (I) generated by the movement of ions through the membrane normalized to the voltage (V) applied to the membrane simply by using a derivative of Ohm's law, $g = I/V$. In electrical terms, conductance is the inverse of resistance ($g = 1/R$). The ionic flux rate through a single ion channel—termed unitary conductance—can be measured by directly performing single-channel current measurements with patch-clamp or lipid bilayer recordings, or indirectly calculated from macroscopic currents using noise-analysis methods for determining the number of channels in a membrane (Neher and

Sakmann 1976; Silberberg and Magleby 1993). Comparisons of ion conductance among different channels are a useful first metric for comparing the ionic flux within their pores and offer a handle for understanding ion permeation paths.

Since conductance is given by the ionic flux, it generally increases with increasing permeant ion concentration. However, a linear relationship between conductance and ion concentration is not obeyed at very high ion concentrations due to intrinsic structural properties of the channel or presence of blockers. Instead, conductance curves obey simple saturation kinetics that follow a Michaelis–Menten relationship (Hille 2001) (Figure 1.6). The typical conductance–concentration saturation curve is comprised of two components: an initial quasi-linear rise at low ion concentrations and a plateau at higher ion concentrations (Latorre and Miller 1983) (Figure 1.6A). The initial linear slope of the curve measures the rate of ion entry into the pore. The plateau portion of the curve represents the maximum conductance—or g_{max}—and is a measure of the exit rate of the permeating ion. Saturation occurs when the steps of ion binding and unbinding are rate limiting. At high ion concentrations, the rate of ion entry increases and it will approach the maximum rates of unbinding (Hille 2001). Therefore, as ion concentrations increase further, the rates of unbinding determine the overall rate of conduction and a plateau at maximum conductance is reached. Intrinsic properties of the ion channel alter this maximum conductance by changing the rate-limiting step of ion exit from the pore.

The absolute upper limit of conductance is set by the diffusion of ions into the mouth of the pore and can be quantified by the phenomena of convergence conductance (Latorre and Miller 1983). The convergence conductance depends on the proportion of the size of the pore to the radius of the permeant ion. The intimate contact between the ion and the

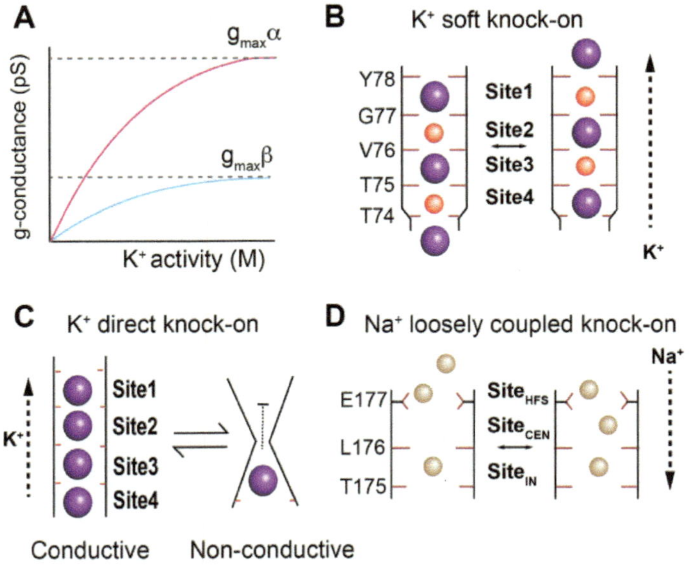

FIGURE 1.6
Ion channel conduction mechanisms. (A) Theoretical graph showing conductance as a function of ionic activity for hypothetical channel α with high maximum conductance (g_{max}) such as KcsA (magenta curve) and for hypothetical channel β with low g_{max} such as K_{ir} (blue curve). The g_{max} for each channel is denoted by a dotted black line. (B) The soft knock-on model for K^+ channel. (C) Transition between conductive selectivity filter (four fully occupied ion-binding sites) and nonconductive (or collapsed) selectivity filter (with one ion-binding site). (D) The loosely coupled knock-on model for Na^+ channel, based on the molecular dynamics simulation analysis of prokaryotic Na_vAb channels. (The illustrations are adapted based on DeMarco, Bekker, and Vorobyov 2019.)

pore walls essential for high selectivity would be antithetical to a large convergence conductance ratio representative of a wide pore radius to small ionic radius (Bezanilla and Armstrong 1972). Paradoxically, K^+ channels can have very high conduction rates while also being highly selective. To circumvent this issue, the specific energetic landscapes embedded by various molecular properties along the permeation pathway have been tailored. Channels will primarily maintain a narrow constriction at the selectivity filter to ensure high ion selectivity—in some instances forcing ion–ion interactions influencing conductance characteristics—but also widen the pore at the base in order to increase the entry rate and promote higher conductance.

A series of mechanisms hypothesized to accommodate the balance of high selectivity and large conductance is centered on the multi-ion theory, postulating the pore's capacity to occupy multiple ions at the same time in a single file. In the multi-ion pore, repulsion between ions increases the exit rate of ions from the pore, enhancing g_{max} (Hille 2001). A model for the barrier-less conduction in K^+ channels was proposed based on the observed configurations of the selectivity filter obtained in different ionic concentrations with X-ray crystallography (Morais-Cabral, Zhou, and MacKinnon 2001; Zhou et al. 2001). In high permeant ion concentrations, the selectivity filter was found in a conductive state, where two ions can bind at the same time in either a water–K–water–K (S2, S4) or K–water–K–water (S1, S3) configuration that follows a soft, or water-mediated, knock-on model (Figure 1.6B). Permeating ions provide counter charges for the 20 electronegative oxygen atoms to stabilize the filter while minimizing the destabilization from same charge repulsion. Under the knock-on model, entry of an additional ion would destabilize this configuration providing the necessary energy to transport the ion creating a mechanism for coupling high conductivity with high selectivity. Molecular dynamics simulations more recently challenged this model with evidence of a water-free selectivity filter that may be fully occupied in a K–K–K–K configuration in support of a direct knock-on model (Kopfer et al. 2014) (Figure 1.6C). At low permeant ion concentrations, the KcsA selectivity filter assumes a collapsed and nonconductive conformation with presumably only one ion bound at one time (Morais-Cabral, Zhou, and MacKinnon 2001). In the collapsed KcsA, a conformational change leads to a lower ion-binding affinity and increases the exit rate of bound ions. Because the nonconductive filter contains only one ion, the second ion-binding event is presumed to rebound the filter from a collapsed to a conductive conformation (Figure 1.6C). This study illustrates yet another possibility of how high selectivity could coexist with high conduction. The presence of multiple high affinity binding sites in the pore allows for selection of certain ions over others, while high conduction only occurs in the occupied selectivity filter due to both electrostatic repulsion. Interestingly, the relationship between ion-binding affinity and selectivity filter conformation as a factor in channel conductance is further supported by studies with MthK, a K^+ channel from the archaea *Methanobacterium thermoautotrophicum*. MthK—which will be discussed as a member of a K^+ channel class that exhibits the largest known conductances—is unlike KcsA because it does not show a collapsed selectivity filter in low concentrations of permeant ions (Ye, Li, and Jiang 2010). Moreover, the MthK selectivity filter was shown to possess higher K^+ binding affinity and consequently a higher probability of binding multiple ions and adopt a conductive state versus a nonconductive one (Boiteux et al. 2020).

The selectivity filter in the Na_V channel likewise follows a multi-ion pore occupancy, but a significantly wider pore than the K^+ channel leads to different permeation mechanisms (DeMarco, Bekker, and Vorobyov 2019). Molecular dynamics simulations of ion permeation in the bacterial voltage-gated Na^+ channel Na_VAb, which harbors a conserved selectivity filter motif to Na_VMs (shown in Figure 1.4E, F), propose a "loosely coupled knock-on" mechanism (Figure 1.6D). In this model, three ion-binding sites impose individual energy barriers and orchestrate to discriminate between ions. Na^+ can move freely between a high-field-strength binding site ($Site_{HFS}$) located toward the extracellular side—designated by a negatively charged glutamate side chain as well as the nearby serine—and a binding site located more toward the internal cavity ($Site_{IN}$)—where it is fully hydrated and coordinated by the backbone carbonyl oxygens pointing into the pore. However, a central binding site ($Site_{CEN}$) located between the external and internal sites is proposed to be the largest energy barrier impeding the permeation of two ions already bound in the selectivity filter. Simulations suggest trapped ions at $Site_{HFS}$ and $Site_{IN}$ can permeate upon an additional ion entering from the extracellular side (Boiteux, Vorobyov, and Allen 2014).

In contrast, attractive forces at ion-binding sites available to multiple ion occupants have been shown to decrease the rate of exit—thus decreasing conductance—as exemplified by the anomalous mole fraction effect. The anomalous mole fraction effect describes changes in permeability ratios of Na^+, K^+ and Ca^{2+} through a common pore when more than one permeant ion type is present (Hess and Tsien 1984; Almers and McCleskey 1984). In certain channels, the conductance passes through a minimum or maximum when plotted against the mole fraction of two permeant ions, and it was understood to occur by allowing the two ions to interact inside the pore and with the pore walls (Nonner, Chen, and Eisenberg 1998) (Figure 1.7). When Hodgkin and Keynes first observed anomalous unidirectional flux ratios in squid axons and offered an explanation by flux coupling, it marked a critical deviation from the expectation that ions pass through independently of one another thereby opening ideas on ion–ion interaction to take hold (Hodgkin and Keynes 1955).

While ion selectivity varies little within a channel subfamily, conductance levels can vary dramatically. Some K^+ channels exhibit a low conductance approximating 10 pS, including inward-rectifying K^+ channels (K_{ir}), K_v, and small-conductance K^+ (SK) channels (Latorre and Miller 1983). In similar conditions, "maxi-K" channels such as the Ca^{2+}-dependent large-conductance K^+ (BK) channels and their archaeal homolog MthK exhibit the largest conductance levels of 100-400 pS that is nearing the diffusion limit ranging from 250 to 300 pS (Zadek and Nimigean 2006; Eisenman, Latorre, and Miller 1986; Latorre and Miller 1983). Since the selectivity filter of different K^+ channels is highly conserved, other factors must account for these varying levels of conductance that can be roughly 10–20 times

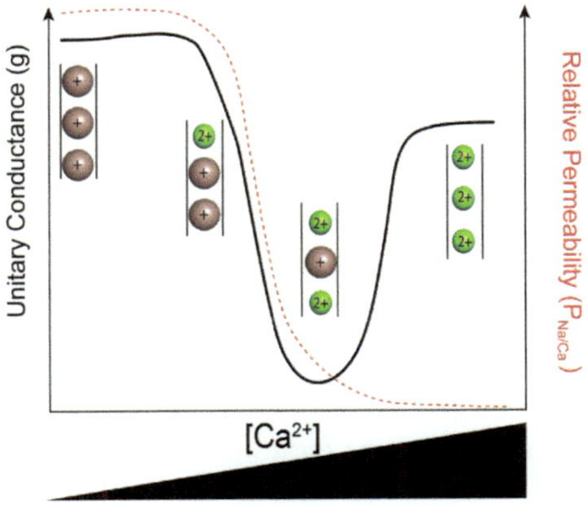

FIGURE 1.7
Anomalous mole fraction. Anomalous mole fraction effect of a cation channel permeable to Na^+ and Ca^{2+} illustrating changes to a channel's unitary conductance (solid black line) and relative permeability (dashed red line) as two competing ions enter the pore. Na^+ concentration remains constant against increasing Ca^{2+} concentrations. Na^+ ions are presented as tan spheres and Ca^{2+} ions as green spheres.

higher in BK/MthK than other K^+ channels (Figure 1.8). One determinant of conductance in K^+ channels is the size of the inner cavity entrance where K^+ ions need to pass through before entering the selectivity filter from the intracellular side. Structures of MthK and BK channels revealed that these channels contain a large aqueous cavity at the inner mouth of the pore measuring 16–20 Å directly under the selectivity filter (Fan et al. 2020; Hite, Tao, and MacKinnon 2017). This cavity is larger than the equivalent in other K^+ channels, such as KcsA and K_v channels, which contain a more modestly sized vestibule (8–12 Å) and correspondingly smaller conductances (10–60 pS) (Doyle et al. 1998; Long, Campbell, and Mackinnon 2005; LeMasurier, Heginbotham, and Miller 2001) (Figure 1.8A–D). The importance of the inner cavity entrance size was supported by studies showing that an increase of side chain volume at the cavity entrance results in a reduction of single-channel conductance in the outward direction, while having no effect on the inward current (Brelidze

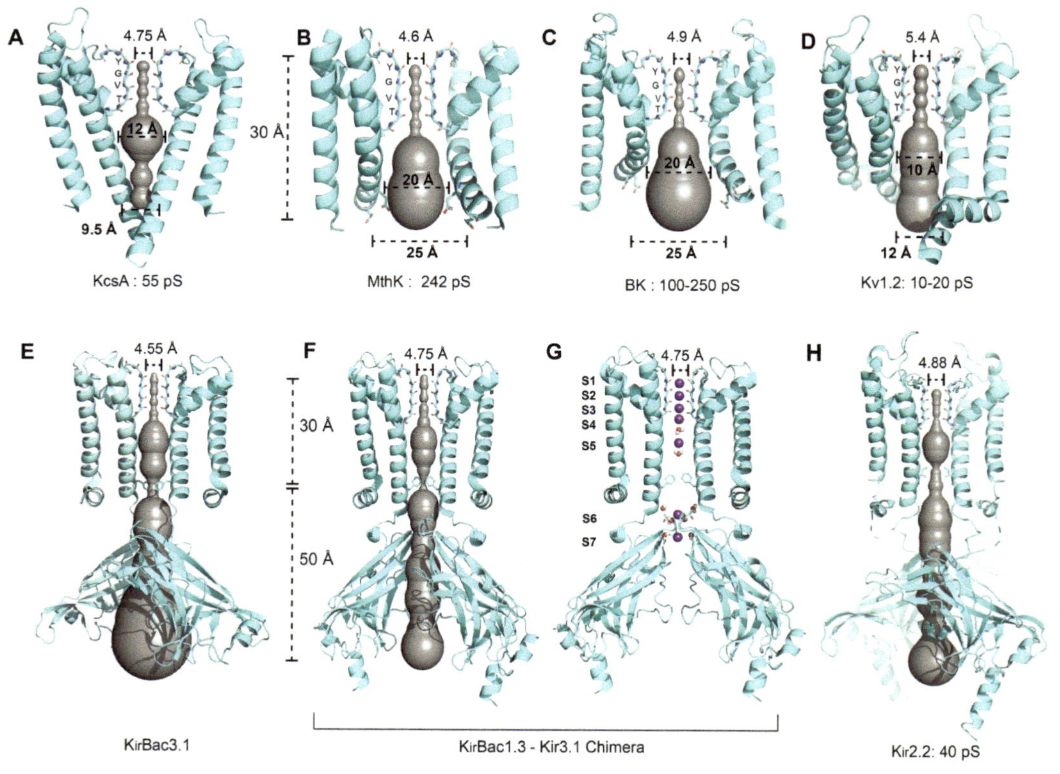

FIGURE 1.8
Pore cavity dimensions and conductance. (A–D) Representation of the pore cavity in K^+ channels: (A) KcsA (PDB: 1K4C), (B) MthK (PDB: 4HYO), (C) the BK channel encoded by the gene *Slo1* (PDB: 6V38) and (D) K_v1.2 (PDB: 2A79). Dashed bars on ion channel structures denote the average width between opposing subunits in the selectivity filter (between carbonyl oxygens), the width of the intermediate cavity, and the width of the cytosolic entrance. Please note that cavity dimensions between opposing subunits extend into the z-axis. (E, F and H) Representation of the pore cavity for (E) K_{ir}Bac3.1 (PDB: 4LP8), (F) a chimera composed of the K_{ir}Bac1.3 transmembrane portion joined to K_{ir}3.1 residues that interface the membrane plus cytosolic domains (PDB: 2QKS), and (H) K_{ir}2.2 (PDB: 3JYC). Note the upper portion of the pore comprising the selectivity filter remains similarly narrow to other K^+ channels of all conductance levels while inner cavities vary in dimensions. (G) Molecular representation of K_{ir}Bac1.3–K_{ir}3.1 showing two opposing subunits illustrating that the formation of a long pore includes the possible formation of two additional K^+ ion-binding sites, not present in KcsA (shown in Figure 1.2). Similar to the chimeric channel, K_{ir}Bac3.1 and K_{ir}2.2 also show additional K^+ ions in the C-terminal portion. Front and back subunits have been removed for clarity in all panels.

and Magleby 2005). The reduction in outward current could be reversed by an increase in intracellular K^+ concentration (Geng, Niu, and Magleby 2011). This suggests that the ions do not encounter a barrier upon entrance into the cavity of these channels, which contributes to the increase in conduction rate. Molecular dynamics simulations provide further support for the importance of the cavity width to conductance as K^+ channel conductance increased with a widening of the entrance to the inner vestibule (Chung, Allen, and Kuyucak 2002). In addition, eight negative charges ring the entrance of BK/MthK to attract K^+ ions to the cavity, which was proposed to elevate the local K^+ concentration, and enhances channel conductance as predicted by the conductance–concentration curve (Nimigean, Chappie, and Miller 2003) (Figure 1.8B).

In addition to cavity size, pore length has been hypothesized to contribute to the single-channel conductance level (Latorre and Miller 1983). This hypothesis is consistent with structural and functional studies of the bacterial inward-rectifying K^+ channels (K_{ir}Bac) and the eukaryotic K_{ir}2.2. K_{ir} channels exhibit a relatively low conductance level of ~10–35 pS at –100 mV for KirBac1.1 and to 40 pS at –80 mV for K_{ir}2.2 (Cheng, Enkvetchakul, and Nichols 2009; Tao et al. 2009). K_{ir} channels are structurally characterized in part by a pore that is ~50–85 Å in length that extends through its large C-terminal domain (Bavro et al. 2012; Tao et al. 2009; Nishida et al. 2007) (Figure 1.8E–H). By comparison, many prokaryotic (KcsA, Mthk) and eukaryotic (Shaker/K_v, BK) channels have a typical pore length of 30 Å (Doyle et al. 1998; Long, Campbell, and Mackinnon 2005; Posson, McCoy, and Nimigean 2013) (Figure 1.8A–D). Nonetheless, channel families such as the Shaker/K_v channels and SK channels also display a low single-channel conductance level (10–20 pS) without a similarly long pore to K_{ir} suggesting the elongated pore harbors additional electrochemical features that strengthen barriers in the energetic landscape. Whether the molecular properties dampening conductance rates are conserved across K^+ channel families remains to be determined. Modifications at the molecular level in pore chemistry, electrostatics, and pore dimensions peripheral to the evolutionarily conserved selectivity filter enables an array of conductance levels. Ultimately, the interplay among a network of ion channels along the same membrane with different ion selectivity and conductance rates cooperate to command the membrane potential for electrical signaling.

Suggested Readings

This chapter includes additional bibliographical references hosted only online as indicated by citations in blue color font in the text. Please visit https://www.routledge.com/9780367538163 to access the additional references for this chapter, found under "Support Material" at the bottom of the page.

Almers, W., and E. W. McCleskey. 1984. "Non-selective conductance in calcium channels of frog muscle: calcium selectivity in a single-file pore." *J Physiol* 353:585–608. doi:10.1113/jphysiol.1984.sp015352.

Bezanilla, F., and C. M. Armstrong. 1972. "Negative conductance caused by entry of sodium and cesium ions into the potassium channels of squid axons." *J Gen Physiol* 60 (5):588–608.

Boiteux, C., D. J. Posson, T. W. Allen, and C. M. Nimigean. 2020. "Selectivity filter ion binding affinity determines inactivation in a potassium channel." *Proc Natl Acad Sci U S A* 117 (47):29968–78. doi:10.1073/pnas.2009624117.

Cheng, W. W., J. G. McCoy, A. N. Thompson, C. G. Nichols, and C. M. Nimigean. 2011. "Mechanism for selectivity-inactivation coupling in KcsA potassium channels." *Proc Natl Acad Sci U S A* 108 (13):5272–7. doi:10.1073/pnas.1014186108.

Derebe, M. G., D. B. Sauer, W. Zeng, A. Alam, N. Shi, and Y. Jiang. 2011. "Tuning the ion selectivity of tetrameric cation channels by changing the number of ion binding sites." *Proc Natl Acad Sci U S A* 108 (2):598–602. doi:10.1073/pnas.1013636108.

Doyle, D. A., J. Morais Cabral, R. A. Pfuetzner, A. Kuo, J. M. Gulbis, S. L. Cohen, B. T. Chait, and R. MacKinnon. 1998. "The structure of the potassium channel: molecular basis of K+ conduction and selectivity." *Science* 280 (5360):69–77. doi:10.1126/science.280.5360.69.

Dwyer, T. M., D. J. Adams, and B. Hille. 1980. "The permeability of the endplate channel to organic cations in frog muscle." *J Gen Physiol* 75 (5):469–92. doi:10.1085/jgp.75.5.469.

Eisenman, G. 1962. "Cation selective glass electrodes and their mode of operation." *Biophys J* 2 (2 Pt 2):259–323. doi:10.1016/s0006-3495(62)86959-8.

Eisenman, G., R. Latorre, and C. Miller. 1986. "Multi-ion conduction and selectivity in the high-conductance Ca++-activated K+ channel from skeletal muscle." *Biophys J* 50 (6):1025–34. doi:10.1016/S0006-3495(86)83546-9.

Fan, C., N. Sukomon, E. Flood, J. Rheinberger, T. W. Allen, and C. M. Nimigean. 2020. "Ball-and-chain inactivation in a calcium-gated potassium channel." *Nature* 580 (7802):288–93. doi:10.1038/s41586-020-2116-0.

Geng, Y., X. Niu, and K. L. Magleby. 2011. "Low resistance, large dimension entrance to the inner cavity of BK channels determined by changing side-chain volume." *J Gen Physiol* 137 (6):533–48. doi:10.1085/jgp.201110616.

Heginbotham, L., Z. Lu, T. Abramson, and R. MacKinnon. 1994. "Mutations in the K+ channel signature sequence." *Biophys J* 66 (4):1061–7. doi:10.1016/S0006-3495(94)80887-2.

Hille, B. 2001. *Ion channels of excitable membranes*. 3rd ed. Sunderland, MA: Sinauer Associates, Inc.

Hite, R. K., X. Tao, and R. MacKinnon. 2017. "Structural basis for gating the high-conductance Ca(2+)-activated K(+) channel." *Nature* 541 (7635):52–7. doi:10.1038/nature20775.

Hodgkin, A. L., and A. F. Huxley. 1952. "Currents carried by sodium and potassium ions through the membrane of the giant axon of Loligo." *J Physiol* 116 (4):449–72. doi:10.1113/jphysiol.1952.sp004717.

Hodgkin, A. L., and R. D. Keynes. 1955. "The potassium permeability of a giant nerve fibre." *J Physiol* 128 (1):61–88.

Kopfer, D. A., C. Song, T. Gruene, G. M. Sheldrick, U. Zachariae, and B. L. de Groot. 2014. "Ion permeation in K(+) channels occurs by direct Coulomb knock-on." *Science* 346 (6207):352–5. doi:10.1126/science.1254840.

Latorre, R., and C. Miller. 1983. "Conduction and selectivity in potassium channels." *J Membr Biol* 71 (1–2):11–30.

Long, S. B., E. B. Campbell, and R. Mackinnon. 2005. "Voltage sensor of Kv1.2: structural basis of electromechanical coupling." *Science* 309 (5736):903–8. doi:10.1126/science.1116270.

Neher, E., and B. Sakmann. 1976. "Single-channel currents recorded from membrane of denervated frog muscle fibres." *Nature* 260 (5554):799–802. doi:10.1038/260799a0.

Noda, M., S. Shimizu, T. Tanabe, T. Takai, T. Kayano, T. Ikeda, H. Takahashi, H. Nakayama, Y. Kanaoka, N. Minamino, et al. 1984. "Primary structure of Electrophorus electricus sodium channel deduced from cDNA sequence." *Nature* 312 (5990):121–7. doi:10.1038/312121a0.

Noskov, S. Y., S. Berneche, and B. Roux. 2004. "Control of ion selectivity in potassium channels by electrostatic and dynamic properties of carbonyl ligands." *Nature* 431 (7010):830–4. doi:10.1038/nature02943.

Pan, X., Z. Li, Q. Zhou, H. Shen, K. Wu, X. Huang, J. Chen, J. Zhang, X. Zhu, J. Lei, W. Xiong, H. Gong, B. Xiao, and N. Yan. 2018. "Structure of the human voltage-gated sodium channel Nav1.4 in complex with beta1." *Science* 362 (6412). doi:10.1126/science.aau2486.

Papazian, D. M., T. L. Schwarz, B. L. Tempel, Y. N. Jan, and L. Y. Jan. 1987. "Cloning of genomic and complementary DNA from Shaker, a putative potassium channel gene from Drosophila." *Science* 237 (4816):749–53. doi:10.1126/science.2441470.

Payandeh, J., T. Scheuer, N. Zheng, and W. A. Catterall. 2011. "The crystal structure of a voltage-gated sodium channel." *Nature* 475 (7356):353–8. doi:10.1038/nature10238.
Shi, N., S. Ye, A. Alam, L. Chen, and Y. Jiang. 2006. "Atomic structure of a Na+- and K+-conducting channel." *Nature* 440 (7083):570–4. doi:10.1038/nature04508.
Starkus, J. G., L. Kuschel, M. D. Rayner, and S. H. Heinemann. 1997. "Ion conduction through C-type inactivated Shaker channels." *J Gen Physiol* 110 (5):539–50. doi:10.1085/jgp.110.5.539.
Thompson, A. N., I. Kim, T. D. Panosian, T. M. Iverson, T. W. Allen, and C. M. Nimigean. 2009. "Mechanism of potassium-channel selectivity revealed by Na(+) and Li(+) binding sites within the KcsA pore." *Nat Struct Mol Biol* 16 (12):1317–24. doi:10.1038/nsmb.1703.
Varma, S., D. Sabo, and S. B. Rempe. 2008. "K+/Na+ selectivity in K channels and valinomycin: over-coordination versus cavity-size constraints." *J Mol Biol* 376 (1):13–22. doi:10.1016/j.jmb.2007.11.059.
Zhou, Y., J. H. Morais-Cabral, A. Kaufman, and R. MacKinnon. 2001. "Chemistry of ion coordination and hydration revealed by a K+ channel-Fab complex at 2.0 A resolution." *Nature* 414 (6859):43–8. doi:10.1038/35102009.

2

Voltage-Dependent Gating of Ion Channels

Baron Chanda and Sandipan Chowdhury

CONTENTS

2.1 Introduction

Electrical signaling is well suited for cell–cell communication and fast processing of complex information in an ever-changing environment. Unlike purely chemical second-messenger pathways, electrical signaling is confined to the plane of a membrane and can travel long distances rapidly. These signaling pathways have been found in all three kingdoms of life and the molecular parallels between distantly related species is quite remarkable. Even in the lowly bacterial cells, "action potential" like electrical signals have been shown to underpin social communication during formation of drug-resistant biofilms (Humphries et al. 2017).

To sense and respond to changes in membrane potential, proteins including ion channels must have charges or dipoles positioned in the transmembrane regions where the electric field is extremely large. The potential gradient within the bilayer can be as high as 10^6 V/m. In a typical voltage-sensitive ion channel, movement of the sensing charges in response to a change in the electric field is coupled to conformational changes in the pore domain, which serves as a gated conduit for the flow of ions (Bezanilla 2000).

DOI: 10.1201/9781003096214-3

The channels in the voltage-gated ion channel (VGIC) superfamily can be very sensitive to membrane potential. Some of these channels increase their activity by many orders of magnitude in response to a few millivolts change in membrane potential. Thus, channels in this family include some of the most tightly regulated allosteric molecules known to date. The overall architecture of members of this superfamily is conserved. Each functional channel has four specialized voltage-sensing domains (VSDs), which surround a central pore domain (Long, Campbell, and Mackinnon 2005a). Each VSD consists of four transmembrane helices where the fourth helical segment contains positively charged residues at every third position. These VSD charges sense and respond to the membrane potential and are the primary drivers of downstream conformational changes that ultimately determine the activity of these channels. Although this architecture is broadly conserved, we should note that some channels like the voltage-gated proton channels (Hv1) lack a specialized pore domain (Sasaki, Takagi, and Okamura 2006; Ramsey et al. 2006) (Figure 2.1).

Since the pioneering studies by Bernstein, Cole, Hodgkin, Huxley and others, voltage-sensitive conductances have been extensively studied because of their primary role in electrical excitability in a variety of cell types which include nerves, muscles, cardiac cells

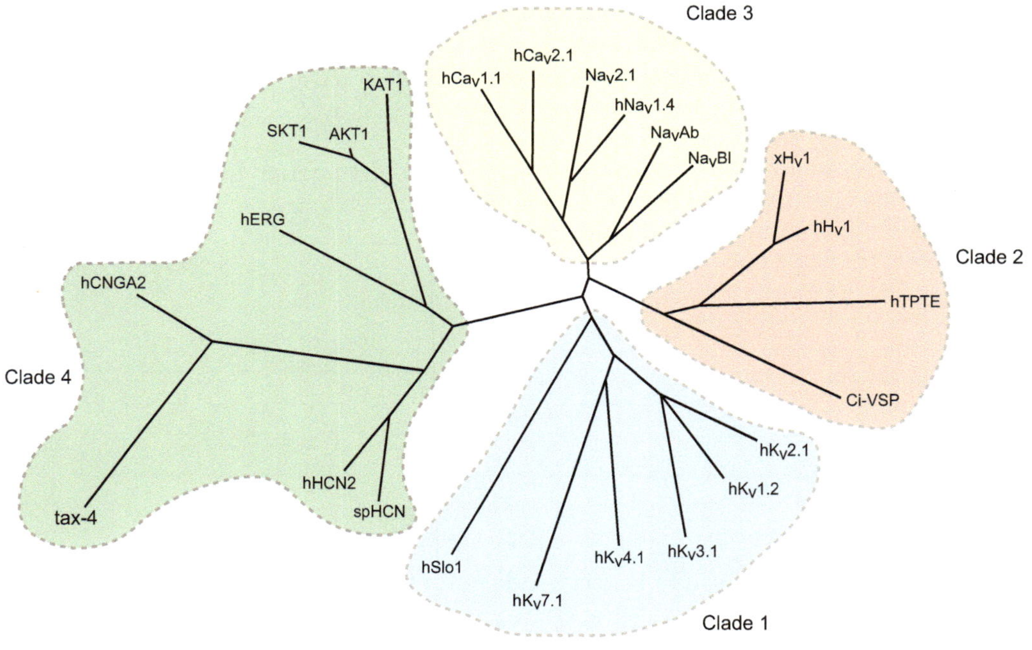

FIGURE 2.1
Phylogenetic tree of voltage-gated ion channels. The cladogram shows the sequence diversity of 24 voltage-sensing domains (VSDs) from different members of the VGIC family, created from a multiple sequence alignment of the VSDs (obtained using MUSCLE). The phylogenetic tree calculation was performed using MATLAB 2012 (MathWorks) and BioNJ. Four distinct clades are directly observable. Clade 1 comprises the different voltage-gated potassium channels members of which form homotetrameric channels. Clade 2 comprises the voltage-sensor-only proteins (VSOPs) – the proton selective H_V channel (human and Xenopus) and the PTEN phosphatase domain containing VSOPs from *Ciona intestinalis* and its human orthologue (hTPTE). Clade 3 comprises the voltage-gated sodium and calcium channels from eukaryotes (which are heterotetrameric; $Na_V2.1$ represents the sodium channel from *Nematostella vectensis*) and prokaryotes (NavAb and NavBl). Clade 4 comprises members of the VGIC family containing a cyclic nucleotide binding domain at their C-termini. Members of this clade include the depolarization-activated hERG channels, hyperpolarization-activated channels from metazoans (hHCN2 and spHCN), the inwardly rectifying plant channels (SKT1, AKT1 and KAT1) and the very weakly voltage-sensitive CNG channels (human and *C. elegans* orthologues).

among others (Bernstein 1902, 1912; Cole and Curtis 1939; Hodgkin and Huxley 1952a, 1952b, 1952c, 1952d, 1952e; Hodgkin and Rushton 1946). Voltage-dependent cation channels, which include the K^+, Na^+ and Ca^{++} channels, are responsible for generation of electrical impulses in many cell types. The influx of Ca^{++} ions through voltage-dependent Ca^{++} channels can also act as a second messenger that initiates various processes such as muscle contraction, exocytosis and cell secretion, proliferation, and migration (Hagiwara 1966; Hagiwara and Nakajima 1966).

Much of our current understanding of the mechanisms of voltage-sensing comes from studies on exemplar VGICs, but it is worth noting that many channels lacking VSD also exhibit voltage-dependent activity. For instance, Kir channels are inwardly rectifying due to voltage-dependent pore block by divalent cations (Lopatin, Makhina, and Nichols 1994). Similarly, K2P channels, which are also lacking in VSD, exhibit strong outward rectification (Schewe et al. 2016). These noncanonical mechanisms will be discussed in detail elsewhere.

2.2 Basic Principles of Voltage Sensing

For more than half a century, voltage-clamp experiments have been the bedrock of electrophysiological methods to analyze biophysical properties of voltage-gated ion channels. In this technique, we measure the amount of current flowing through the membrane expressing channels in response to step changes in potential. From their time-dependent current responses, we can deduce their rates of opening and closing and the fraction of open channels when a new equilibrium is established. The analysis of voltage dependence of open probabilities provides a measure of the relative stability of the open versus closed state of the channel. In the next few sections, we will discuss the quantitative framework for describing the time- and voltage-dependent properties of channel activity and derive fundamental insights into the mechanisms of voltage activation.

2.2.1 Two-State Model of Voltage Gating

From a physical perspective, this voltage-dependent change in channel open probability can be treated as a two-state process wherein the channel exists either in the closed or open state, and voltage biases the equilibrium toward one state or the other.

$$C \rightleftharpoons O$$

Analogous to a chemical reaction, the equilibrium constant for this system is

$$K(V) = \exp\{-\Delta G(V)/RT\} \tag{2.1}$$

where $\Delta G(V)$ is the voltage-dependent energy difference between the two states (Figure 2.2A). In channels that open in response to depolarization, the equilibrium is biased toward the closed state at resting membrane potentials (~−90 mV) and a depolarizing shift in this potential favors channel opening. Furthermore, the energy difference between the two states at a particular voltage is determined by the chemical energy difference between the two states (ΔG_C) and electrical work that is done on this system $\Delta G(V)$:

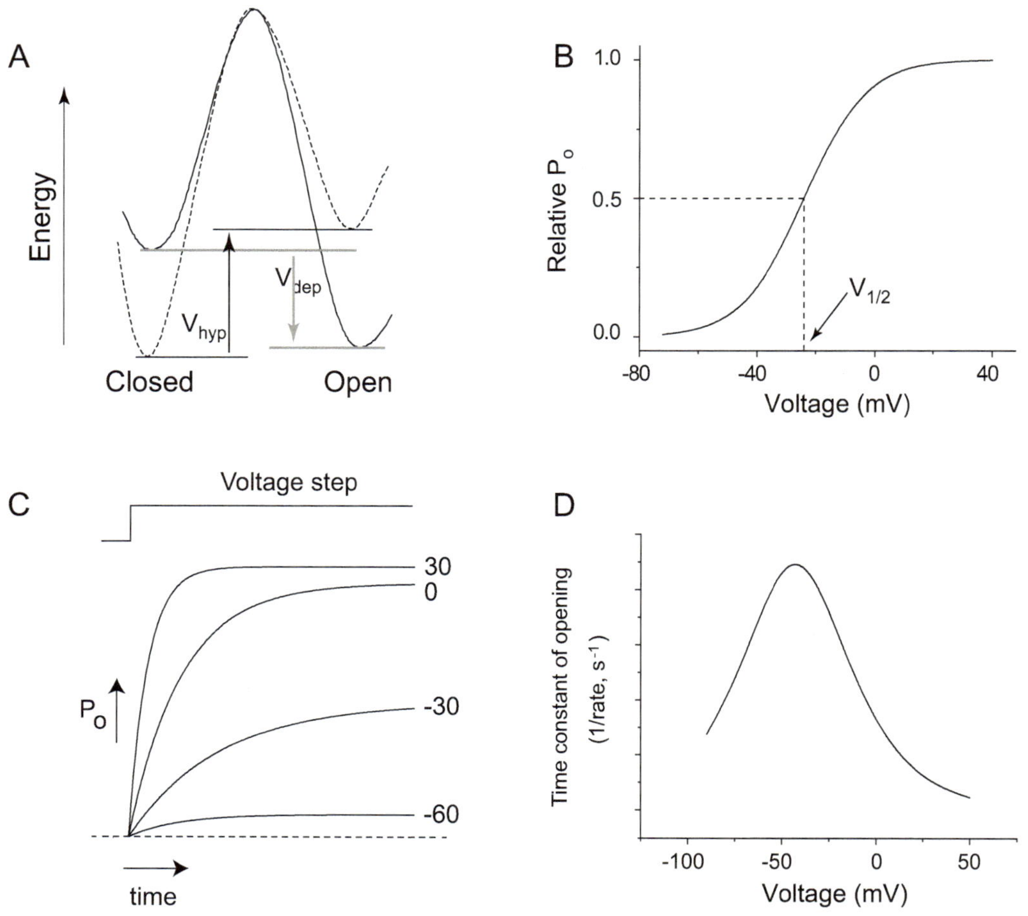

FIGURE 2.2
Steady-state and equilibrium properties for a binary gating model. (A) Energy profile diagram of a closed to open transition for a VGIC. At a hyperpolarizing voltage, V_{hyp} (dotted profile), the closed state is energetically more favorable than the open state. At a depolarizing voltage, V_{dep} (solid profile), the open state is more stable than the closed state. The net energy difference between the closed states at V_{hyp} and V_{dep} are indicated using arrows. (B) The open probability versus voltage curve for a channel that undergoes a binary transition between a closed and an open state. The voltage at which half the channels are open is indicated by $V_{1/2}$. (C) Time-dependent increase in open channel probability of a channel stimulated by voltage jumps to different depolarizing voltages from a constant hyperpolarizing voltage when channels are closed. (D) The time constant of channel opening at different voltages varies nonmonotonically.

$$\Delta G(V) = \Delta G_C - qFV \tag{2.2}$$

where ΔG_C governs the intrinsic conformational bias of the channel in the absence of imposed voltage and q (sometimes referred to as Δq) is the charge difference between the two states, F is Faraday constant, and V is the voltage difference between the inner and outer leaflets of the membrane. The key takeaway here is that a two-state process will be voltage-dependent, only if there is a charge movement associated with the transition.

We can learn about the energetics and kinetics of these processes from measurements of channel activity using voltage-clamp experiments. In a typical experiment, the relative channel open probability is obtained at various voltages and is related to the voltage-dependent equilibrium constant by the following equation:

$$K(V) = \frac{P_O}{1 - P_O} \tag{2.3}$$

$$P_O(V) = \frac{1}{1 + \frac{1}{K(V)}} \tag{2.4}$$

By combining Equations 2.1 and 2.2,

$$K(V) = \exp\{-(\Delta G_C - qFV)/RT\} \tag{2.5}$$

Therefore,

$$P_O(V) = \frac{1}{1 + \exp\{(\Delta G_C - qFV)/RT\}} \tag{2.6}$$

At a particular voltage, ΔG_C will equal qFV and as a result P_O will become 0.5. Thus, we can reparameterize ΔG_C as $qFV_{1/2}$, where $V_{1/2}$ is the voltage at which P_O is 0.5. Thus Equation 2.6 can be rewritten as

$$P_O(V) = \frac{1}{1 + \exp\{qF(V_{1/2} - V)/RT\}} \tag{2.7}$$

The preceding equation is an S-shaped logistic function that is used to describe the open probability versus voltage (P_O–V) relationship for channel opening and is commonly known as the Boltzmann equation. It is noteworthy that, because of the reparameterization, $qFV_{1/2}$ represents the chemical free-energy difference between the closed and open state, and can be obtained by fitting an experimentally deduced P_O–V curve to the Boltzmann equation. A more negative value of $V_{1/2}$ implies that at zero voltage, the open state is intrinsically more stable than the closed state.

In addition to affecting the equilibrium occupancies of the open and closed states, voltage also alters the rates of channel opening and closing. From an Eyring rate theory perspective, channel opening and closing both encounter an activation energy barrier, but these must also be voltage sensitive. Accordingly, the voltage-dependent microscopic rates of channel opening and closing, k_O and k_C respectively, are expressed as

$$k_O = k_O^{\#} \exp(q_O FV) \tag{2.8a}$$

$$k_C = k_C^{\#} \exp(-q_C FV) \tag{2.8b}$$

where $k_O^{\#}$ and $k_C^{\#}$ reflect the respective rates in the absence of voltage (i.e. 0 mV). q_O accounts for the voltage dependence of k_O and can be interpreted as the charge that must move between the closed state and the transition state. Similarly, q_C represents the charge that moves between the open state and the transition state during channel closing. The opposite signs of the exponents in Equations 2.8a and 2.8b are important since it indicates that when voltage is increased, channel opening becomes faster but channel closing becomes slower.

The time-dependent evolution of the open channel probability in response to a depolarization pulse can be expressed as

$$P_O(t) = P_O^{eq}.\left(1 - A.\exp\left\{-\left(k_O + k_C\right)t\right\}\right) \tag{2.9a}$$

In the preceding equation, P_O^{eq} is the equilibrium occupancy of the open state when channels reach a steady-state at the imposed depolarization. A is constant, which is determined by the initial occupancies of the open/closed states (in turn determined by voltage) prior to the depolarization pulse. The important point to note here is that the overall observed rate of channel opening is $\left(k_O + k_C\right)$. Thus, the time evolution of the open state is informed both by the rate constants of channel opening and closing. The reason behind this dependence is that the overall observed rates of channel opening (the time derivative of open channel probability) depend both on the unimolecular (or microscopic) rate constants as well as species fractions, i.e.,

$$\frac{dP_O}{dt} = k_O P_C - k_C P_O \tag{2.9b}$$

When voltage is increased, k_O increases (Equation 2.8a) and thus the positive term of Equation 2.9b increases too. However, with time, as more open channels populate, there will be more kinetic driving force for channel closing (due to the negative term in Equation 2.9b) and some of the channels which have opened, will now begin to close. The contribution of this negative term will depend on the magnitude of microscopic rate of channel closing, k_C. At a certain point in time, when the two terms become equal, the derivative will become zero and equilibrium is achieved. At strong depolarizing voltages, k_C will be negligibly small (Equation 2.8b) and thus the overall rate of channel opening will almost exclusively depend on k_O. The same principles apply to the kinetics of channel closing. This basic formalism can be applied to extract the rate constants of channel opening and closing from time evolution of channel activity in response to step potentials.

As is obvious by now, for a gating transition to be voltage sensitive, it must be associated with movement of charges (Armstrong 1981). Changes in the electric field are sensed either by translocating charges or via movement of dipoles within the membrane field. In the preceding two-state model, this sensing charge is the q, which is the net charge difference between the C and O state. The movement of "sensing" charges (also called gating charges) in response to changes in the electric field generates a current called gating currents. These currents are capacitive in nature because once the translocation of charge from one state to the other is complete, no further charge will be moved. Thermodynamics dictates that for a reversible reaction the total gating charge transported (obtained by integrating gating current over time) during a voltage pulse must be equal to the charge moved when the voltage is restored to its initial value. This equality of ON and OFF charge is a hallmark of gating currents for processes that are completely reversible.

To measure gating currents, it is necessary to eliminate ionic currents, which can be accomplished by either using specific blockers or by replacing the permeant ions. In the absence of ionic current, we are left with capacitive currents, which is due to buildup of charge in the lipid bilayer. The capacitive current, I_C, is the time derivative of the charge and can be written as

$$I_C = C\frac{dV}{dt} + V\frac{dC}{dt} \tag{2.10}$$

where C is the total capacitance and V is the voltage. The first term in Equation 2.10 corresponds to a fixed capacitor, which is due to the charging of the lipid bilayer, whereas the second term describes time-dependent changes in the capacitance of the membrane arising due to conformational transitions in the protein. The capacitance of the membrane itself is practically time invariant while the currents arising due to charge movements in the protein follow special kinetics governed by the energy-landscape of protein dynamics. The definition of capacitance is the ability to store charge and when there is a redistribution of charge in the membrane due to a gating transition, its capacity to store charge must also change. Since the proteins are sensitive to voltage over a relatively narrow, specified voltage-range the contribution of the second term to *Ic* will saturate at extreme potentials once the transition is completed and the charges within the protein reach their final position. In contrast, the first term will grow linearly with increasing potential as more free charges from the bath will keep accumulating across the lipid bilayer. This difference in the voltage dependence of the two components can be utilized to extract gating charges associated with voltage-sensing gating transitions (see Chapter 7 for more experimental details).

The time course of gating charge (Q) movement can be obtained by integrating the gating currents over time. For a two-state model, the time course of gating charge in response to a step depolarization corresponds to

$$Q(t) = N.q.P_O^{eq}\left(1 - A.\exp\left\{-\left(k_O + k_C\right)t\right\}\right) \tag{2.11}$$

where N is the number of channels and q is the gating charge per channel. P_O^{eq}, A and k_O, k_C have the same definitions as in Equation 2.9. Similarly, the voltage dependence of the gating charge in a two-state model is

$$Q(V) = \frac{N.q}{1 + \exp\left\{qF\left(V_{1/2} - V\right)/RT\right\}} \tag{2.12}$$

Except for a multiplicative factor, the expressions describing $Q(V)$ and $Q(t)$ are identical to $P_O(V)$ and $P_O(t)$, respectively. For a two-state model, the voltage dependence and time dependence of ionic currents and charge movement will be completely superposable. This is a hallmark of a two-state model but is rarely observed in practice. While gating behavior of most channels is more complicated and requires consideration of multiple transitions, the two-state model, nonetheless, serves as an excellent pedagogical tool to understand the behavior of a voltage-dependent process.

2.2.2 Multistate Models of Voltage Gating

In the 1950s, using the newly developed voltage-clamp technique, Hodgkin and Huxley (1952a, 1952b, 1952c, 1952d, 1952e) carried out a series of groundbreaking experiments that led to a detailed quantitative description of an action potential in the squid giant axon. They discovered that the sodium and potassium conductances in the squid giant axon are regulated independently and depend on membrane voltage rather than membrane currents. Furthermore, the time course of these currents did not correspond to monoexponential kinetics as expected in a simple two-state model (Equations 2.9 and 2.11). Nonetheless, the two-state model served as a starting point to develop their phenomenological model, which recapitulated the major features of excitability.

Their model, which is widely known as the Hodgkin–Huxley model, has two separate equations to describe the sodium and potassium conductances during an action potential. The potassium current (I_K) is represented as

$$I_K = n^4.g_K\left(V - V_K\right) \tag{2.13}$$

where g_K is the conductance of the conductance of the potassium channels, V is the applied voltage, V_K is the reversal potential and n refers to voltage-dependent probability of a gating particle to exist in either a permissive or nonpermissive state. Hence, the voltage dependence of this gating particle will be similar to the two-state particle shown in Equation 2.9, and we can rewrite the time dependence of the potassium current as

$$I_K(t) = n_\infty\left(1 - A.\exp\{-(k_O + k_C)t\}\right)^4 g_K\left(V - V_K\right) \tag{2.14}$$

The preceding equation is a powered exponential function that accounts for the steep voltage dependence and the pronounced lag phase observed in the potassium currents. As the power increases, the lag phase becomes prominent, and in the case of potassium currents, the time course was well fit to power of 4. Based on this empirical relationship, Hodgkin and Huxley presciently inferred that the increase in potassium conductance is regulated by four independent gating particles, all of which must move to a permissive state for ion flux. We now know that the potassium channel is a tetramer where a voltage-sensing domain in each subunit regulates the opening and closing of the channel.

Hodgkin and Huxley also described the time-dependent changes in sodium conductance along the same lines with four independent gating particles. However, to account for inactivation of sodium currents, they reassigned one of the probability terms to inactivation. In the following equation, m is probability of the three particles that contribute to channel opening to be in a permissive state and h is the probability of the inactivation particle to be in a nonpermissive state:

$$I_{Na} = m^3.h.g_{Na}\left(V - V_{Na}\right) \tag{2.15}$$

The kinetics of the inactivation particle is slower than those of the activation particles, which gives rise to the characteristic sodium channel transients. Remarkably, 50 years later, studies probing the structural dynamics of the voltage-sensing domains found that the voltage-dependent activation of the first three homologous domains of the sodium channel contributes to channel opening, whereas slow moving domain IV is a rate-limiting step for entry into an inactivated state (Chanda and Bezanilla 2002).

Some of the predictions of the Hodgkin–Huxley model were put to further tests when gating currents from sodium channels in the squid giant axon were first measured by Armstrong and Bezanilla (1973). To record these small transient currents, which till that point was viewed using oscilloscopes and chart recorders, they had to develop specialized electronics. They found that the sodium channel gating current is not monoexponential as predicted but instead exhibits a fast rising phase and a multiexponential decay. Furthermore, they found that the entry into inactivation slows the recovery of the gating charge when the voltage is returned to the original value, indicating that the activation and inactivation are coupled. These and other observations formed the basis of the coupled inactivation model, which postulates that the inactivation is not due to movement of the gating particle itself but follows upon channel opening (Armstrong 1992, 1981). This notion of inactivation as a linked reaction is widely accepted and has been validated by multiple lines of evidence.

Discovery of methods to make gigaohm resistance seals and the development of the patch-clamp technique by Neher and Sakmann ushered a new era whereupon it became possible to reliably record activity of single channels from native tissues with tremendous precision (Hamill et al. 1981). These measurements enabled the development of detailed kinetic models that were both parsimonious as well as mechanistically insightful. A detailed discussion of these models is provided in Chapter 7, but suffice it to say, these kinetic models of activation were further constrained by gating current recordings. The key feature of these models is that the pore opening and voltage sensing are distinct but linked processes. According to the model proposed by Zagotta, Hoshi and Aldrich for the Shaker potassium channel, the overall process of channel opening involves two sequential "voltage-sensing" transitions occurring independently in different subunits of the protein (Hoshi, Zagotta, and Aldrich 1994; Zagotta, Hoshi, and Aldrich 1994; Zagotta et al. 1994). Once all four voltage sensors are fully activated, a single concerted transition gates the channel open. The charge movement that accompanies the voltage-sensing steps are largely confined to a specialized voltage-sensing domain within each subunit of the protein and occur early in the activation pathway. The opening of the channel pore exhibits very little intrinsic voltage dependence and occurs late in the activation pathway (Figure 2.3).

The voltage dependence of channel gating depends on the amount of charge associated with voltage-dependent transitions, but how do we estimate the gating charge per channel (referred to as Q_{max})? In principle, all that is required is measurement of the maximum gating charge and an independent estimate of the number of channels in the same cell or patch of membrane. However, precisely estimating the number of active channels in a single cell is experimentally nontrivial. It was not until the 1990s that methods were developed and successfully implemented to calculate Q_{max}. Almers (1978) proposed a strategy that involves measuring the voltage-dependence of channel opening at voltages where channel opening probabilities are infrequent. This approach, referred to as the limiting slope analysis, allows researchers to calculate the charge per channel by simply using *Po* measurements. The slope at limiting voltages is calculated from a semilog plot of ln P_O versus V and can be described mathematically as

$$Q_{max} = \lim_{V \to -\infty} \frac{\partial \ln P_O}{\partial V} \tag{2.16}$$

Under these limiting conditions, the channel will spend most of its time in the first closed state, and to open it must traverse through all the intermediate states and hence the voltage

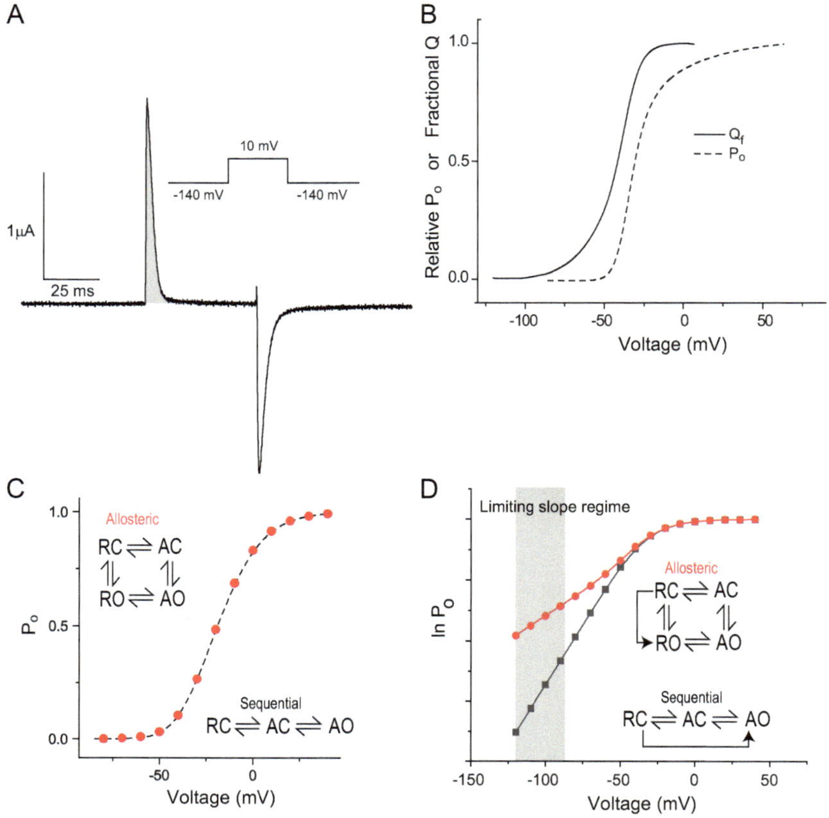

FIGURE 2.3
Gating charge movement in voltage-dependent ion channels. (A) An example of a gating current trace for the Shaker potassium channel (in the background of the W434F and inactivation-removing mutations) obtained in response to a depolarizing pulse. The shaded area represents the gating charge transferred during the depolarization. (B) Comparison of the normalized gating-charge displacement (Q_f) and relative open-probability curves for the Shaker potassium channel (inactivation removed mutant). (C and D) Open probability versus voltage relationships for channels which activate either allosterically (red dots) or strictly sequentially (dashed line). RC corresponds to resting voltage sensor and closed pore, RO corresponds to resting voltage sensor and open pore, AC corresponds to activated voltage sensor and closed pore, and AO corresponds to activated voltage sensor and open pore. In allosteric model, the channel consisting of a voltage sensor and pore can exist in all four possible states whereas in a strictly sequential scheme, some of the states such as RO are forbidden. Note that P_O versus V relationships (C) will appear identical, but the differences between the two models will become evident in $\ln P_O$ versus V plots (D). The limiting slope of the $\ln P_O$ versus V curves, measured in the shaded voltage range, correspond to the charge movement between the states indicated by the arrows. In the allosteric model, the limiting slope represents the charge moved between the RC and RO states, whereas in the sequential model, the limiting slope equals to the charge moved between the RC and AO states, which corresponds to the total gating charge associated with channel activation.

dependence of opening at these voltages reflects the sum total of all voltage-dependent transitions in the gating pathway. The underlying premise is that the channel activation is fully sequential and that the voltage sensor must always activate before the pore can open. However, in an allosteric scheme, where pore openings can occur even in absence of voltage-sensor activation (albeit with very low probability), the limiting slope method cannot be applied to calculate charge per channel.

2.2.3 Model-Free Methods for Estimating Free Energy of Channel Gating

The gating models described until now not only provide us a quantitative description of the time- and voltage-dependent behavior of channel activity but can also give mechanistic insights into the structural features. More sophisticated and structurally detailed models have been built since then by using orthogonal information from voltage-clamp fluorometry and above all by solving high-resolution structures of voltage-gated channels in various conformations. While the structures of the channel in various conformations are visually satisfying and provide a physical view of channel gating, they cannot be used to directly make quantitative predictions. Experimentally, the effects of mutations or perturbations can be manifested both in terms of equilibrium and kinetic properties. As a first approximation, it is fair to assume that interactions at equilibrium determine the structures of macromolecules, and changes in these interactions underlie conformational rearrangements that ultimately dictate function. In this section, we will review model-agnostic methods to estimate the free energy of channel gating and how these enable estimates of site-specific interactions.

Let us begin by considering the Zagotta–Hoshi–Aldrich model for voltage-dependent gating:

$$(R_1 \underset{}{\overset{K_1}{\rightleftharpoons}} R_2 \underset{}{\overset{K_2}{\rightleftharpoons}} C)_4 \underset{}{\overset{L}{\rightleftharpoons}} O$$

K_1, K_2 and L indicate the voltage-independent components of the equilibrium constants for each of the transitions. The net free energy of activation in this case is

$$\Delta G_C = -RT \ln\left\{\left(K_1 K_2\right)^4 L\right\} \tag{2.17}$$

In principle, it is not necessary to develop detailed kinetic models to estimate the net free energy of activation. We should also point out that these kinetic models have been developed for only a limited set of exemplar ion channels but are missing for the vast majority of the channels. A more direct approach to obtain G_c exploits an important thermodynamic principle that the free-energy change in a system will be equal to the work done on the system by the external force under reversible conditions (Chowdhury and Chanda 2012). The work done, in turn, depends on the external force and its conjugate displacement. For instance, in the case of mechanical work on an ion channel, the external force is surface tension or pressure whose conjugate displacements are surface area or volume expansions. In a similar way, when the work done is electrical in nature, the external force is voltage and its conjugate displacement is charge. Thus, for voltage-dependent ion channels, the free-energy change, G, can be estimated from the QV curve. Specifically, G is given by the integral of the QV curve (Figure. **2.4**a):

$$\Delta G = \int_{-\infty}^{\infty} Q\,dV \tag{2.18}$$

Integration of this equation by parts allows us to separate this G into two components:

$$\Delta G = \lim_{V \to -\infty} Q_{max} FV - \int_{0}^{Q_{max}} V\,dQ \tag{2.19}$$

The first ($Q_{max}FV$) is the electrical component (where Q_{max} is the maximum number of gating charges transferred during channel activation); the second component, which is the

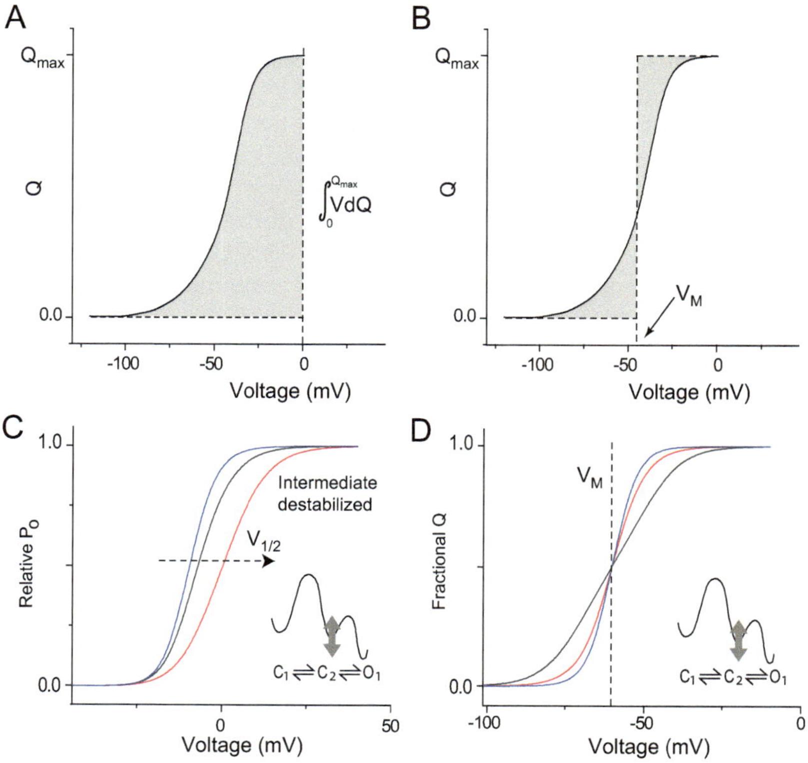

FIGURE 2.4
Model-free measurements of free energy of voltage-dependent activation. (A and B) A typical charge–voltage (*QV*) plot of a voltage-gated ion channel where the y-axis is the charge per channel and x-axis is the voltage. Q_{max} is the maximum charge per channel. The shaded area under the *QV* curve and ordinate axis (shown in A) represents the chemical components of the total free-energy change associated with voltage-dependent activation of the channel. The dashed line in (B) represents the median voltage axis (V_M), which is the voltage where the two shaded areas on either side of the V_M axes are equal. When this condition is satisfied, then the area under the curve depicted in (A) can be calculated by simply taking the product of Q_{max} and V_M. (C and D) A comparison of how the stabilities of intermediate states alter the P_O–V and *QV* curves. (C) shows P_O–V for a sequential model (inset). These curves were generated by changing the stability of the intermediate C_2 state, while keeping the net energy difference between the initial and final states of the system (C_1 and O_1, respectively) fixed. Fitting these data to a Boltzmann function and using the fit parameters $zFV_{1/2}$ to calculate the free energies will give three different values even though the net free energy remains unchanged. (D) *QV* plots with the same three state schemes as in (C). Changes in the stability of the intermediate state C_2 changes the shape of the *QV* curve but does not affect the value of V_M of activation. Thus, the parameter V_M is solely governed by the difference in the energy between the initial and final states of the system.

area between the QV curve and the Q (ordinate) axis, reflects the chemical component of the free-energy change, G_c. It can be shown that G_c is simply

$$\Delta G_C = Q_{max} F V_M \tag{2.20}$$

where V_M (median voltage for activation) is the voltage that divides the QV curves into two parts of equal areas (Figure 2.4b). The preceding free-energy relationship allows us to accurately and directly estimate the free-energy associated with channel gating without any prior knowledge of the underlying kinetic mechanisms (Figure 2.4c and 2.4d).

The ability to obtain accurate measurements of the free energy of channel gating makes it possible to use thermodynamic tricks to learn more about the molecular forces that determine structure. One such method is the generalized interaction energy analysis (GIA), which has been used to measure pairwise interaction energies between different sites or side chains (Chowdhury, Haehnel, and Chanda 2014). This method is based on the double mutant cycle analysis pioneered by Alan Fersht and his colleagues to study Barnase–Barstar interactions (Carter et al. 1984). Simply stated, the method involves comparing the free energies of the double mutations with linear summation of the free energies of the individual mutants. For instance, if we consider two sites, A and B, in an ion channel, the free energy of interaction between the two sites (ΔG_{A-B}) can be obtained by using the following equation:

$$\Delta G_{A-B} = \left(\Delta G_{WT} - \Delta G_{A}\right) - \left(\Delta G_{B} - \Delta G_{AB}\right) \tag{2.21}$$

ΔG_A and ΔG_B correspond to the free energy of gating of single mutants at the *A* and *B* positions. ΔG_{AB} and ΔG_{WT} are median voltage values for the *AB* double mutant and *WT* channel, respectively. If the sites are not interacting, the ΔG_{A-B} is zero otherwise and it corresponds to the net interaction energy between these sites. These methods have been recently used to map allosteric interaction pathways underlying voltage-dependent opening in voltage-gated potassium ion channels (Fernandez-Marino et al. 2018).

2.3 Biophysical Methods to Probe Voltage-Sensing Mechanisms

The advent of molecular cloning stimulated mechanistic thinking in terms of the underlying chemical structures and their role in ion channel function. In this section, we will briefly summarize the key biophysical approaches to study the mechanisms of voltage-dependent gating.

2.3.1 Gating Charge per Channel

To understand the mechanism of voltage sensing, it is necessary to first identity the voltage-sensing charges. This is akin to identifying what the ligand-binding site is in a ligand-gated ion channel. While in the VGICs, the positive-charged residues (arginines and lysines) in the S4 segment are likely to be the gating charges; there is no a priori reason to assume that these are the only voltage-sensing residues. Moreover, not all charged residues in the S4 segment may contribute to voltage sensing. Only if a mutation of a charged residue reduces the gating charge per channel, then that residue is considered a bona fide voltage sensor. As discussed earlier, gating charge per channel (Q_{max}) can be estimated by using the limiting slope method when the channel activation is completely sequential. A more general and direct method is to measure the total gating charge per cell and then independently determine the number of channels in that cell. In experiments on the Shaker potassium channel, the total number of channels per cell were estimated by either using a radioligand binding measurements or by noise analysis (see Chapter 8 for details on noise analysis) (Aggarwal and MacKinnon 1996; Seoh et al. 1996; Schoppa et al. 1992).

More recently, fluorescence from ion channels tagged with a fluorescent protein has been used to obtain a relative estimate of gating charge per channel (Gamal El-Din et al. 2008).

In these experiments, fluorescence-gating charge (F-Q) curves were generated by plotting the surface fluorescence with respect to gating currents for every cell. By comparing the slope of the F-Q plot of mutants with that of the wild-type channel, it is possible to obtain a relative estimate of the gating charge per channel for the mutants.

2.3.2 Substituted Cysteine Accessibility Method (SCAM)

One of the early approaches to probe conformational change in a voltage-gated ion channel is the substituted cysteine accessibility method (SCAM). In a membrane protein, each residue is either accessible to the surrounding water molecules or buried in the lipid bilayer or protein interior. Water-exposed cysteine side chains form ionized thiolate (RS–), which reacts a billion times faster with MTS derivatives than the nonionized thiol groups (RSH). Therefore, reactivity to MTS derivatives is a measure of water accessibility of a cysteine at a particular site and, depending on their location in the structure, changes in cysteine accessibility in response to a stimulus can be a probe of local conformation. Cysteine reactivity to MTS reagents is typically assayed by measuring the effect on this modification on channel function. Akabas and Karlin first developed this method to probe conformational changes in acetylcholine receptor (Akabas et al. 1992). Yang and Horn (1995) were the first to apply this approach on voltage-gated ion channel to show that the S4 voltage-sensing segment of a sodium channel moves in response to a change in membrane potential. Since then, SCAM has been widely adopted to probe conformational changes in various other ion channels, including in the Shaker potassium channel.

2.3.3 Voltage-Clamp Fluorometry (VCF)

Another approach to probe voltage-dependent conformational change is to introduce a fluorescent reporter at various sites in the ion channel. Fluorescence emission is highly sensitive to the local environment; solvent, pH, quenching groups and electric field can influence the emission spectra and intensity of the fluorescence signal. In voltage-clamp fluorometry (VCF), fluorescence measurements are carried out in voltage-clamp configuration that allows simultaneous tracking of function and structure. Unlike SCAM, VCF can also provide information about the kinetics of conformational change near the labeling site. Isacoff and colleagues were the first to implement this approach to measure voltage-induced conformational changes in the Shaker S4 segment (Mannuzzu, Moronne, and Isacoff 1996). Similar measurements on the S4 segments of the voltage-gated sodium channel revealed the roles of S4 of domain III and IV in sodium channel inactivation (Cha, Ruben, et al. 1999). Over the past two decades, VCF and a related PCF (patch-clamp fluorometry) have been powerful biophysical approaches to probe conformational dynamics and correlate it with the functional state of the channel (Zheng and Zagotta 2000). A more detailed review of this approach will be provided in Chapter 11.

2.3.4 Gating Pore Currents

Starace and Bezanilla discovered that the mutations of some of the charged residues to histidines made the Shaker potassium channel permeable to protons (Starace, Stefani, and Bezanilla 1997). They were able to demonstrate that the proton permeation pathway resides in the voltage-sensing domain (VSD) and that it is voltage-dependent. Subsequently, it was shown that neutralizing the S4 charges can elicit gating pore currents that are both voltage-dependent and can pass cations as large as guanidium ions (Tombola, Pathak,

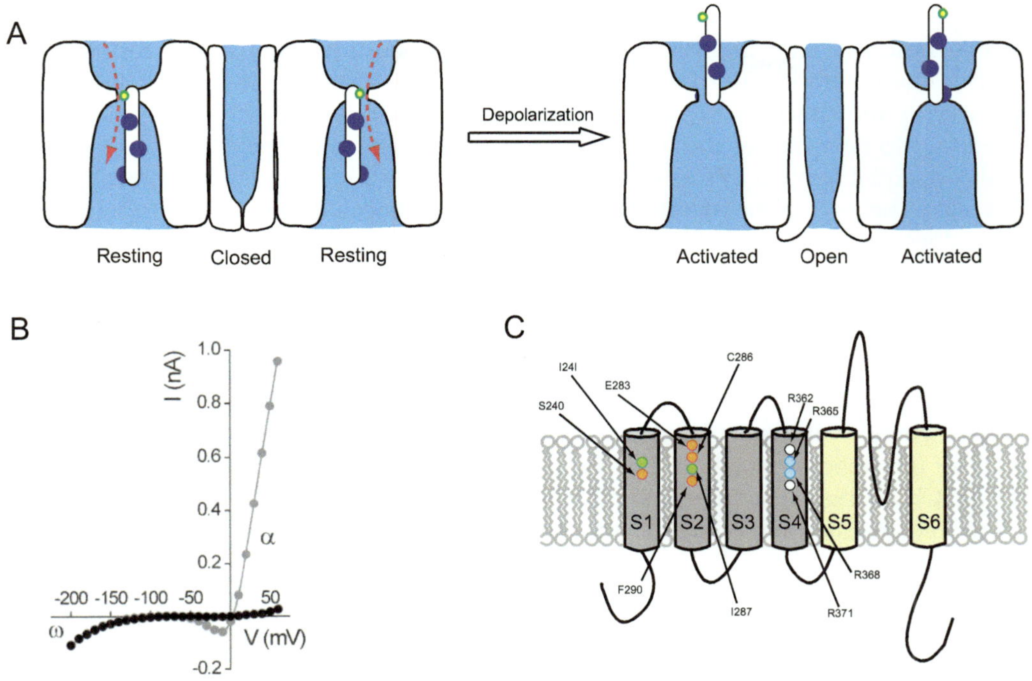

FIGURE 2.5
Gating pore currents. (A) At hyperpolarizing voltages, when the channel is closed and the voltage sensor is in resting conformation, the first gating charge is positioned near the gating septum, which separates the external and internal water-filled crevices in the voltage sensor. Mutation of this gating charge to a small/neutral amino acid allows small free cations in solution to sneak through the gating septum giving rise to gating pore currents. Activation of the voltage sensor by depolarization positions the last gating charge near the gating septum, thereby occluding flux through the gating pore pathway and making it state-dependent. (B) Current–voltage plots of the R1 mutant of the Shaker potassium channel in the absence and presence of Agitoxin. Agitoxin specifically blocks the flux of potassium ions through the principal pore (constituted by S6 helices) but does not inhibit the flux of the ions through the gating pore. (Data from Tambola et al. 2005.) (C) Sites in the Shaker potassium channel, which, upon mutation, give rise in state-dependent gating pore currents. (Adapted from Chanda and Bezanilla, Neuron 2008).

and Isacoff 2005). Gating pore currents are a consequence of a remarkable feature of the voltage-sensing domain wherein the electric field is focused over a very narrow region referred to as the gating septum, which separates the intracellular side from the extracellular side (Ahern and Horn 2004). In absence of charged S4 residues, ions and protons can sneak through the gating septum at a measurable rate that is detected as gating pore currents. These small currents can be detected only when the primary pore currents are abolished by a pore blocker or in the background of a nonconducting mutation. Depending on the position of the mutated residue, the gating pore currents are observed either in the activated state or in resting conformation. These voltage-dependent VSD currents become a very useful tool to monitor the conformational state of the voltage sensor (Figure 2.5).

2.3.5 Thermodynamic Mutant Cycle Analysis of Interaction Energies

Thermodynamic mutant cycle analysis is a widely used technique to determine interaction energies between two sites within a protein molecule or between two interacting molecules. Since energies are additive, a change in free energy due a double mutation should correspond to the sum of the change in free energy due to each of the single mutations, unless these two sites interact. In that case, the nonadditive component corresponds to the free energy of the pairwise interaction between the two sites. Mutant cycle analysis is a powerful and conceptually straightforward tool to probe interaction energies, but it is important to be aware of the limitations of this approach. The principle of mutant cycle analysis can be applied to free energies and not to any functional or structural property. The reason being that the free energy of a system is a thermodynamic state function whose value does not depend on the path but only on the initial and the final states. Structural and functional parameters, by contrast, are path functions rather than state functions. For instance, consider the conductance-voltage curve, which is a functional property of a voltage-dependent ion channel but has been used as a surrogate to estimate the free energy of channel activity by fitting it to a single Boltzmann function under the two-state approximation (Zandany et al. 2008). For most ion channels, this approximation is not valid and the free-energy value calculated from G-V curves in such cases is dependent on the energetics of intermediate states and hence, it no longer represents the free energy of channel gating (Chowdhury, Haehnel, and Chanda 2014).

For voltage-gated ion channels, the free energy of channel gating can be obtained by measuring the charge-voltage curves and using the median voltage method (Equation 2.20). As described earlier, thermodynamic mutant cycle analysis based on median voltage-estimates (GIA) can be used to probe pairwise interaction energies between various residues and their contribution to channel gating. To apply this approach, one also needs to calculate the charge per channel. If the mutations do not involve charge residues, it can be assumed that the charge per channel is unchanged and use wild-type values. While this approach is broadly applicable to any voltage-gated ion channel, we should note that in many cases it is challenging to record gating currents due to lack of sufficient expression.

2.3.6 Structural Approaches

Structure-based approaches can provide molecular-level descriptions of the structural changes during the voltage-gating cycle and in recent decades have revolutionized our understanding of the mechanisms of voltage gating (Doyle et al. 1998; Long, Campbell, and Mackinnon 2005a). High-resolution structures set the stage for building detailed mechanistic models and designing new experiments to test these models. Developments in membrane protein crystallography and more recently cryo-EM technology has ushered in a resolution revolution in the field (Liao et al. 2013). Representatives of at least one member of every family of VGICs are now available. Structure of VGICs have been solved in multiple conformations representing various intermediates during the gating cycle (Clairfeuille et al. 2019; Guo et al. 2016; Shen et al. 2018; Shen et al. 2019; Long et al. 2007; Wisedchaisri et al. 2019). In some cases, the structures have no obvious correlate with any known functional state. This lack of correlation is not necessarily surprising because kinetic models are built on the principle of parsimony, and it is not possible to rule out more complex gating schemes that involve additional conformational states that may be captured in some studies but not others.

Voltage-clamp fluorometry can bridge some of the divide between the various structures and functional states (Tombola, Pathak, and Isacoff 2006; Cowgill and Chanda 2019). FRET and related methods provide subnanometer level descriptions of relative conformational changes during the gating cycle (Posson et al. 2005). The primary advantage of this approach is that these measurements are carried out in functional channels in a native lipid environment. In some cases, it is possible to correlate specific conformational change with a kinetic transition or entry into a functional state.

2.3.7 Computational Approaches

Computational approaches have become increasingly important even though at present it is not possible to run molecular simulations that drive the channel through a complete gating cycle. In a technical tour de force, Shaw and his group used a special-purpose supercomputer to simulate the voltage-driven deactivation of a voltage-sensing potassium channel by applying extraordinarily large electric fields to hasten the process (Jensen et al. 2012). However, most academic labs have limited access to these resources and much of the effort in the field has focused on enhanced sampling methods to bridge molecular dynamics (few microseconds) and experimental (tens of milliseconds and beyond) time scales. Computational approaches have been used to identify the possible interaction sites during the gating process and also track how conformational energy is transmitted from the sensing or modulator domain to pore gates by using allosteric network analysis (Kasimova, Lindahl, and Delemotte 2018).

To simulate voltage-dependent gating, it is necessary to apply an electric field across the lipid bilayer. Computationally there are two ways this can be achieved. In one case, charge imbalance is created by introducing ionic gradients across the lipid bilayer (Sachs, Crozier, and Woolf 2004). While the forces experienced by the atoms in the transmembrane regions are more realistic, given the small size of the periodic cell even a single charge translocation can significantly alter the membrane potential. An alternate approach involves the application of a uniform electric field throughout the entire periodic cell, oriented such that it is perpendicular to the plane of the membrane (Roux 1997). In bulk solution, the ions reorient to neutralize most of the electric field such that the steepest gradient is over the low dielectric membrane. A detailed discussion of the various computational approaches will be covered in Chapter15.

2.4 Voltage-Sensor Motions

The first structures of voltage-sensing ion channels revealed that the voltage-sensing domain consisting of S1–S4 helix is a four-helix bundle domain where the charged S4 residues either interact with water molecules inside the crevices or with lipid headgroups (Long, Campbell, and Mackinnon 2005a; Long et al. 2007). Much of the early studies characterizing the mechanism of voltage sensing were carried out on the Kv channels, but more recently high-resolution structures of two-pore channels and sodium channels have provided new insights (Guo et al. 2016; Shen et al. 2019; Xu et al. 2019). Here, we will present some of the key features of the voltage-sensing domain and emerging views regarding the movements of the voltage sensor. First, the voltage-sensing domain forms a water-filled permeation pathway for movement of voltage-sensing charges (Krepkiy, Gawrisch, and Swartz 2012). However, the permeation pathway is not contiguous but interrupted by the charge transfer center, which is capped on the extracellular side by conserved phenylalanine and negatively charged residues on the intracellular side (Tao et al. 2010). Note that the charge transfer center has also been referred to as the gating septum by researchers in the field. The charge transfer center facilitates the movement of the voltage-sensing charges between the extracellular and intracellular water-filled crevices. These structural features ensure that the electric field is focused within the voltage-sensing domain and most of that drop occurs over a distance of 4 Å (Ahern and Horn 2004; Starace and Bezanilla 2004) (see Box 2.1). The first four arginine residues on the S4 helix carry most of the gating charges (Seoh et al. 1996; Aggarwal and MacKinnon 1996). Some of these charges are stabilized by interaction with acidic residues in other helices, but the first charge has been shown to interact with acidic headgroups of the surrounding lipids (Tiwari-Woodruff et al. 1997; Long et al. 2007; Xu, Ramu, and Lu 2008). The composition of the surrounding lipid can have a significant influence on the voltage dependence of gating (Figure 2.6).

The nature and extent of the voltage-sensing motion underlying the activation of depolarization-activated channels have been extensively investigated (Swartz 2008; Bezanilla 2008; Tombola, Pathak, and Isacoff 2006). Before the high-resolution structures became available, there were two competing models of voltage sensing. In the transporter model, the water-filled crevices in the voltage-sensing domain were envisioned to undergo a conformational switch from an outside-facing conformation to an inside-facing state upon membrane hyperpolarization (Starace and Bezanilla 2004; Cha, Snyder, et al. 1999). The

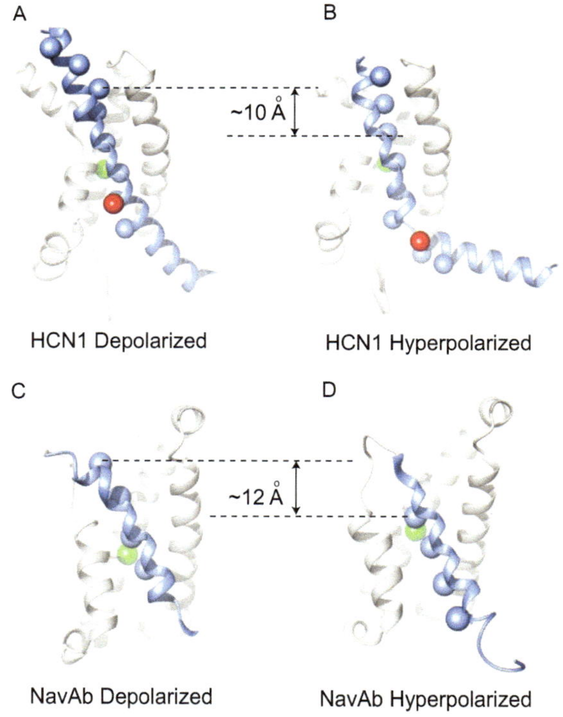

FIGURE 2.6
Structures of the voltage sensors of hyperpolarization- and depolarization-gated channels in activated and resting conformations. (A and B) The structure of the voltage sensor of the HCN1 channel in a depolarized state (A) and in a metal-bridge-locked hyperpolarized state (B). The S4 helix is in blue, while S1, S2, S3 are in light gray. The blue spheres represent the charged residues on the S4 segment and the red sphere indicates a conserved Serine residue that facilitates the bending of the S4 helix between the two conformations. The conserved phenylalanine residue representing the gating charge transfer center is shown as a green sphere. There is a vertical displacement of ~10Å in the S4 segment between the two conformations. (C and D) The structure of the voltage sensor of the NavAb channel in an activated conformation (C) and in a metal-bridge-locked resting state (D). The color scheme is the same as before. There is a vertical displacement of ~11.5Å in the S4 segment between the two conformations.

proposed movement of the S4 segment is limited in this model and the charge transport occurs due to a change in the position of the gating septum. The alternate model, by contrast, posits that the S4 helix undergoes a helical screw translational motion to transfer gating charges from the outside to inside past the gating septum (Yang, George, and Horn 1996; Baker et al. 1998). Most voltage-sensing domains are in activated conformation at 0 mV and since the structures are solved in absence of an electric field, voltage-sensing domains in these structures are only captured in activated conformation. By chemically trapping the voltage sensor in down conformation, researchers have successfully solved a number of structures of the voltage-sensing domain in resting state (Guo et al. 2016; She et al. 2018; Shen et al. 2018; Clairfeuille et al. 2019; Xu et al. 2019; Shen et al. 2019; Wisedchaisri et al. 2019). These resting-state structures of the voltage-sensing domain in two-pore channels and sodium channels are in general consistent with helical screw motion in the voltage-sensor domain.

For many years, it was implicitly assumed that similar voltage-sensing motions occur in the hyperpolarization-activated ion channels, but the primary difference between them and their depolarization-activated channels is the inversion of the coupling such that the pore is open when the voltage sensor is in down conformation (Mannikko, Elinder, and Larsson 2002; Latorre et al. 2003). Structural studies reveal the voltage-sensing S4 helix in hyperpolarization-activated channels is at least one helical turn longer (Lee and MacKinnon 2017). Unexpectedly, when this S4 helix moves to a down position, it breaks in the middle with the lower half forming a structure that resembles the S4–S5 linker in domain-swapped channels (Lee and MacKinnon 2019). Functional measurements show that the polarity of residue at the breakpoint is a critical determinant of the direction of rectification; hydrophobic residues at that position favor depolarization activation, whereas the hydrophilic residues favor hyperpolarization activation (Kasimova et al. 2019). Sequence analysis of the protein database shows that the polarity of the amino acids residues is also directly correlated with the helix-breaking propensity in membrane proteins. While the generality of these concepts remains to be established, the emerging view is that the voltage-sensor motion in the hyperpolarization-activated channels involves an additional conformational switch that it not observed in their depolarization-activated counterparts.

2.5 Coupling of Voltage-Sensor Motion to Pore Opening

One of the key questions in the field is: How are the voltage-sensor motions coupled to opening and closing of the pore gates in the VGIC superfamily? Before we get into that, let us take a moment to consider the overall structure of the voltage-sensing domain and pore in a functional VGIC. They are either arranged in a domain-swapped or non-domain-swapped architecture. The main characteristic of domain-swapped architecture is that the VSD is juxtaposed to the neighboring pore domain and the structured S4–S5 linker connects these two domains, whereas in the non-domain-swapped channels, the VSD is placed right next to its own pore domain and is connected to its pore domain by a short unstructured linker (Long, Campbell, and Mackinnon 2005a; Whicher and MacKinnon 2016). Thus, there are two possible pathways for communication between the voltage sensor and the pore: one involving the structured S4–S5 linker and the other via the transmembrane interfaces between the S4 and S5 segments. For historical reasons, the allosteric pathways involving the structured S4–S5 linker are delineated as canonical pathways,

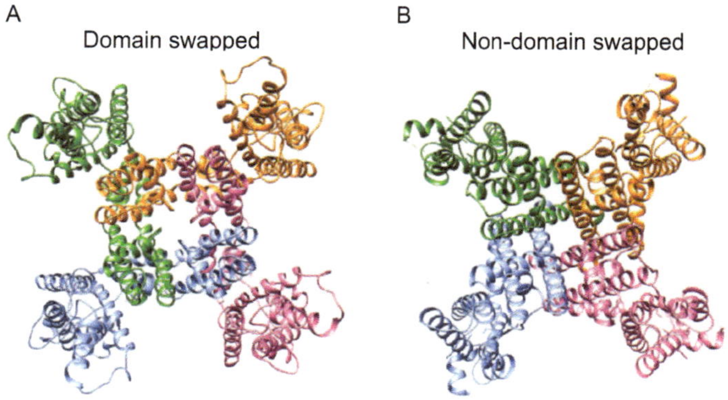

FIGURE 2.7
Arrangements of the voltage sensor and pore in VGICs. (A) Top view of the Kv1.2/2.1 paddle chimera in lipid nanodiscs showing a domain-swapped arrangement of the VSD and the pore. The four subunits are colored differently. The voltage sensor of a subunit forms intimate contacts with the pore domain (specifically the S5 helix) of the neighboring domain in a clockwise direction. (B) Top view of the HCN1 channel showing a non-domain-swapped arrangement of the VSD and the pore. In this case, each VSD is proximal to the pore domain of its own subunit.

whereas those involving direct communication between the membrane spanning S4 and S5 segments are loosely referred to as noncanonical pathways (Figure 2.7).

One of the strongest evidence regarding the role of the S4–S5 linker in the voltage-sensor pore coupling came from studies by Lu and coworkers who discovered that the bacterial potassium channel can be made voltage-dependent by fusing the voltage-sensing domain of the Shaker potassium channel (Lu, Klem, and Ramu 2001). They also found that the S4–S5 linker and distal parts of the S6 are crucial for conferring voltage dependence. Subsequent studies using site-directed mutagenesis and other approaches further supported the crucial role of the linker region and distal part of S6 in electromechanical coupling (Lu, Klem, and Ramu 2002). The structures of the domain-swapped voltage-gated ion channels show that the S4–S5 linker is a short helix that sits parallel to the membrane and is also in direct contact with the distal region of the S6 segment. One can envision a physical mechanism wherein a movement of the S4 voltage-sensing helix will be transmitted via the linker to the S6 helix that forms part of the pore gates (Long, Campbell, and Mackinnon 2005b).

A characteristic structural feature of the non-domain-swapped channels is that the S4–S5 linker helix is replaced by a short unstructured loop (Whicher and MacKinnon 2016). Insertion of a break or even deletion of this loop does not appear to have a significant effect on voltage gating (Lorinczi et al. 2015; Flynn and Zagotta 2018). Even in the Shaker channel as well as in the KCNQ channels, it has been shown that the interactions between S4 and S5 helices are involved in mediating voltage-sensor pore coupling (Li-Smerin, Hackos, and Swartz 2000; Fernandez-Marino et al. 2018; Hou et al. 2017, 2020). These interaction pathways mediated by direct interactions between the S4 and S5 helices are referred to as the noncanonical coupling pathway and are likely to be important in electromechanical coupling of non-domain-swapped voltage-gated ion channels.

Despite much progress in the past decade, our understanding of the mechanisms of electromechanical coupling remains far from complete. Although multiple structures of depolarization-activated ion channels in the resting and activated conformations are now available, many details regarding the mechanisms of electromechanical coupling remain to be worked (see Box 2.2). For instance, we do not have a clear sense as to the relative energetic contributions of the two major electromechanical coupling pathways in VGICs. Further investigation of other members of the VGIC family beyond the handful of prototypical channels will help build a more comprehensive view of the mechanisms of electromechanical coupling. Finally, compared to the depolarization-activated ion channels, our understanding of the forces and physical mechanisms of EM coupling in hyperpolarization-activated channels remains very limited.

2.6 Concluding Remarks

It is becoming increasingly evident that other processes involving membrane signaling also exhibit voltage dependence, which in some instances have physiological significance. Many G protein-coupled receptors (GPCRs), which are primarily involved in chemical signaling, are also regulated by voltage, which is likely to be mediated via sodium ion binding pocket deep in the transmembrane regions. While the voltage-sensing mechanisms discussed in this chapter focused on the VGICs, many of these principles and approaches are broadly applicable to many other membrane proteins and voltage-sensing molecules in living cells.

Acknowledgments

We are grateful to the National Institutes of Health for continuously funding our research. Due to space constraints, we were unable to cover many interesting aspects of voltage gating, but we hope that some of them will be introduced in more specialized sections of this book series. We also owe a debt to the giants in the field whose original writings inspired some of the material in this chapter.

Suggested Readings

This chapter includes additional bibliographical references hosted only online as indicated by citations in blue color font in the text. Please visit https://www.routledge.com/9780367538163 to access the additional references for this chapter, found under "Support Material" at the bottom of the page.

Aggarwal, S. K., and R. MacKinnon. 1996. "Contribution of the S4 segment to gating charge in the Shaker K+ channel." *Neuron* 16 (6):1169–77.

Almers, W. 1978. "Gating currents and charge movements in excitable membranes." *Rev Physiol Biochem Pharmacol* 82:96–190.

Armstrong, C. M. 1981. "Sodium channels and gating currents." *Physiol Rev* 61 (3):644–83. doi:10.1152/physrev.1981.61.3.644.

Armstrong, C. M., and F. Bezanilla. 1973. "Currents related to movement of the gating particles of the sodium channels." *Nature* 242 (5398):459–61.

Bezanilla, F. 2000. "The voltage sensor in voltage-dependent ion channels." *Physiol Rev* 80 (2):555–92. doi:10.1152/physrev.2000.80.2.555.

Carter, P. J., G. Winter, A. J. Wilkinson, and A. R. Fersht. 1984. "The use of double mutants to detect structural changes in the active site of the tyrosyl-tRNA synthetase (Bacillus stearothermophilus)." *Cell* 38 (3):835–40. doi:10.1016/0092-8674(84)90278-2.

Chanda, B., and F. Bezanilla. 2002. "Tracking voltage-dependent conformational changes in skeletal muscle sodium channel during activation." *J Gen Physiol* 120 (5):629–45. doi:10.1085/jgp.20028679.

Chanda, B., and F. Bezanilla. 2008. "A common pathway for charge transport through voltage-sensing domains." *Neuron* 57 (3):345–51. doi:10.1016/j.neuron.2008.01.015.

Chowdhury, S., and B. Chanda. 2012. "Estimating the voltage-dependent free energy change of ion channels using the median voltage for activation." *J Gen Physiol* 139 (1):3–17. doi:10.1085/jgp.201110722.

Chowdhury, S., B. M. Haehnel, and B. Chanda. 2014. "A self-consistent approach for determining pairwise interactions that underlie channel activation." *J Gen Physiol* 144 (5):441–55. doi:10.1085/jgp.201411184.

Guo, J., W. Zeng, Q. Chen, C. Lee, L. Chen, Y. Yang, C. Cang, D. Ren, and Y. Jiang. 2016. "Structure of the voltage-gated two-pore channel TPC1 from Arabidopsis thaliana." *Nature* 531 (7593):196–201. doi:10.1038/nature16446.

Hagiwara, S. 1966. "Membrane properties of the barnacle muscle fiber." *Ann N Y Acad Sci* 137 (2):1015–24.

Hamill, O. P., A. Marty, E. Neher, B. Sakmann, and F. J. Sigworth. 1981. "Improved patch-clamp techniques for high-resolution current recording from cells and cell-free membrane patches." *Pflugers Arch* 391 (2):85–100. doi:10.1007/BF00656997.

Hodgkin, A. L., and A. F. Huxley. 1952e. "A quantitative description of membrane current and its application to conduction and excitation in nerve." *J Physiol* 117 (4):500–44.

Humphries, J., L. Xiong, J. Liu, A. Prindle, F. Yuan, H. A. Arjes, L. Tsimring, and G. M. Suel. 2017. "Species-independent attraction to biofilms through electrical signaling." *Cell* 168 (1–2):200–9 e12. doi:10.1016/j.cell.2016.12.014.

Islas, L. D., and F. J. Sigworth. 2001. "Electrostatics and the gating pore of Shaker potassium channels." *J Gen Physiol* 117 (1):69–89.

Jensen, M. O., V. Jogini, D. W. Borhani, A. E. Leffler, R. O. Dror, and D. E. Shaw. 2012. "Mechanism of voltage gating in potassium channels." *Science* 336 (6078):229–33. doi:10.1126/science.1216533.

Kasimova, M. A., D. Tewari, J. B. Cowgill, W. C. Ursuleaz, J. L. Lin, L. Delemotte, and B. Chanda. 2019. "Helix breaking transition in the S4 of HCN channel is critical for hyperpolarization-dependent gating." *Elife* 8. doi:10.7554/eLife.53400.

Lee, C. H., and R. MacKinnon. 2017. "Structures of the human HCN1 hyperpolarization-activated channel." *Cell* 168 (1–2):111–20 e11. doi:10.1016/j.cell.2016.12.023.

Liao, M., E. Cao, D. Julius, and Y. Cheng. 2013. "Structure of the TRPV1 ion channel determined by electron cryo-microscopy." *Nature* 504 (7478):107–12. doi:10.1038/nature12822.

Long, S. B., E. B. Campbell, and R. Mackinnon. 2005a. "Crystal structure of a mammalian voltage-dependent Shaker family K+ channel." *Science* 309 (5736):897–903. doi:10.1126/science.1116269.

Lorinczi, E., J. C. Gomez-Posada, P. de la Pena, A. P. Tomczak, J. Fernandez-Trillo, U. Leipscher, W. Stuhmer, F. Barros, and L. A. Pardo. 2015. "Voltage-dependent gating of KCNH potassium channels lacking a covalent link between voltage-sensing and pore domains." *Nat Commun* 6:6672. doi:10.1038/ncomms7672.

Lu, Z., A. M. Klem, and Y. Ramu. 2001. "Ion conduction pore is conserved among potassium channels." *Nature* 413 (6858):809–13. doi:10.1038/35101535.

Mannikko, R., F. Elinder, and H. P. Larsson. 2002. "Voltage-sensing mechanism is conserved among ion channels gated by opposite voltages." *Nature* 419 (6909):837–41. doi:10.1038/nature01038.

Mannuzzu, L. M., M. M. Moronne, and E. Y. Isacoff. 1996. "Direct physical measure of conformational rearrangement underlying potassium channel gating." *Science* 271 (5246):213–16. doi:10.1126/science.271.5246.213.

Seoh, S. A., D. Sigg, D. M. Papazian, and F. Bezanilla. 1996. "Voltage-sensing residues in the S2 and S4 segments of the Shaker K+ channel." *Neuron* 16 (6):1159–67.
Starace, D. M., E. Stefani, and F. Bezanilla. 1997. "Voltage-dependent proton transport by the voltage sensor of the Shaker K+ channel." *Neuron* 19 (6):1319–27.
Xu, Y., Y. Ramu, and Z. Lu. 2008. "Removal of phospho-head groups of membrane lipids immobilizes voltage sensors of K+ channels." *Nature* 451 (7180):826–9. doi:10.1038/nature06618.
Yang, N., and R. Horn. 1995. "Evidence for voltage-dependent S4 movement in sodium channels." *Neuron* 15 (1):213–18.
Zagotta, W. N., T. Hoshi, and R. W. Aldrich. 1994. "Shaker potassium channel gating. III: evaluation of kinetic models for activation." *J Gen Physiol* 103 (2):321–62.
Zheng, J., and W. N. Zagotta. 2000. "Gating rearrangements in cyclic nucleotide-gated channels revealed by patch-clamp fluorometry." *Neuron* 28 (2):369–74. doi:10.1016/s0896-6273(00)00117-3.

3

Ligand-Dependent Gating Mechanism

William N. Zagotta

CONTENTS

3.1 Introduction

Ligand-gated ion channels are the chemosensors of the brain. Their essential role is to transduce changes in the concentration of a chemical into changes in membrane potential. In this role, they are responsible for the cell's response to neurotransmitters and neuromodulators (e.g., acetylcholine (ACh) and glutamate (Glu)), intracellular second messengers (e.g., Ca^{2+} and cAMP), cellular metabolites (e.g., ATP) and signaling lipids (e.g., PIP_2). The changes in membrane potential they produce may generate (or inhibit) action potentials, release a hormone, contract a muscle or activate a lymphocyte. Ligand-gated ion channels are the most rapid link between a cell and its environment.

Ligand-gated channels are best classified by the structural family in which they belong. Figure 3.1 shows a list of some of the major ligand-gated channel families with some representative members. Later chapters will discuss these channels in more detail. Structural families are grouped by the type of ligand they bind. The 2-TM P-loop containing channels bind some intracellular ligands such as ATP and G-protein beta-gamma subunits, while the 6-TM P-loop channels (voltage-gated family) bind different intracellular ligands such as Ca^{2+} and cAMP. Both P-loop families contain members that are regulated by PIP_2. Extracellular ligand-gated channels are separated into three families: the glutamate receptor family which binds glutamate; the cys-loop family which binds ACh, γ-aminobutyric acid (GABA), glycine and serotonin; and the trimeric ligand-gated channel family which binds ATP (extracellular) and protons. Within each family the channels can vary in many other properties such as ion selectivity, regulation by other factors (e.g. voltage), and the cell and location on the cell in which they act.

DOI: 10.1201/9781003096214-4

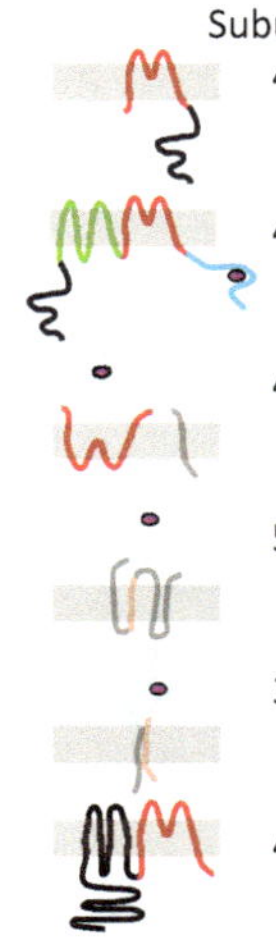

Subunits	
4	2-TM P-loop containing channels (K_{ATP}, KIR, GIRK)
4	6-TM Voltage-gated channels (BK, CNG, HCN, TRP, KCNQ)
4	Glutamate Receptor (AMPA, Kainate, NMDA)
5	Cys-loop Channels (AChR, GABAa, Glycine Receptor, 5-HT_3 Receptor)
3	Trimeric ligand-gated channels (ASICs, ENaC, P2X)
4	IP3 Receptor, Ryanodine Receptor

FIGURE 3.1
A list of some of the major ligand-gated channel families with some representative members.

3.2 Energetics of Ligand Gating

Despite the differences between the ligand-gated channel families, ligand gating shares several mechanistic features across all channels. The first, and maybe the most important, is that ligands almost always regulate the channels by changing the probability that the channel is open, not by changing the current that flows through an open channel. In other words, ligands effect gating not permeation. This is illustrated for a single cyclic nucleotide-gated (CNG) channel in Figure 3.2 (Sunderman and Zagotta, 1999). In the absence of ligand, these channels are closed almost all the time (open probability about 1 ´ 10^{-6}). However, upon addition of the agonist cGMP to the intracellular face of the channels, their open probability increases. The open probability increases steadily with increasing cGMP concentration until eventually it saturates at a probability near one. However, the single-channel current is unaffected by the cGMP concentration as seen in the amplitude histograms in Figure 3.2 (right). This behavior, seen in almost all ligand-gated channels, reflects the fact that the binding of ligand to the channel regulates a conformational change between the closed and open states of the channel, and that the open conformation is the same for different ligand concentrations. For most channels, ligand binding promotes channel opening and the ligand is referred to as an agonist (e.g., cGMP on CNG channels). For some channels, however, the ligand inhibits opening (promotes closing) and is referred to as a reverse agonist (e.g., ATP on K_{ATP} channels). For the rest of this chapter, we will consider only ligand-activated channels, but the same principles apply to ligand-inhibited channels.

This effect of ligand binding on the open probability can be viewed from an energy perspective. The progress of the conformational change from closed to open can be described by a plot of the free energy as a function of the reaction coordinate for the closed to open conformational change as shown in Figure 3.3A. The low points on the free energy profile represent the stable closed and open conformations of the channel, and the high point represents the high energy transition state for the conformational changes between closed and open. According to Boltzmann's law, the ratio of probability of a channel being in the

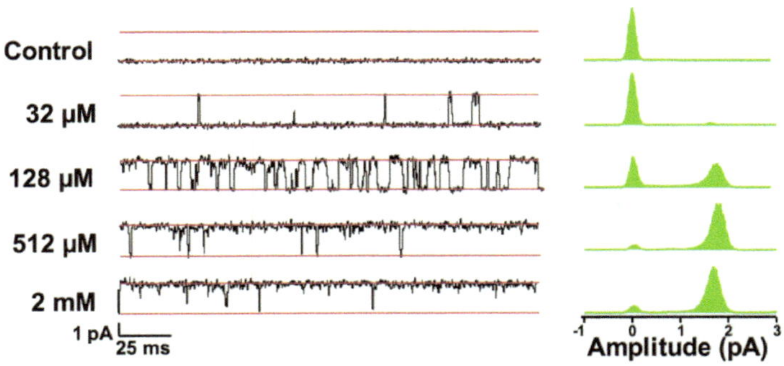

FIGURE 3.2
Single CNGA1 channels recorded in an inside-out patch in the presence of the indicated concentrations of cAMP. Amplitude histograms are shown on the right. (From Sunderman and Zagotta, 1999.)

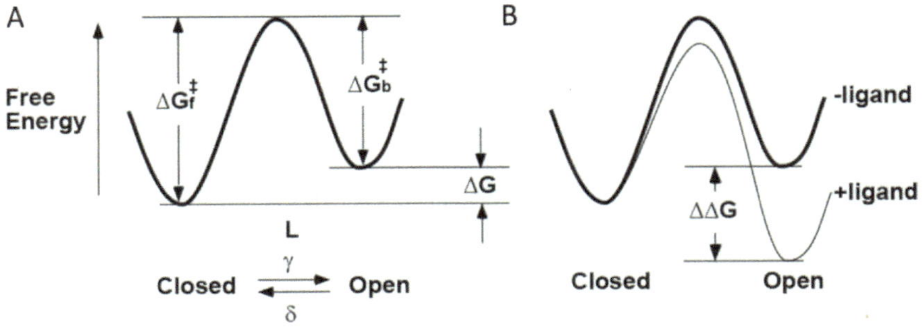

FIGURE 3.3
Energetics of ligand gating. (A) Hypothetical free energy profile of a closed to open transition. (B) Effect of ligand on the free energy profile.

open state (P_O) to the probability of it being in the closed state (P_C) *at equilibrium* is an exponential function of the free energy difference between the states (ΔG):

$$\frac{P_O}{P_C} = L = e^{-\frac{\Delta G}{kT}} \tag{3.1}$$

where L is the equilibrium constant for the conformational change, k is Boltzmann's constant and T is absolute temperature. In other words, when the energy difference between the states is on the order of thermal energy (kT), the channel will exist in both conformations at equilibrium. At the single-molecule level, this appears as the channel chattering between closed and open conformations. If ΔG is positive, the channel spends most of its time closed. If ΔG is negative, the channel spends most of its time open. A ligand changes the open probability by simply changing this free energy profile (Figure 3.3B). Most ligands decrease ΔG (stabilizing the open state) and therefore increase the probability of the channel being open.

The rates of opening (γ) and closing (δ) depend on the free energy difference between the closed state and the transition state $\left(\Delta G_f^!\right)$, and the open state and the transition state ($\Delta G_b^!$), respectively, as given by Eyring rate theory:

$$\gamma = Ae^{-\frac{\Delta G^{!}_{f}}{kT}} \tag{3.2}$$

$$\delta = Ae^{-\frac{\Delta G^{!}_{b}}{kT}} \tag{3.3}$$

where A represents the frequency factor (the rate if there was no transition state) (Figure 3.3A). The frequency factor can vary widely for different reactions, but for a first-order reaction it is typically 10^{-11} to 10^{-12} s^{-1}. Note that by calculating the ratio of the rate constants before and after some perturbation, the frequency factor cancels out and one can calculate the change in $\Delta G^{\ddagger}$ ($\Delta \Delta G^{\ddagger}$) produced by the perturbation. Because the transition state generally has a structure intermediate between the open and closed states, an activating ligand generally partially stabilizes the transition state, speeding channel opening relative to no ligand. If the stabilization of the transition state is less than the stabilization of the open state (relative to the closed state), then the ligand also slows channel closing relative to no ligand. The speeding of channel opening and slowing of channel closing by the agonist increases the open probability of the channel, the hallmark of a ligand-activated channel.

Another general principle of ligand gating (indeed all allosteric modulation) is that ligands exhibit state-dependent binding. For a ligand that promotes channel opening, the ligand binds with higher affinity to the open state than to the closed state. This is a simple consequence of thermodynamics: if ligand binding promotes opening, then opening promotes ligand binding. This means that channels, like all allosteric proteins, actually have two (or more) binding affinities. As discussed later, it is very difficult to measure the affinities of the ligand for a particular state separate from the conformational transitions between the states.

3.3 Steady-State Properties of Ligand Gating

Ion channels behave as molecular switches, opening or closing a channel pore in response to changes in the concentration of a ligand. To understand this behavior, let's start by considering a two-state model with an unbound closed state and an agonist-bound open state:

$$C \overset{K_A}{\underset{}{\overset{A}{\rightleftharpoons}}} O$$

Scheme 3.1

For this model, it can easily be shown that, at equilibrium,

$$P_O = \frac{[A]K_A}{1+[A]K_A} = \frac{[A]}{K_D+[A]} \tag{3.4}$$

where $[A]$ is the free concentration of the agonist, K_A is the association equilibrium constant (in units of M^{-1}) and K_D is the dissociation equilibrium constant (in units of M). This equation is the Langmuir isotherm used to describe the saturating concentration dependence of any bimolecular interaction. The equation displays a hyperbolic relationship between

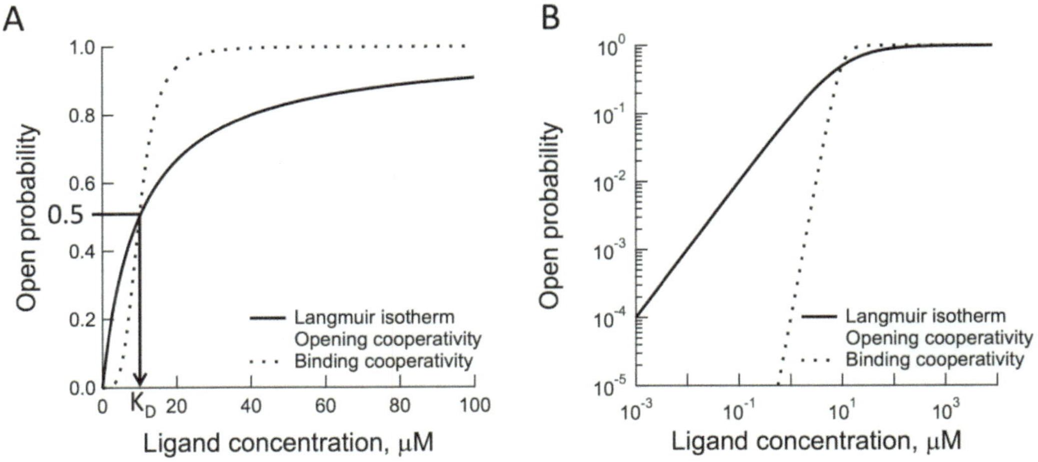

FIGURE 3.4
Theoretical steady-state dose–response curves for a simple two-state model, a model with opening cooperativity and a model with binding cooperativity. (A) Linear axes. (B) Log–log axes.

P_O (or fraction bound) and the free agonist concentration $[A]$ (Figure 3.4A, solid line). The apparent affinity of the interaction, the concentration that produces half-maximal opening (or binding) is simply the K_D. When plotted on log–log axes, this equation shows a limiting slope at low agonist concentrations of one (x-fold change in $[A]$ produces x-fold change in P_O) (Figure 3.4B).

3.4 Cooperativity

Channels almost always exhibit more complex behavior than predicted by a simple Langmuir isotherm. One of the complexities arises from the fact that most channels are multimers, composed of three to five subunits (Figure 3.1). Because some or all of the subunits usually bind ligand, channels have the potential to exhibit cooperativity. Cooperativity is a loosely used term that usually means that the dose–response relation or the binding curve has a limiting slope of greater than one. There are two kinds of cooperativity that are found in channels: opening cooperativity and binding cooperativity. Most channels exhibit opening cooperativity, meaning that the binding of more than one agonist is required to open the channel (more precisely, opening of the channel is superlinearly dependent on the number of agonist molecules bound). A simple example is shown in Figure 3.5A where the agonist binds independently and identically to each of four subunits, but the channel only opens when all four agonists are bound. Because the multiple configurations of binding one, two or three agonists are considered identical, the equilibrium constants in the model are weighted accordingly. For example, the equilibrium constant for binding the first agonist is $4K_A$ because any one of four subunits can bind agonist for the first transition, but only one subunit can unbind to make the reverse transition. For this model, the equilibrium probability of being open by binding n agonists is given simply by the nth power of the probability of binding to a single subunit:

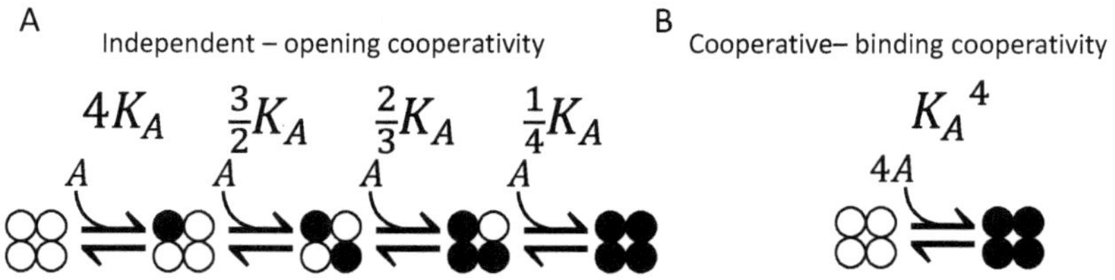

FIGURE 3.5
Models for gating of tetrameric channels. (A) Model with opening cooperativity. (B) Model with binding cooperativity.

$$P_O = (P_A)^n = \left(\frac{[A]}{K_D + [A]}\right)^n \quad (3.5)$$

This model for ligand gating is similar to the independent gating particles proposed by Hodgkin and Huxley (1952) for voltage-gated channels. Opening cooperativity makes the opening of the channel more steeply dependent on the agonist concentration, behaving more like a molecular switch (Figure 3.4, dotted trace). This is particularly apparent in log–log dose–response plots where the limiting slope at low agonist concentrations is four, with an x^4-fold change in P_O for an x-fold change in $[A]$ (Figure 3.4B, dotted trace). Note, however, that, because the binding is independent, binding curves would still exhibit a limiting slope of one (x-fold change in $[A]$ produces x-fold change in bound A).

The second kind of cooperativity is binding cooperativity. Binding cooperativity means that the bind of one agonist affects the binding of subsequent agonists. Binding cooperativity can arise from two general mechanisms. One way is if the binding of agonist produces a local conformational change (induced fit) that affects the binding of the next agonist, usually to a neighboring subunit. This mechanism was first proposed by Koshland, Nemethy and Filmer (KNF) in 1966 (Koshland et al., 1966). In the context of the model in Figure 3.5A, the KNF mechanism means that the equilibrium constants no longer conform to the values predicted for independent binding (for a tetramer: $4K_A$, $3K_A/2$, $2K_A/3$, $K_A/4$). If the equilibrium constants get successively bigger, then the binding exhibits positive cooperativity, and if the equilibrium constants get successively smaller, then the binding exhibits negative cooperativity (only the KNF mechanism will produce negative cooperativity). The other general mechanism for binding cooperativity is that the binding of each agonist promotes a quaternary rearrangement in the protein that increases the binding affinity of all the sites. This mechanism was proposed by Monod, Wyman, and Changeux (MWC) in 1965 (Monod et al., 1965) and will be discussed in more detail later. For both mechanisms, binding cooperativity causes binding curves to exhibit a limiting slope greater than one (assuming positive cooperativity).

The most extreme form of binding cooperativity is produced by the Hill model (Figure 3.5B) in which all of the agonists bind at once. This model has both opening and binding cooperativity. The equation that describes the equilibrium open probability for the model in Figure 3.5B is the Hill equation:

$$P_O = \frac{[A]^n}{K_{0.5}^n + [A]^n} \quad (3.6)$$

where $K_{0.5}$ is the concentration that produces half-maximal activation and n is the Hill coefficient. Notice that Equation 3.6 is similar to Equation 3.5 except that $K_{0.5}^n + [A]^n$ replaces $(K_D + [A])^n$. While the cooperative opening model and the cooperative binding model predict the same limiting slopes in the dose–response relationship (Figure 3.4B), the slope at intermediate ligand concentrations is steeper for cooperative binding (Figure 3.4A). Furthermore, the binding curve for the Hill model also has a steep slope (for binding, the limiting slope at low agonist concentration is n). This is the hallmark of binding cooperativity.

3.5 Separate Ligand Binding and Opening Transitions

Another complexity in the ligand activation of channels is that, unlike in Scheme 3.1, for channels the ligand binding step is separate from the channel opening step. The ligand binding step involves a docking of the ligand with a binding site on the channel, perhaps with a concurrent induced fit conformational change in the binding site. The channel opening step, however, involves a conformational change in the protein that opens a pore region located at a distance from the ligand binding site. This ability of the ligand to produce a change at a distance is the property of allostery, and this conformational change is frequently referred to as the allosteric transition. In fact, ligand-gated channels were one of the first allosteric proteins to be studied. In 1957, working on acetylcholine-gated (AChR) channels, Del Castillo and Katz were the first to write the binding step and the conformation change as two separate steps:

$$C \underset{}{\overset{K_A}{\underset{A}{\rightleftharpoons}}} C \overset{L}{\rightleftharpoons} O \qquad \text{Scheme 3.2}$$

where K_A is the association equilibrium constant for binding, and L is the equilibrium constant for opening of the fully bound channel. The equilibrium open probability for Scheme 3.2 is given by the following equation:

$$P_O = \frac{[A]K_A L}{1+[A]K_A+[A]K_A L} \tag{3.7}$$

One of the big differences between Scheme 3.1 and Scheme 3.2 is that the maximum open probability (at saturating agonist concentration) for Scheme 3.2 is not one, but is given by the following equation:

$$P_{O\,max} = \frac{L}{1+L} \tag{3.8}$$

Plots of Equation 3.7 for different values of L are shown in Figure 3.6. When L is large (>>1), $P_{O\,max}$ is near one. However, when L is small, $P_{O\,max}$ is less than one.

The separation of the binding step from the opening step has two important ramifications. The first is that it is very difficult to differentiate effects of perturbations, such as mutations, on ligand binding versus channel opening. When L>>1 (as it frequently is), then changes in L cause a shift in the dose–response relation indistinguishable from changes in K_A (Figure 3.6). This can also be seen by normalizing P_O by $P_{O\,max}$, a common practice in ion

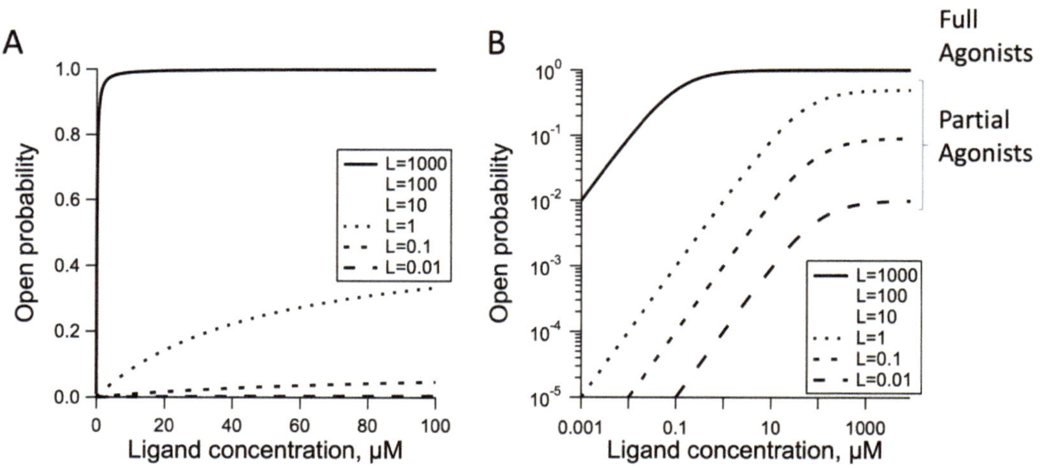

FIGURE 3.6
Theoretical steady-state dose–response curves for a model with separate binding and opening transitions with different values of *L*. (A) Linear axes. (B) Log–log axes.

channels because P_O is usually not measured directly but measured indirectly from the conductance, which is proportional to P_O.

$$\frac{P_O}{P_{O\,max}} = \frac{[A]K_A(1+L)}{1+[A]K_A(1+L)} = \frac{[A]K_{app}}{1+[A]K_{app}} \tag{3.9}$$

where

$$K_{app} = K_A(1+L) \tag{3.10}$$

Equation 3.9 shows that plots of $P_O/P_{O\,max}$ versus $[A]$ will conform to a Langmuir isotherm (Equation 3.4) with an apparent association equilibrium constant K_{app} (Equation 3.10). Since K_{app} depends on both K_A and L, it is impossible from such a plot to tell if a change in K_{app} resulted from a change in K_A or a change in L. Surprisingly measuring ligand binding instead of channel opening does not get around this problem. The probability of the agonist being bound for Scheme 3.2 is given by the same equation as for $P_O/P_{O\,max}$ (Equation 3.9) with the same apparent association equilibrium constant K_{app} (Equation 3.10). This paradox has been the cause of numerous errors where a mutation of a channel (or other protein) that caused a change in the apparent affinity of a ligand was erroneously thought to be in the binding site for the ligand (for review, see Colquhoun, 1998).

One way to differentiate mutations in the ligand binding site from mutations that affect channel opening is by measuring the ligand dependence of the alteration. If a mutation has a different energetic effect on two different ligands that bind to the same site, then it is highly likely that the mutant residue resides in the binding site and interacts with the ligand in a portion where the ligands differ. For example, mutation of D604 in the CNGA1 channel causes a decrease in the apparent affinity for cGMP but an increase in the apparent affinity for cAMP (Varnum et al., 1995). From this, it was concluded that D604 interacts directly with the cyclic nucleotide in the portion of the purine ring where cGMP and cAMP differ (Figure 3.7). This conclusion was later confirmed with X-ray crystallography

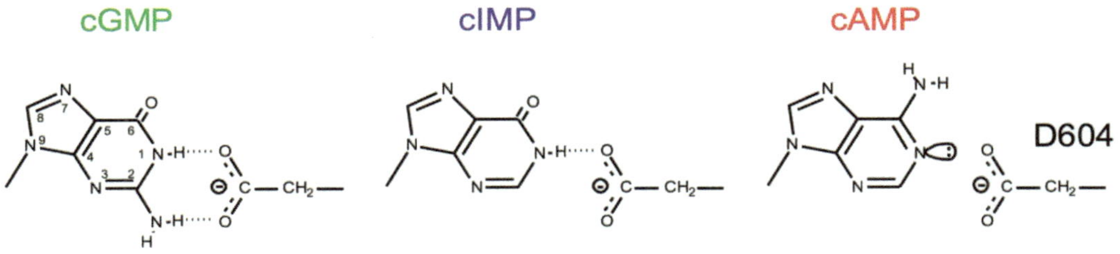

FIGURE 3.7
Predicted interactions of D604 of CNGA1 with the purine ring of cGMP, cIMP and cAMP. (From Varnum et al., 1995.)

on a related channel (Flynn et al., 2007). Since the effects of the mutation were primarily on the $P_{O\,max}$, it appears that the interaction of the cyclic nucleotide with D604 is one that forms primarily during the allosteric transition (see later).

3.6 Partial Agonists

The second ramification of the separation of the binding transition from the opening transition (Scheme 3.2) is that sometimes the ligand does not fully activate the channel even at saturating concentrations. For some channels or ligands or mutants, L is small (<10) so that $P_{O\,max}$ is appreciably less than one (Figure 3.6 and Equation 3.8). These ligands are referred to as partial agonists to differentiate them from full agonists where $P_{O\,max}$ is near one. Desensitization or inactivation can also cause a decreased open probability at saturating ligand concentrations as discussed in Chapter 5.

Partial agonists are a very powerful tool for measuring the properties of the allosteric conformational change. At saturating ligand concentrations, one can isolate just the conformational transitions in the fully liganded channel. Figure 3.8 shows an example of partial agonists for CNG channels. While the open probability in the presence of saturating cGMP is about 0.95, it is about 0.5 for the related agonist cIMP, and only about 0.01 for cAMP. The value of L can then be calculated using Equation 3.8 ($L_{cGMP} = 20$, $L_{cIMP} = 1$, $L_{cAMP} = 0.01$). Finally, the free energy difference between the closed and open states ($\Delta\, G$) can be calculated for each of the cyclic nucleotides using Equation 3.1 ($\Delta\, G_{cGMP} = -1.8$ kcal/mol, $\Delta\, G_{cIMP} = 0$ kcal/mol, $\Delta\, G_{cAMP} = 2.7$ kcal/mol). With the assumption that the different ligands induce a similar conformational change in the protein (albeit with different stabilities), the differences in these $\Delta\, G$ values should reflect structural differences between how the ligands interact with the binding site during the allosteric transition.

Partial agonists offer another way to differentiate alterations in ligand binding from alterations in channel opening. In the context of Scheme 3.2, if a mutation changes $P_{O\,max}$, it must reflect changes in L and therefore changes in the allosteric transition. This is the basis for phi analysis for mapping regions of the channel involved in the allosteric transition (Auerbach, 2003). In general, regions that effect L have been identified throughout the structure of channel proteins including the ligand binding site, the pore and almost everywhere in between. Conversely changes in apparent affinity without changes in $P_{O\,max}$ for a partial ligand must reflect changes in K_A, presumably due to alterations in the ligand binding site itself.

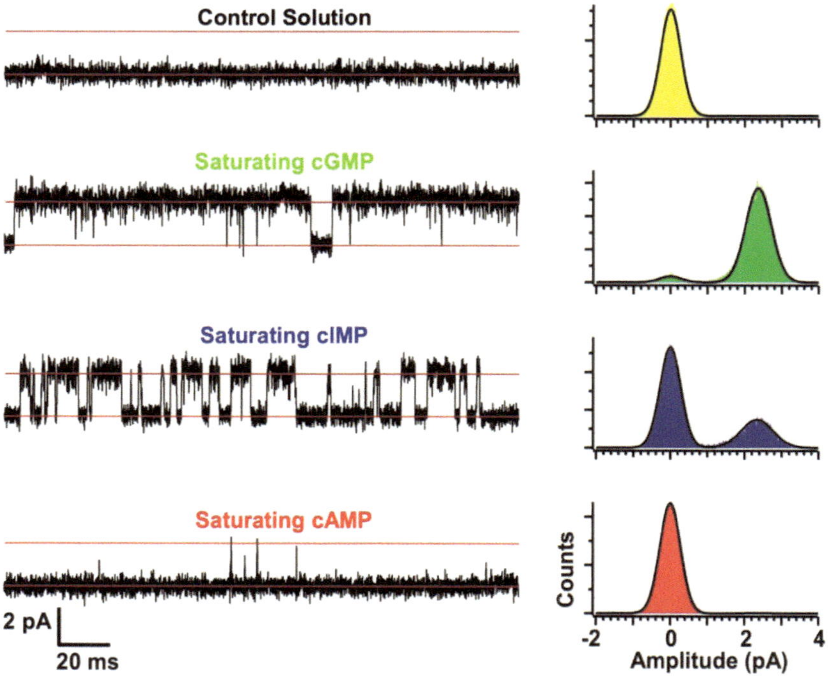

FIGURE 3.8
Single CNGA1 channels recorded in an inside-out patch in the presence of cGMP, cIMP and cAMP. Amplitude histograms are shown at the right. (From Sunderman and Zagotta, 1999.)

3.7 MWC Model

A more general model for ligand-dependent activation that incorporates separate binding and opening events is shown in Figure 3.9A. This model explicitly considers the possibility that the channel can open both without ligand bound (the apo state) and with ligand bound. The opening from the apo state has an equilibrium constant of L_0, while the opening from the ligand bound state has an equilibrium constant of fL_0. For a ligand that activates the channel, f is greater than one, and opening from the ligand bound state is more favorable by a factor of f than opening from the apo state. This is equivalent to saying that the binding of the ligand stabilizes the open state relative to the closed state ($\Delta\Delta G$) by $-RT\ln(f)$ kcal/mol. This $\Delta\Delta G$ must come from new or strengthened interactions (or weakened negative interactions) between the ligand and the channel binding site during the allosteric transition.

To preserve microscopic reversibility, a more favorable opening with ligand bound also indicates that ligand binding to the open state (fK_A) is more favorable than ligand binding to the closed state (K_A). In other words, the simple cyclic allosteric model in Figure 3.9A illustrates that ligand-gated channels have at least two different affinities K_A and fK_A. The ligand dependence of the open probability of this model will still conform to a Langmuir isotherm with an apparent affinity intermediate between K_A and fK_A depending on the value of L_0 (Figure 3.9B). For low values of L_0 (<1/f) the apparent affinity is similar to K_A, while for large values of L_0 (>1) the apparent affinity is close to fK_A.

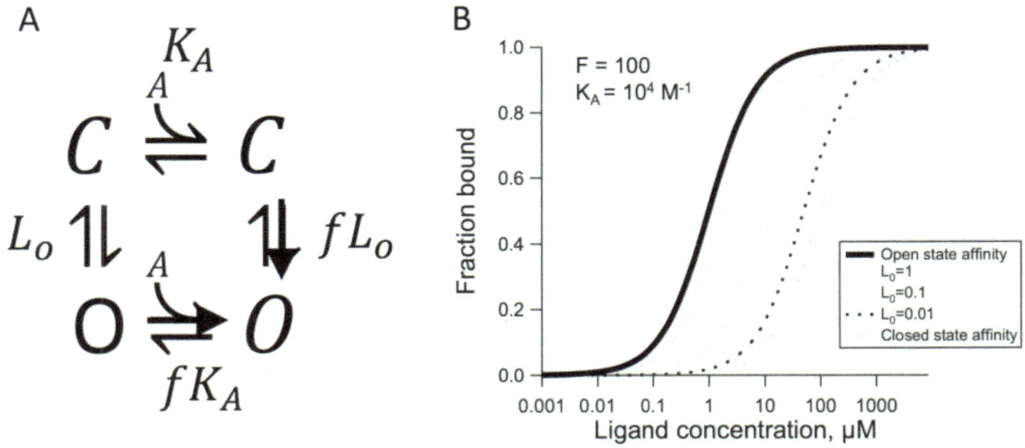

FIGURE 3.9
Separate but coupled binding and opening transition. (A) Cyclic model. (B) Theoretical steady-state binding curves for different values of L_0 compared to the closed state and open state binding curves.

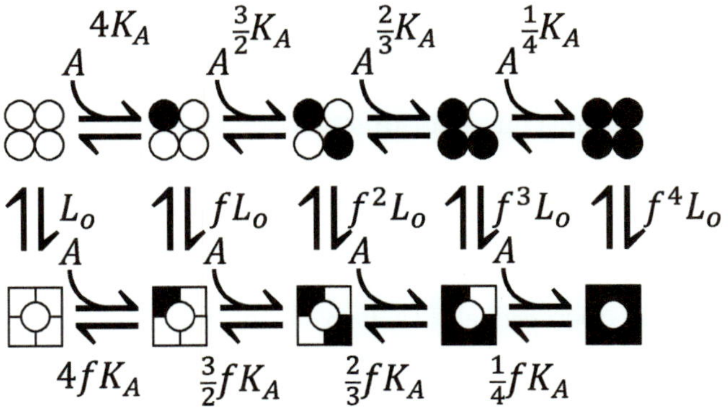

FIGURE 3.10
MWC model for a tetrameric ion channel.

Perhaps the most widely used model for allostery in multimeric proteins in general, and ion channels in particular, is the MWC model (Monod et al., 1965) (Figure 3.10). This model is essentially a combination of the independent binding model in Figure 3.5A and the cyclic allosteric model in Figure 3.9A. It supposes that the protein can exist in two different conformational states; for channels these are closed and open. Within each state, the ligands bind independently and identically. Importantly, the binding of each ligand promotes the opening allosteric transition by a factor of *f*. Therefore, opening with one ligand bound is *f*-fold more favorable than opening with no ligands, opening with two ligands is *f*-fold more favorable than opening with one ligand, etc. This produces opening cooperativity – the channel opens more favorably when multiple ligands are bound. Similarly, ligand binding to the open state is *f*-fold more favorable than ligand binding to the closed state. The binding of ligands to the low affinity closed state promotes a transition to the high affinity open state. Therefore, the MWC model also exhibits binding cooperativity.

3.8 Macroscopic Gating Kinetics

The preceding discussion describes the steady-state properties of ligand-gated ion channels. To understand the molecular mechanisms and physiological role of the channels, it is also important to understand their kinetic properties. Kinetics describes the time course of reactions. For ion channels, the time course of the open probability is simply proportional to the time course of the current at a constant voltage (voltage clamp). Electrophysiology is perhaps the most sensitive assay for any protein, able to measure time courses that span eight orders of magnitude (10 μs to 1000 s) and channel numbers that range from thousands of channels to just a single channel.

Once again, let's start by considering a two-state model:

$$C \underset{\beta}{\overset{A\alpha}{\rightleftharpoons}} O \qquad \text{Scheme 3.3}$$

where α is the bimolecular rate constant for ligand binding (units of $M^{-1}s^{-1}$), and β is the unimolecular rate constant for ligand unbinding (units of s^{-1}). The equilibrium dissociation constant K_D is given by the following equation:

$$K_D = \frac{\beta}{\alpha} \tag{3.11}$$

It can be shown that the time course for this two-state model can be described by the following equation:

$$P_O(t) = P_O(\infty) + \left(P_O(0) - P_O(\infty)\right)e^{-t/\tau} \tag{3.12}$$

where $P_O(0)$ is the initial open probability, $P_O(\infty)$ is the equilibrium open probability and τ is the time constant of relaxation (the time for the open probability to change by *e*-fold of the total change). For Scheme 3.3:

$$P_{O\infty} = \frac{[A]}{[A] + \beta/\alpha} \tag{3.13}$$

and

$$\tau = \frac{1}{\alpha[A] + \beta} \tag{3.14}$$

In words, what Equation 3.12 says is that, in a two-state model, the open probability will relax to equilibrium with a single-exponential time course with a time constant that depends equally on both the opening and the closing rate constants. For a given [*A*], the time constant will be the same whether the channels are opening from a low probability or closing from a high probability. According to Equation 3.14, the time constant will decrease hyperbolically with the increasing agonist concentration, with an affinity of β/α (K_D) (Figure 3.11A). Plots of 1/τ versus [*A*] will be linear with a slope of α and a y-intercept of β (Figure 3.11B). In this way one can get the individual rate constants from the ligand dependence of τ.

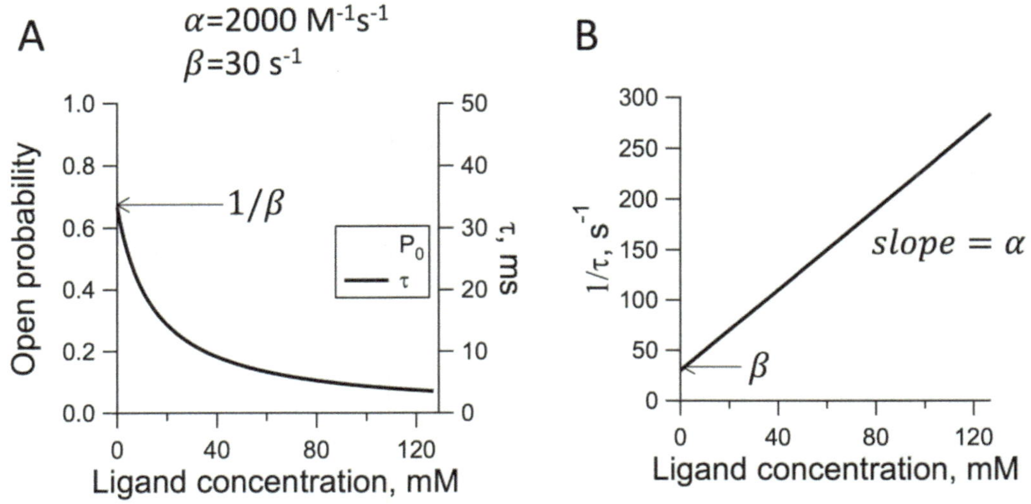

FIGURE 3.11
Macroscopic kinetics for a two-state model. (A) Concentration dependence of the open probability and the time constant. (B) Linear dependence of $1/\tau$ on the ligand concentration.

The time course of the open probability for a three-state model is described by a double exponential function with two time constants.

$$P_O(t) = P_O(\infty) + F_1 e^{-t/\tau_1} + F_2 e^{-t/\tau_2} \tag{3.15}$$

where the time constants τ_1 and τ_2 depend on the rate constants between the states. The amplitudes of the exponentials, F_1 and F_2 depend on both the rate constants and the initial conditions (the probability of being in each state at time zero). In general, any model with N states will be described by $N - 1$ macroscopic time constants. The time constants only depend on the rate constants (and ligand concentration for a ligand-gated channel). Importantly, each time constant, in general, will be a function of some or all of the rate constants. There is not a one-to-one association between the time constants and the transitions. Therefore, it is a mistake to assign any particular time constant to a particular transition except under special circumstances.

3.9 Single-Channel Gating Kinetics

The rate constants for particular transitions can be determined more directly from an analysis of single-channel currents. An example is shown for AChR channels in Figure 3.12A (Purohit et al., 2007). Single-channel currents record the moment a channel makes a conformational change from closed to open and the moment it makes a conformational change from open to closed. Therefore, the durations of the open and closed events reflect the rates of the transitions. Since single-channel behavior (indeed any single-molecule behavior) is stochastic, the rate constants are determined from histograms of the open and closed durations (Figure 3.12B). These duration histograms are exponentially distributed

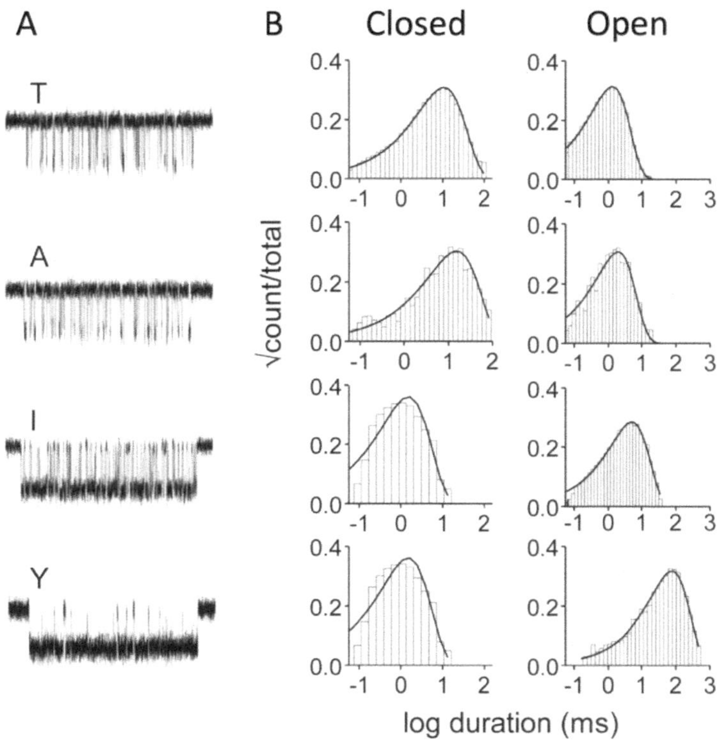

FIGURE 3.12
Single-channel recordings of AChR channels containing T254, T254A, T254I or T254Y in the pore region of the α subunit. (A) Single-channel traces recorded with saturating concentrations of the partial agonist choline. (B) Closed and open duration histograms fit with single-exponential functions. (From Purohit et al., 2007.)

where the time constants are a function of a subset of the opening and closing rate constants. They are frequently plotted on the square root of the number of events versus the log duration axes where each exponential component of the histogram produces a separate peak at the time constant for that component (Figure 3.12B). In general, the open and closed distributions will be multiexponential where the number of time constants is equal to the number of open and closed states, respectively. Once again, these time constants are generally a function of multiple rate constants, so cannot be assigned to a particular transition except under special circumstances.

One of those special circumstances is seen in some ligand-gated channels at saturating concentrations of ligand. As predicted by Scheme 3.2, at saturating $[A]$ some ligand-gated channels will transition between just a single closed state and a single open state:

$$C \underset{\delta}{\overset{\gamma}{\rightleftharpoons}} O \qquad \text{Scheme 3.4}$$

Under these conditions, the closed durations will be single-exponentially distributed with a time constant (τ_c) dependent on only the opening rate constant (γ):

$$\tau_C = \frac{1}{\gamma} \tag{3.16}$$

And the open durations will be single-exponentially distributed with a time constant (τ_o) dependent on only the closing rate constant (δ):

$$\tau_O = \frac{1}{\delta} \tag{3.17}$$

Therefore, under this condition, the rate constants can be calculated directly from the time constants. This behavior is observed in AChR channels at saturating concentrations of either the full agonist acetylcholine or the partial agonist choline (Figure 3.12) (Purohit et al., 2007).

3.10 Phi Analysis

Measuring the effects of mutations on the opening and closing rate constants is the basis for a very powerful method for studying the allosteric transition called phi (Φ) analysis (also called Brönsted analysis, linear free energy relationship (LFER) analysis or rate-equilibrium free energy relationship (REFER) analysis). This method has been most rigorously applied to AChR channels (Purohit et al., 2007). Phi analysis involves measuring the effect of mutations at a particular site on the energy of the transition state relative to the effect on the energy of the open state. From the rate constants of the opening and closing transitions, one can calculate the energy change of the transition state ($\Delta G^{!}_{f}$) using Equation 3.2

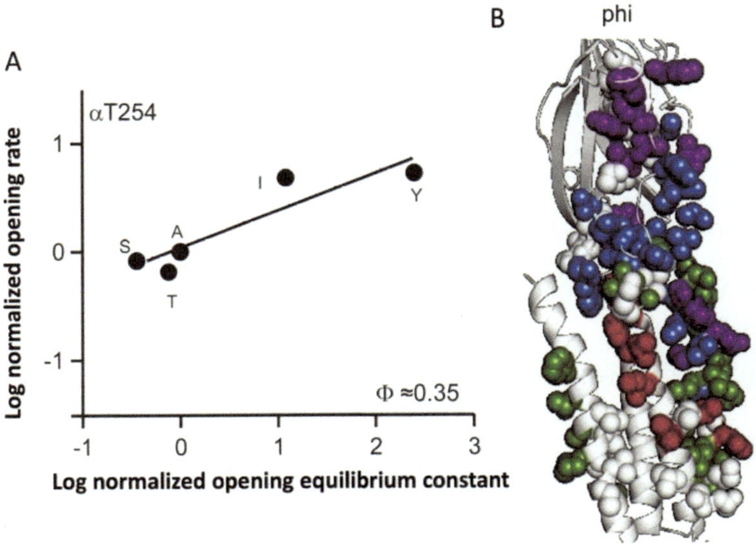

FIGURE 3.13
Phi analysis of AChR. (A) Plot of log normalized opening rate versus log normalized equilibrium constant for a number of mutations of T254 in the pore region of the α subunit. The slope of the plot (Φ) is 0.35. (From Purohit et al., 2007.) (B) Map of the Φ values for a number of mutations in the α subunit plotted on the structure of AChR. Similar Φ values are colored as follows: purple (0.85–1), blue (0.7–0.85), green (0.5–0.7) and red (0.25–0.35). (From Auerbach, 2013.)

and the energy change of the open state (ΔG) using Equation 3.1. A plot of $\Delta G_f^!$ versus ΔG (or simply log γ versus log L) for a number of mutations at a particular site is frequently linear with a slope Φ (Figure 3.13A). Φ indicates the fraction of the energetic effect of the mutations on the open state that has occurred at the time of the transition state. To the extent that the transition state conformation is intermediate between the open state and the closed state, Φ indicates the open state-like nature of the transition state, with a Φ of one indicating similarity to the open state and a Φ of zero indicating similarity to the closed state.

Different amino acids in the channel will have different Φ values. These Φ values can be mapped onto the three-dimensional structure of the channel (Figure 3.13B). One interpretation of these Φ maps is that they represent the sequence of events that occur during the allosteric transition between binding of the ligand and opening of the channel. For the AChR channel, Φ values near the acetylcholine binding site are near one, while Φ values in the transmembrane segments are near zero (Purohit et al., 2007). This would seem to indicate a conformational wave associated with ligand gating that starts in the ligand binding site and ends at the gate in the transmembrane segments. Interestingly, the Φ values tend to cluster, suggesting that some regions of the channel move as a unit, referred to as a gating module or a nanotectonic plate. While other interpretations are possible, the broad range of regions of the channel involved in the transition and their non-equivalency is clear from these maps.

Suggested Readings

Auerbach, A. 2003. "Life at the top: the transition state of AChR gating." *Sci STKE* 2003:re11. doi:10.1126/stke.2003.188.re11.

Auerbach, A. 2013. "The energy and work of a ligand-gated ion channel." *J Mol Biol* 425:1461–75. doi:10.1016/j.jmb.2013.01.027.

Colquhoun, D. 1998. "Binding, gating, affinity and efficacy: the interpretation of structure-activity relationships for agonists and of the effects of mutating receptors." *Br J Pharmacol* 125:924–47.

Del Castillo, J., and B. Katz. 1957. "Interaction at end-plate receptors between different choline derivatives." *Proc R Soc Lond B Biol Sci* 146:369–81.

Flynn, G. E., K. D. Black, L. D. Islas, B. Sankaran, and W. N. Zagotta. 2007. "Structure and rearrangements in the carboxy-terminal region of SpIH channels." *Structure* 15:671–82. doi:10.1016/j.str.2007.04.008.

Hodgkin, A. L., and A. F. Huxley. 1952. "A quantitative description of membrane current and its application to conduction and excitation in nerve." *J Physiol* 117:500–44.

Koshland, D. E., Jr., G. Nemethy, and D. Filmer. 1966. "Comparison of experimental binding data and theoretical models in proteins containing subunits." *Biochemistry* 5:365–85.

Monod, J., J. Wyman, and J. P. Changeux. 1965. "On the nature of allosteric transitions: a plausible model." *J Mol Biol* 12:88–118.

Purohit, P., A. Mitra, and A. Auerbach. 2007. "A stepwise mechanism for acetylcholine receptor channel gating." *Nature* 446:930–3. doi:10.1038/nature05721.

Sunderman, E. R., and W. N. Zagotta. 1999. "Mechanism of allosteric modulation of rod cyclic nucleotide-gated channels." *J Gen Physiol* 113:601–20.

Varnum, M. D., K. D. Black, and W. N. Zagotta. 1995. "Molecular mechanism for ligand discrimination of cyclic nucleotide-gated channels." *Neuron* 15:619–25.

4

Mechanosensitive Channels and Their Emerging Gating Mechanisms

Sergei Sukharev and Andriy Anishkin

CONTENTS

4.1 Introduction

Cells constantly generate and experience mechanical stresses of different scales for many reasons. For example, cells generate mechanical stress as part of progression through the cell cycle, cytoskeletal contractility and motility, tissue differentiation and growth, as well as through metabolic activity that creates an osmotic imbalance. Cells also perceive external forces characteristic of their environments, which are converted into intracellular signals that trigger cell-specific responses. This conversion is implemented by several types of primary mechanotransducers.

DOI: 10.1201/9781003096214-5

In this chapter we focus specifically on mechanosensitive (MS) ion channels and their basic mechanisms, though MS channels are not the only type of mechanotransducers in cells. Other mechanotransducers include molecules that connect the cytoskeleton to focal adhesions. Talin, for instance, responds to linear stress by partially unfolding and exposing cryptic binding sites that recruit other proteins, thus leading to remodeling of stress-bearing complexes (Wolfenson et al., 2019). Kinases associated with focal adhesions also respond to forces mediating cell remodeling and differentiation. Bacterial osmosensory kinase EnvZ perceives hydration stresses – apparently through changes in ionic strength and cytoplasmic macromolecular crowding – forcing a more compact "active" conformation (Foo et al., 2015). G protein-coupled receptors have also been shown to mediate stretch responses in membranes (Yasuda et al., 2008).

MS channels are distinctive mechanotransducers because they reside in membranes separating different compartments and directly respond to stresses in the lipid bilayer or in associated cytoskeletal or extracellular elements. They use the energy of preexisting ionic or solute gradients to amplify and convert mechanical forces into electrical or chemical signals. Because they are directly opened by force without any preceding chemical steps, MS channels are the fastest mechanotransducers. Mammalian auditory channels, for instance, respond to hair cell bundle displacement within 10 microseconds (Abeytunge et al., 2021), allowing us to hear a wide range of frequencies and enjoy music. Tactile receptors in the skin are also characterized by exceptionally short latency and a variety of adaptive behaviors producing transient (phasic) or sustained (tonic) signals (Handler and Ginty, 2021), allowing us to discern details of surface softness and texture. Bacterial mechanosensitive channels acting as fast release valves reduce osmolyte gradients and hydrostatic pressure in small cells subjected to abrupt osmotic shock within ~30 milliseconds (Cetiner et al., 2017), thereby preventing osmotic damage.

In contrast to voltage-gated channels, MS channels do not represent a cohesive structural family with a characteristic membrane topology and conserved signature sequences exemplified by voltage-sensor domains. MS channels are found in all clades of organisms and belong to several structurally unrelated families. Instead of being structurally unified, they are functionally unified by a significant increase of open probability in response to applied external force. The force reaches the gate either through the lipid bilayer or attached cytoskeletal or extracellular elements. In all cases, the channel complexes "comply" with the applied force in order to undergo an opening transition. More precisely, the energy input of the external force biases the probability distribution between the resting and conductive states.

In the next section we will provide a brief historic overview of the field emphasizing the extreme diversity of MS channels and challenges this poses. In the following sections, we will present some unifying principles and details of force-dependent gating of specific channels.

4.2 Diversity of MS Channels

Progress in the field is based on the rich functional and structural data obtained in a variety of organisms such as *Escherichia coli* and other bacteria, *Chlamydomonas reinhardtii, Caenorhabditis elegans, Drosophila melanogaster, Arabidopsis thaliana, Danio rerio,*

Mus musculus, *Homo sapiens* and several others. The field benefits from the fact that many MS channels are homomultimeric protein complexes, and heterologous expression in convenient cell systems often preserves key electrophysiological traits. Some channels also permit functional reconstitution from pure protein and lipid, a system that not only defines the "minimal" set of components needed for function but also allows investigators to probe the role of the lipid environment (Moe and Blount, 2005). Some mechanosensory complexes, such as the auditory channel, involve many different protein components and do not function as pure or individually expressed proteins. Functional cooperation between different types of MS channels has also been observed in sensory systems, which may suggest partial redundancy and thus complicate interpretations of knock-out or knock-down experiments. Fast accumulation of structural information with the cryo-EM technique in the past few years drives precise structure-based experiments and increases the predictive power of molecular dynamics simulations. Besides mechanistic questions about novel classes of MS channels (Piezo, OSCA/TMEM, MSA, SWELL) and newly identified members of existing families (MscS-like Flycatcher), an equally important part of the field now addresses multiple functional roles of MS channels in normal physiology and pathology, cell differentiation, tissue and organ development, and cancer progression (Syeda, 2021).

4.2.1 The Auditory Mechanotransduction Channel

Studies of the hearing process, the finest example of mechanosensitivity, focus on cochlear auditory hair cells with apical bundles of stereocilia interconnected with tip links. The ionic nature of the deflection-induced hair cell receptor potential was first reported by Corey and Hudspeth (1979), establishing that the primary auditory mechanotransducer must be an ion channel. The channel was located at the tips of stereocilia near the points of tip link insertion (Beurg et al., 2009). Its cationic conductance ranges from 100 to 200 pS and the channel, when active, passes the 611 Da FM1-43 dye into the tip of the stereocilium. The conductance and kinetics of adaptation in these transduction channels change with the position in the cochlea according to the frequency to which they are most sensitive (Ricci et al., 2003). Molecular identification of transduction channels has been the major challenge. The tetraspan membrane protein of hair cell stereocilia (TMHS) and transmembrane channel-like proteins TMC1 and TMC2 have been cloned through analysis of mutations causing deafness (Kawashima et al., 2011; Longo-Guess et al., 2005). TMC1 now appears to be a pore component of the auditory channel complex co-assembling with several other proteins at the tips of stereocilia (Pan et al., 2018). TMIE, a small TMC-associated protein was also predicted to be a part of the pore (Cunningham et al., 2020). TMC1 and TMC2 proteins share homology with representatives of the TMEM16/OSCA family (see Section 4.2.5), which function as dimers of 10 TM subunits (Figure 4.1a). The recent cryo-EM study has visualized the *C. elegans* TMC1 complex and identified two tightly associated proteins: TMIE, a single-membrane span protein; and a Ca^{2+} and integrin-binding protein CALM-1 attached to the cytoplasmic side of the complex (Jeong et al., 2022). The ion-conductive pathway in TMC1 is identified as the slit between helices 6, 7 and 8, partially exposed to lipids and covered by TMIE. Tip links, the critical extracellular elements connecting adjacent stereocilia, are formed by dimers of protocadherin 15 and cadherin 23 interconnected by Ca bridges (Sotomayor et al., 2012). Bundle displacement produces stress in the tip links and thus opens the channels. More about the function of the auditory transduction channel will be discussed in Section 4.5.1.

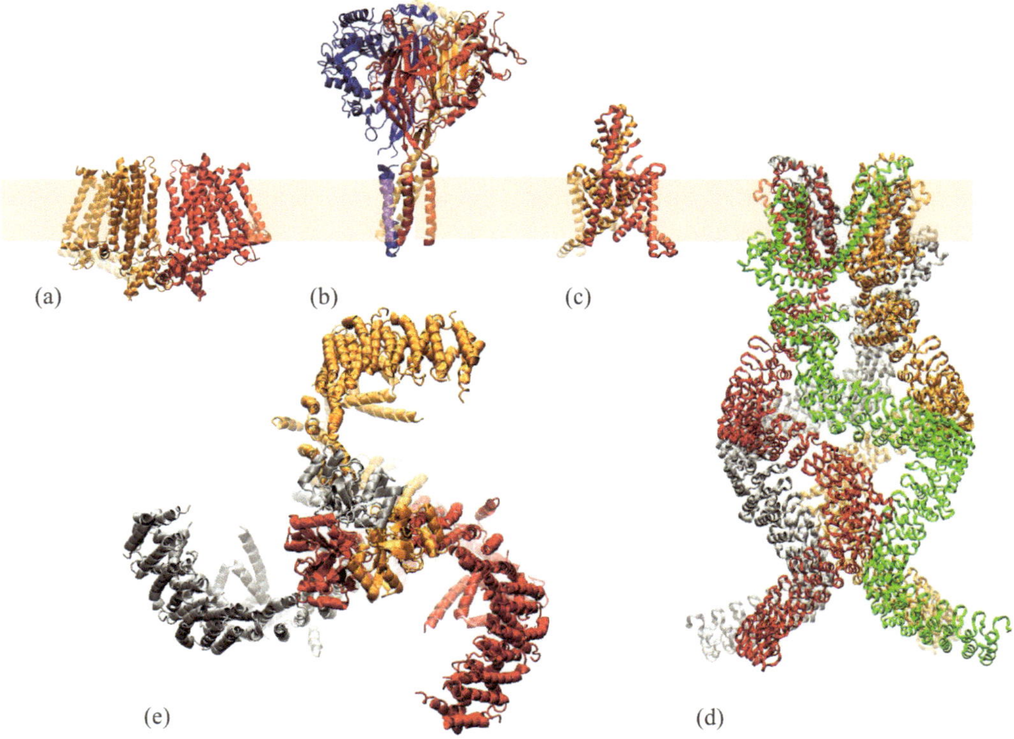

FIGURE 4.1
Structures of representatives from five major families of mechanosensitive channels. (a) OSCA1.2 mechanosensitive channel (PDB ID 6MGW) representing the TMEM16/OSCA/ TMC family. (b) Acid-sensitive ASIC channel (PDB ID 6AVE), a member of DEG/MEC/ENaC family. (c) TRAAK mechanosensitive channel (PDB ID 4WFE) representing the family of two-pore potassium (K2P) channels. (d) NompC mechanosensitive channel (PDB ID 5VKQ) representing the transient receptor potential (TRP) family. (e) Piezo1 channel shown as a top view (PDB ID 6B3R) from the Piezo family.

4.2.2 Phenomenological Patch-Clamp Studies of MS Channels in Non-Sensory Cells

The first direct observation of activation of cationic channels in chicken myoblasts by stretching a patch membrane by suction was reported by Guharay and Sachs (1984). Soon after, MS channel activities were reported in bacteria (Martinac et al., 1987), yeast (Gustin et al., 1988), amphibians (Zhang et al., 2000) and plants (Cosgrove and Hedrich, 1991). The attempts to apply pharmacology to the characterization of mechano-activated currents revealed that gadolinium (Gd^{3+}) ions, amiloride, aminoglycoside antibiotics, ruthenium red and *Grammostola* spider peptide GsMTx4 block different subsets of MS channels (Hamill and McBride, 1996; Bae et al., 2011). However, no universal agent that could reliably block and "tag" the enigmatic MS channel molecules for isolation, the way tetrodotoxin binds to neuronal NaV channels, has been reported. Cloning and structural studies now drive searches for pharmacological agents directed toward specific classes of MS channels.

4.2.3 The DEG/ENaC/MEC Family

The first molecular candidates for the role of MS channels came from genetic screens in the nematode *C. elegans* (Chalfie and Sulston, 1981), which revealed 18 mec genes (mechanosensory abnormal) necessary for sensing light touch mediated by the worm's six specialized "touch" neurons. Twelve mec genes encode for components responsible specifically for mechanotransduction, which include tubulins (mec-7, mec-12), extracellular matrix elements (mec-5, mec-9), paraoxonase-like (mec-6) proteins, cholesterol recruiting stomatin-like (mec-2) proteins, and mec-4 and mec-10 channel-like proteins from the DEG/ENaC channel family (Arnadottir and Chalfie, 2010). These channel proteins have 2-TM topology with cytoplasmic N- and C-termini and a large extracellular domain that mediates subunit trimerization (Figure 4.1b). Electrophysiological recordings in these miniature worm neurons were challenging, but they have shown that out of these two genes mec-4 is indispensable for transduction (Arnadottir et al., 2011). In *C. elegans* and *Drosophila* mechanotransduction channels are predicted to be connected with the microtubular structures inside the cell and with the cuticle outside, apparently to collect the stress from a larger "receptive field." Other members of the family, acid-sensitive ion channel (ASIC-1) and DEG-1 in particular, are essential for mechanotransduction in other mechanosensory neurons in *C. elegans* (Geffeney and Goodman, 2012). In *Drosophila*, bristle receptors and multidendritic sensory neurons rely on the presence of ENaC-like Pickpocket (Ppk) proteins (Adams et al., 1998).

4.2.4 Bacterial Channels

Preparations of giant bacterial spheroplasts enabled exploration of bacteria with patch-clamp electrophysiology and immediately led to the discovery of prokaryotic MS channels (Martinac et al., 1987). Bacterial MS channels proved highly amenable to experimentation in terms of abundance, biochemistry, ease of reconstitution and initial crystallographic analysis. Patch-clamp surveys of the *E. coli* cytoplasmic membrane and of reconstituted proteo-liposomes (Berrier et al., 1996; Sukharev et al., 1993) revealed three major phenotypical classes of MS channels differing in conductance and activating pressure: large (MscL, 3 nS), small (MscS, 1 nS) and mini (MscM, 100–300 pS). The large-conductance mechanosensitive channel MscL was identified and cloned through biochemical fractionation and reconstitution (Sukharev et al., 1994). The mechanosensitive channel of small conductance MscS was cloned by homology to the potassium efflux protein KefA (Levina et al., 1999), which turned out to be a K-dependent MS channel now called MscK (Li et al., 2002). Subsequently, analysis of the *E. coli* genome revealed four more MscS paralogs: YnaI, YbdG, YbiO and YjeP. The mini-channel activities were attributed to the products of ynaI, yjeP and ybdG genes (Edwards et al., 2012). Together, these bacterial channels constitute an adaptive system that adjusts turgor pressure in the cell by releasing small metabolites from the cytoplasm in the event of osmotic downshift. The two structurally unrelated channels, MscS and MscL, are the major release valves rescuing bacteria from lysis, and thus only the deletion of both genes produced an osmotically fragile *E. coli* phenotype (Levina et al., 1999).

MscL and MscS were purified, functionally reconstituted with lipids, and shown to gate directly by tension in the lipid bilayer (Moe and Blount, 2005; Sukharev, 2002). Structures of a MscL homolog from *Mycobacterium tuberculosis* and *E. coli* MscS were solved crystallographically (Steinbacher et al., 2007) and became model systems for mechanistic studies of channel gating by tension. While MscL-type channels are confined primarily to bacteria and archaea, many MscS homologs have been found in essentially all organisms with walled cells including fission yeast, alga, flagellates and higher plants (Balleza and Gomez-Lagunas, 2009). MscS-like channels are present not only in the plasma membranes of fungi, protists and plants, but also in internal membranes and membranes of organelles where they stabilize plastids, mitochondria and contractile vacuoles (Hamilton et al., 2015). Structures and mechanisms of MscL and MscS will be discussed in Section 4.5.

4.2.5 MS Channels in Plants

The first patch-clamp surveys of plant protoplasts prepared from guard, epidermal and mesenchymal cells revealed a variety of currents activated with pressure and carried by Cl^-, K^+ or Ca^{2+} (Cosgrove and Hedrich, 1991; Qi et al., 2004). Epidermal channels permeable to Ca^{2+} and Mg^{2+} were blocked by micromolar Gd^{3+}, the ion that abolished gravitropic responses in plant roots (Ding and Pickard, 1993). Genomic data shows that the higher plant *Arabidopsis thaliana* possess ten homologs of bacterial MscS (MSL1-10) (Hamilton et al., 2015). Two of these channels, MSL2 and MSL3, were localized to plastids. Elimination of these two genes lead to morphological abnormalities and swelling of plastids, which experience hyperosmotic stress due to sugar production. A similar regulation of shape and division of chloroplasts by MSC1, a MscS homolog in unicellular alga *Chlamydomonas reinhadtii* (Fujiu et al., 2011), confirmed the critical role of MscS-like channels in the maintenance of endosymbiotic organelles. MSL9 and MSL10 were located in the plasma membrane of Arabidopsis root cells, whereas MSL8 is expressed specifically in pollen where it assists in normal germination and pollen tube growth (Hamilton et al., 2015). Flycatcher (FLYC1), a MscS homolog in the carnivorous Venus flytrap plant, is located in the specific indented regions of prey-sensing hairs that trigger the action potential and the closure of the trap (Procko et al., 2021).

A. thaliana also possesses five two-pore domain potassium channels (TPK1-5) with the mechanosensitive TPK1 expressed in osmotically driven guard cells. A single Piezo channel (PZO1) and two unique MCA (1 and 2) Ca^{2+}-permeable channels were shown to perceive the hardness of the soil and assist in deeper root penetration (Hamilton et al., 2015). A family of novel OSCA channels (Figure 4.1a) is involved in mediating Ca^{2+} influx under hyperosmotic stress (Murthy et al., 2018). OSCA channels belong to a broad TMEM16/TMEM63 family of eukaryotic channels and transporters that also includes TMCs. Interestingly, the genomes of *A. thaliana* and other land plants completely lack transient receptor potential (TRP)-like channels. In contrast to land plants, the green unicellular alga *Chlamydomonas reinhardtii* possesses several TRP channels; one of them, TRP11 is located at the base of the flagella and mediates the reversal of flagella beating upon cell collision with a hard object (Fujiu et al., 2011).

4.2.6 Two-Pore Potassium (K2P, TPK) Channels

When the first genomic databases for eukaryotic organisms were completed, new classes of channels such as two-pore potassium (K2P) channels were discovered. The TWIK channel (tandem of P domains in a weak inwardly rectifying K+ channel) was described first (Lesage et al., 1996). K2Ps constitute a family of ubiquitous leakage channels that set the resting potential in excitable and non-excitable cells. Among them, TREK 1, 2 (TWIK-related K^+) and TRAAK (TWIK-related arachidonic acid-stimulated K^+) channels were activated by stretch, producing robust fast-adapting currents (Honore et al., 2006). TREK1 in particular is unique in that it responds to many different types of stimuli such as pH, temperature, polyunsaturated fatty acids, phospoinositide lipids and general anesthetics. For this reason, it is called multimodal (Honore, 2007).

The successful reconstitution of human TRAAK and zebrafish TREK-1 (Brohawn et al., 2014) demonstrated that gating of these channels is driven directly by tension in the lipid bilayer. The solved crystal structures of TRAAK (Figure 4.1c) reveal an unusual connectivity of the pore interior with the aliphatic core of the membrane. Lipids entering the inner cavity through "lateral openings" were proposed to act as a mechanically activated "gate" that closes the pore in the absence of tension (Brohawn et al., 2012). Other studies implicated neither the lipids nor the cytoplasmic ends of pore-lining helices in the gating of TREK-1 but rather a "C-type gate" working through rearrangements in the selectivity filter (Bagriantsev et al., 2011).

K2P channels are highly selective for K^+ and are therefore hyperpolarizing and inhibitory. As a part of the "leakage" channel population, K2P's roles include regulation of excitability by adjusting the resting potential setting receptor sensitivity thresholds, and neuroprotection from adverse factors such as ischemia and concussion (Honore, 2007). Not surprisingly, inhibitory TREK channels are pharmacological targets for general anesthetics and antidepressants.

4.2.7 TRP Channels

The highly diverse transient receptor potential (TRP) family of channels serve multiple functions related to the sensation of temperature, nociception and pain, taste, and olfaction among others. It also participates in regulation of intracellular Ca^{2+}, determination of the redox status of the cell, and osmo- and mechanosensation (Gees et al., 2012). Several representatives have been suggested to be mechanosensitive; they belong to the subfamilies of TRPC (canonical), TRPA (ankyrin domain-containing), TRPV (vanilloid), TRPN ("no mechanoreceptor potential") and TRPP (related to polycystic disease). The yeast vacuolar channel TRPY1, resembling the putative ancestor of the TRP family, is directly sensitive to membrane tension (Zhou et al., 2003). The osmosensitive TRPV4 channel activated by cell swelling also activates in patches under applied tension (Loukin et al., 2010). TRPN1 (NompC), initially identified in *Drosophila*, is implicated in sensing bristle deflections, vibrations by the specialized ciliated cells in the chordotonal organ as well as in proprioception. Its cryo-EM structure (Figure 4.1d) shows an extended N-terminal cytoplasmic domain made of ankyrin repeats (Jin et al., 2017). Function of TRPN1 in *Drosophila* critically depends on the presence of this N-terminal domain where ankyrin repeats directly associate with microtubule (Liang et al., 2013). Molecular dynamics simulations suggested that not the pulling but rather the pushing of ankyrin domains into the membrane domain results in channel activation (Wang et al., 2021). TRP-4, a TRPN1 homolog in *C. elegans*, was shown to be the transduction channel generating fast electrical responses in ciliated mechanosensory neurons (Geffeney and Goodman, 2012).

In mammals, Ca^{2+}-permeable channels TRPP1 (aka PC2, Pkd2 or polycystin 2) are located at the base of the primary cilium on the apical side of kidney epithelia. Co-assembled with another membrane protein Pkd1, TRPP1 channels fulfill the function of flow sensors responding to cilia deflection with the influx of Ca^{2+} (Nauli et al., 2003). Pkd2 was also shown to play a role in the early stages of vertebrate development setting left–right symmetry of the body (Delmas, 2004). Many TRPV channels exhibit multimodal activation by a variety of stimuli, thus making them "integrators" of mechanical stimuli with temperature, osmolarity and the presence of anionic lipids, polyunsaturated fatty acids or Ca^{2+}.

4.2.8 Volume-Regulated Anion Channels

Bestrophin-1 (BEST1) was initially identified as a site of mutation linked to Best vitelliform macular dystrophy. The protein was found to function as a Cl-selective channel activated by elevated Ca^{2+} and swelling (Kunzelmann, 2015). Mutations in this protein deregulate the volume of pigment cells in the retina that support rods and cones, and remove reactive oxygen species. The structure of BEST1 revealed a pentamer of 4TM subunits with five cytoplasmic helices forming the hollow inner vestibule (Figure 4.2a). BEST1 homologs are found in bacteria, fungi, worms, insects and vertebrates where they may function as sensors that couple intracellular Ca^{2+} concentration with cell volume.

The second family of volume-regulated anion channels (VRACs) is exemplified by LRRC8 proteins (LRRC8A-E). These channels mediate regulatory volume decrease by extruding inorganic anions and small organic osmolytes from cells in the event of hypoosmotic swelling. In humans, functional VRACs represent a heterogenic family of channels that are formed by co-assembly of five different paralogous subunits. LRRC8A (SWELL1) is the subunit that is required in all functional complexes (Syeda et al., 2016). The structures of homohexameric SWELL1 have been determined using cryo-EM (Saotome et al., 2018) and revealed that the hexamers are in fact trimers of dimers. The transmembrane barrel of SWELL1 is formed out of six 4TM subunits with the fold and outer loops similar to that of innexins (invertebrate connexins). Importantly, the C-terminal half of each subunit carries 16 leucine-rich repeats (LRRs) that fold into a hollow cage-like structure in the cytoplasm (Figure 4.2b). SWELL1 activates under a moderate osmotic gradient across the membrane and decreased ionic strength in the cytoplasm. Cell osmotic swelling is obligatorily accompanied by a reduction of concentration of all intracellular components including ions, small osmolytes and macromolecules. In Figure 4.2 we illustrate BEST1 and LRRC8A architecture together with bacterial MscS; all of them feature hollow cytoplasmic domains structurally connected to the gate. It seems plausible that not only the decrease of ionic strength but also the reduction of crowding pressure in the cytoplasm (a poorly explored parameter) could be a decisive activating factor, whereas an increase of crowding pressure due to dehydration should drive deactivation. Responsiveness to 3D pressure would undoubtedly unify these molecules in the category of mechanosensitive channels (see Section 4.3 below).

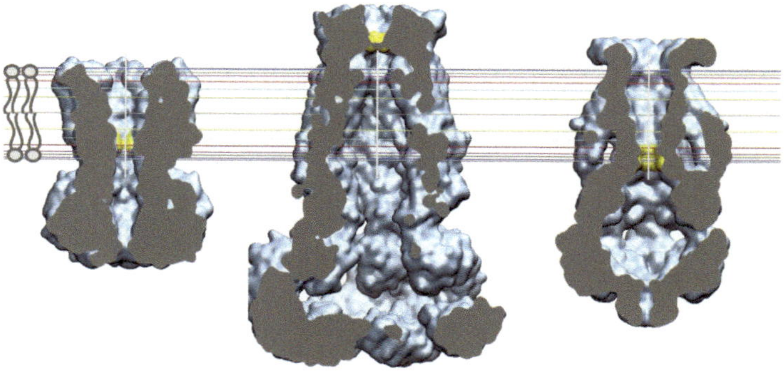

(a) (b) (c)

FIGURE 4.2
The swelling-activated channels characterized by hollow cytoplasmic vestibules with relatively narrow entrances impermeable to larger cytoplasmic components. All three structures are shown as vertical cross-sections of space-filled models to show internal volumes of the cytoplasmic domains. (a) Bestrophin-1 (PDB ID 6n24), (b) SWELL1 (PDB ID 6djb) and (c) bacterial MscS (PDB ID 6PWP). The horizontal lines delineate the position of the lipid membrane.

4.2.9 Piezo Channels

The Piezo family was discovered in 2010 by the Patapoutian group (Coste et al., 2010). Utilizing SiRNA knock-down technology combined with electrophysiology, they correlated the presence of transient mechano-activated currents in neuroblastoma cells with the product of the Fam 38A gene (now called Piezo1) previously associated with hereditary xerocytosis. The gene was found to be widely expressed in the bladder, colon, kidney, lung and skin. Piezo1 homologs are found in most eukaryotic clades, including protists, plants and metazoans. While most genomes contain a single homolog of Piezo1, vertebrates have a second paralog, Piezo2, that is highly expressed in the sensory dorsal root ganglion (DRG) neurons. Both Piezo1 and Piezo2 were shown to be nonselective Ca-permeable cationic channels of 20-45 pS that generate robust decaying currents in response to pressure steps applied to the patch membrane or to pocking the cell surface with a blunt glass probe. In patch-clamp experiments, membrane tension appears to be the main parameter that drives opening, with a midpoint estimated between 2 and 6 mN/m (Cox et al., 2016; Lewis and Grandl, 2015). In addition to tension, Piezo1 robustly activates by share stress created by fluid flow parallel to cell surface (Maneshi et al., 2018).

Both Piezo1 and Piezo2 proteins are ~2500 amino acids long, varying with the splice variant. The structures of Piezo1 and Piezo2 were solved by several groups (Guo and MacKinnon, 2017; Saotome et al., 2018; Wang et al., 2019), which revealed a similar trimeric architecture for these channels (Figure 4.1e). Each monomer is predicted to have 38 transmembrane helices organized as four-helix repeats. The C-terminal regions of each subunit contribute to the central pore that is covered by an extracellular cap and by intracellular central and lateral plug segments. The larger N-terminal parts of each monomer form membrane-embedded "arms" so the trimer resembles a curved three-blade propeller (Figure 4.1e). The trimeric complex thus imposes local curvature to the membrane forming a membrane dome bulging inside that is ~10 nm tall (out-of-plane) and has a ~24 nm in-plane diameter. Membrane tension is predicted to flatten the dome and produce the opening transition that can be associated with up to a 120 nm^2 in-plane expansion of the curved membrane-protein complex (Guo and MacKinnon, 2017) providing exceptional tension sensitivity (see discussion of the MS channel energetics in the next section).

Despite similar structure and gating properties, Piezo1 and Piezo2 are expressed in different sets of cells: Piezo1 are found primarily in nonsensory tissues, whereas Piezo2 function in sensory neurons and endings. Piezo1 and Piezo2 are expressed together only in some DRG neurons, articular chondrocytes (Lee et al., 2014), and aortic and carotid baroreceptors (Zeng et al., 2018), where apparently coordinated action of the two receptors is required. Piezo1 and Piezo2 knock-outs in mice are lethal, whereas tissue-specific knockouts produce severe sensory abnormalities as well as developmental defects. Mutations in Piezos are associated with at least 18 hereditary diseases including arthrogryposis, stomatocytosis and lymphatic dysplasia. For this reason, the channels have become important pharmacological targets. So far, the channels are nonspecifically blocked by Gd^{3+} ions and ruthenium red, whereas *Grammostola* venom peptide GsMTx4 acts as gating modulator shifting the tension-activation curve to the right in a concentration-dependent manner (Gnanasambandam et al., 2017). Flow-induced Piezo1 currents are not affected by GsMTx4, but effectively blocked by amphipathic Aβ peptides whose release is associated with brain trauma (Maneshi et al., 2018). Pharmacological screens have identified two specific Piezo1 activators Yoda1 and Jedi 1,2, whereas Dooku1 acts as a specific blocker. The action sites have been recently identified by cryo-EM studies. See Volume II, Chapter 26 on Piezo channels.

4.3 The Ways External Forces Are Conveyed to the Channel and the Energetics of Gating

There are several paradigms on how force can be applied to a membrane-embedded channel and how it may drive opening (Figure 4.3). The force may originate from cytoskeletal or extracellular tethers. In this case, the tether displacement accompanying the transition multiplied by linear force ($f \times \Delta l$) would be the "work" input that biases the energy landscape of the system toward the open state. This mode of gating is called *force from filaments* and implies a transition along the vector of the force (1D) which can either push or pull (Figure 4.3a).

Alternatively, external force can be conveyed through the surrounding lipid bilayer under stress (*force from lipid*; Figure 4.3b) and the energy-biasing input in this two-dimensional (2D) case will be tension multiplied by the in-plane area change associated with the transition ($\gamma \times \Delta A$). The recent structural characterization of curvature-imposing Piezo channels has introduced another principle of gating by tension, which involves flattening a curved membrane segment embedding the channel (*curvature-assisted gating*; Figure 4.3c) (Guo and MacKinnon, 2017; Young et al., 2022). In physics terms, this mechanism is similar to the 2D case, but the flattening of the membrane "dome" drastically increases the effective in-plane area change of the complex (ΔA), thus sensitizing the transition. The curvature-imposing arms of the Piezo channel likely act as levers passing the force from the flattened structure to the pore constriction (gate).

Channels can also be sensitive to the bulk pressure in the cytoplasm. Examples include the volume-activated channels described earlier. The entropic forces generated in the cytoplasm can be a result of increased or decreased concentration of impermeable cytoplasmic components acting on a hollow intracellular domain and can be unified as osmotic or crowding force (*osmosensitive gating*; Figure 4.3d). In this case the energy factor perturbing the equilibrium between the states is the pressure times volume change ($p \times \Delta V$) associated with the 3D transition.

Another mechanical factor acting on protruding extracellular domains of channel proteins pinned to the membrane is shear stress created by fluid flow (*shear stress or "sail" model*; Figure 4.3e), which reflects the action of viscous drag force. Gating induced by flow has been reported for epithelial sodium channels (ENaCs) (Carattino et al., 2004) and for Piezo channels (Maneshi et al., 2018).

In order to change the equilibrium between two functional states, a molecule should *comply* with the external force and change its geometry in that specific direction. This basic mechanochemical principle applies to MS channels irrespective to the dimensionality of the system (Corey and Hudspeth, 1983; Markin and Sachs, 2004). In all cases, a simplified two-state diagram illustrating the dependence of system energy on a spatial parameter (reaction coordinate) can represent the sensor molecule distributed between the inactive (closed) or active (open) states separated by energy G_0, with probabilities P_c and P_o, obeying the Boltzmann equation (Equation 4.1):

$$\frac{P_o}{P_c} = e^{-(G_0 - W_e)/kT} \tag{4.1}$$

In the absence of an external stimulus (W_e), the intrinsic energy bias G_0 shifts the equilibrium toward the resting state (P_c is high), while the open state is accessible with a low probability. The applied external force or tension makes the expanded (active) conformation

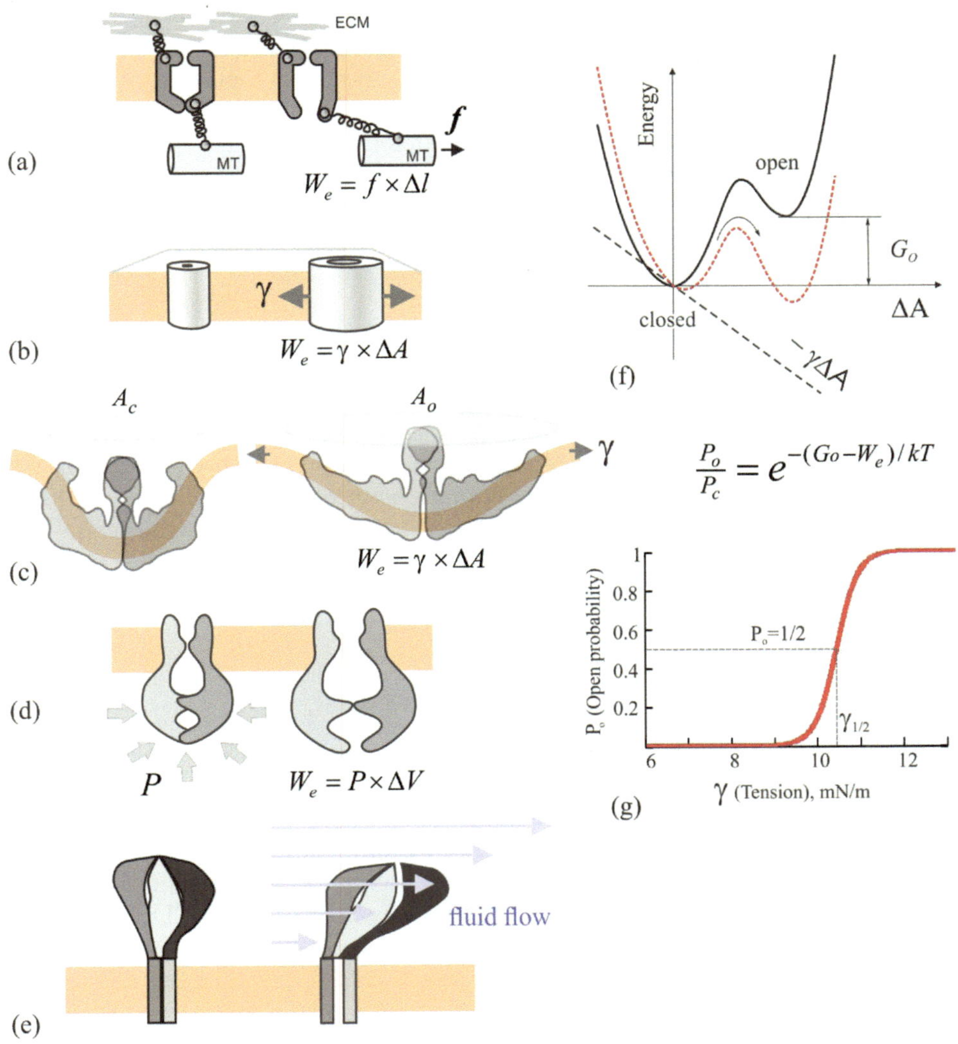

FIGURE 4.3
The five general ways a mechanical stimulus can reach the channel to produce a gating event. (a) Gating by linear force applied to the channel gate by movement of a microtubule through a gating spring; the spatial parameter Δl is the linear displacement of the channel gate in the direction of force and *We* is the work produced by the external force (*f*). (b) Gating of membrane-embedded channel by two-dimensional tension (γ). The work in this case is equal to the product of tension and in-plane area change (ΔA) of the channel complex associated with opening. (c) Gating of the curvature-generating Piezo channel. The curved "blades" of the channel bend the segment of surrounding membrane creating a dome-like structure. Tension is predicted to flatten the dome that opens the central gate. The lateral expansion of the structure due to the flattening is larger than could be expected from simple stretching, which provides higher tension sensitivity. (d) Gating driven by osmotic or crowding stress. In this case, the opening of the channel is caused by an increase of cytoplasm hydration. The hollow sensor domain of the channel has openings that allow the passage of solvent and small molecules but are impermeable to large osmolytes. The domain changes its volume with the gating transition, imparting sensitivity to osmotic (or crowding) pressure (*P*) in the cytoplasm, which is proportional to *P* and ΔV. (e) The "sail" model of channel gating by shear stress generated by fluid flow parallel to the membrane surface. (f) A two-well energetic diagram representing the energy of a tension-gated channel with expansion (ΔA) chosen as the transition coordinate. The open and closed states are separated by G_o, the energy gap between the two states in the absence of tension. Tension introduces a linear term $-\gamma\,\Delta A$, which changes the energy profile and lowers the energy of the open state. Obeying the Boltzmann distribution, the population of channels in the open state will increase. (g) A dose–response curve for a tension-activated channel plotted using Equation 4.2 with the experimental parameters $G_o = 50$ kT and $\Delta A = 20\ \text{nm}^2$ determined for MscL.

more favorable. For the 3D case, the cartoon (Figure 4.3d) depicts a channel whose cytoplasmic domain is compacted by osmotic or crowding pressure, which keeps the pore closed. Dilution of the cytoplasm relieves pressure, and the cytoplasmic domain expands, leading to pore opening.

An example of a two-well energy profile (Figure 4.3f) depicts energy as a function of a 2D expansion of the molecule driven by tension added as the linear energy term $\gamma \times \Delta A$. As the magnitude of tension increases, it offsets the intrinsic energy bias more strongly, thus redistributing the population toward the open state. Tension also decreases the height of the transition barrier, thus increasing the rate of opening. Rearrangement of Equation 4.1 with the two-state condition ($Po = 1 - P_c$) leads to Equation 4.2, which produces an S-shaped dose–response curve for a simple tension-activated transition, as depicted in Figure 4.3g. The maximal slope of the open probability (Po) on force is determined solely by the spatial parameter of the transition, Δl, ΔA or ΔV. The larger the spatial parameter of the sensor molecule, the steeper the activation curve. The midpoint of the dose–response curve, i.e., the "equipartitioning" tension at which half of the channels are open ($P_c = P_o = 0.5$) as seen from Equation 4.1, would be achieved at the tension that compensates the intrinsic bias exactly ($G_o - \gamma \Delta A = 0$). The actual dose–response curve of the mechanotransducer, defined by the position and slope of the dose–response curve, is determined by both the intrinsic bias G_o and the spatial parameter of the transition (Equation 4.2).

$$P_o = \frac{1}{1 + e^{(G_0 - W_e)/kT}} \tag{4.2}$$

All energies in the Boltzmann formalism are normalized to kT, the parameter characterizing the thermal energy of the surroundings. It defines the threshold above which the energy change will have noticeable effect. Changing the probability ratio e-fold (2.72 times) requires a change of the biasing energy by $1\ kT = 4.1 \times 10^{-21}$ J (k is the Boltzmann constant and T is the absolute temperature). This is equal to the work produced by a 1 pN (piconewton) force over a distance of 4.1 nm; similarly, a 1 kT work in 2D and 3D cases will amount to 4.1 mN/m $\times$ nm^2, or 4.1 atm $\times$ 10 nm^3, respectively.

A word of caution should be added to every consideration of dimensionality of the mechanical stimulus. Embedded proteins can be differently affected by intrinsic pressure profiles characteristic of membranes of different compositions. The same conformational transition in a given protein might translate into a different effective expansion area depending on the lateral pressure profile in a particular membrane. Also, membranes and proteins behaving like soft elastic bodies may interconvert and redirect forces arriving from the outside. For instance, the linear (1D) force exerted on the membrane at the tip of a stereocilium by the tip link causes "tenting" of the membrane (Kachar et al., 2000), which in turn creates 2D tension in the region near the insertion point of the tip kink. This poses a dilemma regarding the true stimulus driving the transduction channel: Is it a linear force or 2D tension? More detailed structural knowledge of the components constituting the system will permit a better description of force distribution.

Gating sensitivity may critically depend on the location and the structural context. For instance, the channel can be sensitized when it is positioned at the flexible base of a sensory hair or cilium. In addition to membrane tension, multimodal channels can be allosterically modulated by chemical/physical factors such as increased cytoplasmic Ca^{2+}, presence of phosphorylated inositide lipids, unsaturated fatty acids, elevated temperature or lowered pH.

4.4 Transient Responses: Adaptation, Desensitization and Inactivation

Like neuronal voltage-gated channels, which produce conductance spikes and inactivate (Hille, 2001), most MS channels generate current transients in response to a stepped mechanical stimulation. The mechanism behind these transient responses may not always be the same. The simplest model implies a viscoelastic stress relaxation in the medium around the channel such that the stimulus reaching the gate decays with time due to passive rearrangement of soft stress-bearing elements (Sachs, 2010). This could explain the transient activation of MS channels in *Xenopus* oocytes, where uncoupling of the cortical cytoskeleton from the patch membrane removes adaptation (Zhang et al., 2000).

In contrast to adaptation, which usually means a rightward shift of the activation curve with a prolonged stimulus, inactivation refers to an intrinsic process that renders the channel both nonconductive and insensitive to the stimulus. This can be viewed as uncoupling of the activation gate from the stimulus followed by gate reclosure, which some authors call "desensitization." Alternatively, a return to a nonconductive state can be due to closure of a dedicated inactivation gate. This appears to be the case with Piezo1 where activation was proposed to expand the pore constriction formed by M2493 side chains of the inner helices (Guo and MacKinnon, 2017), whereas inactivation was attributed to the hydrophobic collapse of the pore four helical turns above the main gate (Zheng et al., 2019). Independently, inactivation properties of both Piezo1 and Piezo2 channels were tied to the C-terminal extracellular cap domain (Wu et al., 2017).

The ~100 ms desensitization of TREK-1 also appears to be intrinsic to the channel complex (Honore et al., 2006) since it is unaffected by patch excision or cytoskeleton-depolymerizing agents. Consistently, purified and reconstituted TREK-1 and TRAAK channels exhibit adaptive currents in liposomes (Brohawn et al., 2014). The intrinsic mechanism of gate uncoupling that produces silent inactivation in the bacterial tension-gated channel MscS from the closed state will be described later.

The fast (~1 ms) adaptation of the auditory channel currents in vertebrate hair cells was shown to depend on both membrane potential and the influx of Ca^{2+} at the apical side. The slower components of hair bundle adaptation reset the activation curve according to the new resting position of the tip of the stereociliary bundle and are attributed to tension adjustments in the tip link through the slippage and restoring pulling action of myosin motors (Vollrath et al., 2007).

4.5 Experimental Parameters of MS Channel Gating

In this section we will consider only three examples. The first is the auditory transduction channel that gates by linear force applied through a tip link. The two bacterial channels considered next, MscL and MscS, are both gated by tension in the surrounding lipid bilayer. The large-conductance channel MscL shows an essentially two-state gating mechanism, whereas the smaller-conductance channel MscS exhibits a complex adaptive behavior that involves both adaptation and complete inactivation. The gating of Piezo channels is described in Volume II, Chapter 26.

4.5.1 The Auditory Transduction Channel

A basic theory for the gating for force-gated ion channels, which involves an elastic gating spring pulling on a movable channel gate, came from two biophysical measurements: the kinetics of hair cell receptor currents evoked by the stereociliary bundle deflection (Corey and Hudspeth, 1983) and concomitant changes of bundle compliance that matched the activation range of those receptor currents (Markin and Hudspeth, 1995). Figure 4.4a represents the configuration of the stereocilia bundle at rest and upon deflection by force applied to the tip; panels b and c show two ways that a channel at the tip of a stereocilium could feel tension. In one, the elastic element is intracellular and connected directly to the channel; in the other, the tension is conveyed through the lipid membrane. When the bundle moves in the direction of the tallest stereocilium, all stereocilia pivot around their basal insertions, which tensions the tip links. The ratio of shear between adjacent stereocilia to the horizontal displacement of the tip of the stereociliary bundle (γ) varies with bundle geometry but is roughly 0.1–0.2. Panel d shows a simplified equivalent mechanical scheme of one interconnected pair made of model elements, including the gating spring (k_g) attached to the channel gate whose probability of being open (P_o) depends on tension and presence of Ca^{2+}. Parameter d is the gate "swing" associated with the channel opening and $\Delta\ x_g$ represents calcium-induced relaxation of a hypothetical release element proposed as an additional contributor to the compliance increase on opening. The motor present in the system maintains resting tension in the tip link. The parallel elastic (ks) and viscous ($\Xi\ s$) elements represent the pivoting stiffness of the stereocilium and fluid drag of the bundle. Panels e–g show data obtained by Cheung (Cheung and Corey, 2006) using optical tweezers to simultaneously apply force and measure bundle movement. The receptor current (panel e) increases with deflection and then saturates, exhibiting a midpoint at a deflection near 50 nm. The force–deflection relationship (panel f) is linear at the left end of the graph, where all channels are closed, and at the right end, where they are open, indicating about the same stiffness of the bundle; however the middle part, where the channels gate, shows flattening and the stiffness in that region (panel g) has a minimum. The decrease of stiffness can be understood as a slackening of the gating spring associated with the opening or swing of the channel's gate in the direction of force. Fitting multiple experimental curves suggested the ranges of model parameters. In amphibian hair cells, the effective stiffness of the gating spring kg was on the order of 700 μN/m, which collectively describes all the elastic elements that are in series with the transduction channel combined in a single element. To reach the midpoint of the activation curve ($P_o = 0.5$), the gating spring needs to be extended by about $50 \times \gamma \approx 7$ nm under a force of about 5 pN. Channel opening is associated with an effective gate swing of ~4 nm (Cheung and Corey, 2006).

One should be careful interpreting the "swing" of the gate literally. It could be nearly as shown in Figure 4.4b if the channel gate is indeed directly attached to actin filaments through an elastic gating spring protein. The gating spring is not the tip link, however, crystal structures of protocadherin 15 and cadherin 23 forming the tip link and molecular dynamics simulations have shown that these filamentous proteins are too stiff to satisfy the properties of the gating spring (Sotomayor et al., 2012). Other components present in stereocilia need to be considered, and some of them may have the requisite stiffness and extension. In an alternative arrangement (Figure 4.4c), the channels may not be attached to the tip links or gating springs, but may reside in the membrane near the point of tip link insertion and may experience tension transmitted through lipids when the force applied to the tip link causes "tenting" of the membrane (Powers et al., 2014).

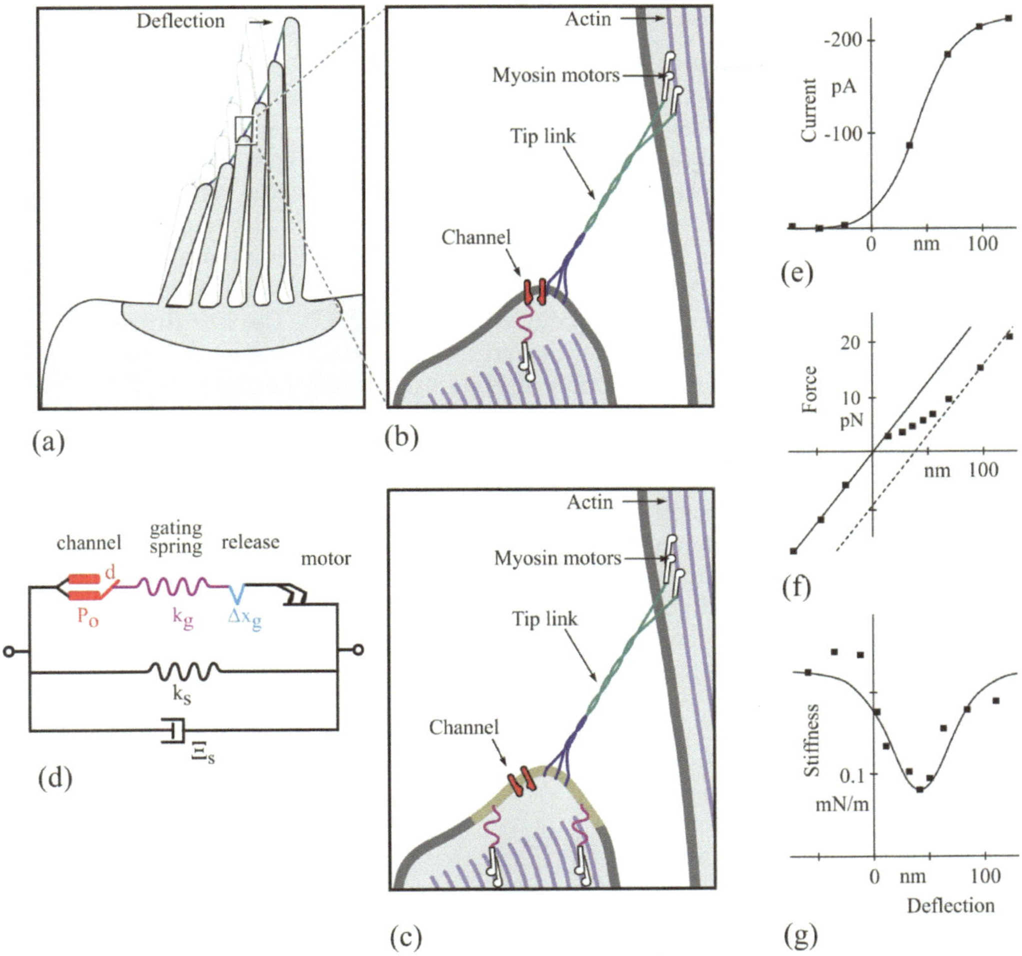

FIGURE 4.4
Gating of the inner ear transduction channel. (a) The arrangement of the stereociliary bundle at the apical side of a hair cell. The bundle is interconnected by cadherin filaments called tip links. Deflection of the tip of the bundle produces stress of the tip links. (b) A hypothetical arrangement in which the transduction channel residing at the tip of the lower stereocilium is directly connected to the tip link (made of protocadherin 15 and cadherin 23) and the elastic element is attached to the actin below the channel. Tension in the tip link and the elastic element is maintained by the myosin motors. (c) The alternative arrangement in which the transduction channel resides in the membrane and experiences tension produced in the bilayer by the linear force exerted via the tip link near the point of insertion. (d) The equivalent mechanical scheme depicting the serial connection of the channel with its gate (P_o); the gating spring (k_g), which combines mechanical compliances of tip link and all connected elements inside stereocilia; and the hypothetical "release" element that provides a $Ca2^+$-dependent extension, which can be associated with the channel and the motor. The two parallel elements *ks* and Ξ *s* represent the pivoting stiffness and drag, respectively. (e) The receptor current–deflection curve for a bullfrog hair cell. (f) The force–deflection relationship showing linear regimes at both ends and a transition regime in the middle. (g) The stiffness–deflection curve with a minimum corresponding to the point of steepest current rise caused by transduction channel gating. (Panels d–g are redrawn from Cheung and Corey 2006.)

4.5.2 Bacterial MS Channels as Models for Gating by Membrane Tension

The mechanosensitive channels of small and large conductance, MscS and MscL, are responsible for rescuing bacteria in the event of a sudden osmotic downshock (Levina et al., 1999). They act as tension-driven osmolyte release valves that increase membrane permeability to small metabolites that provide the release rate commensurate with the rate of osmotic swelling (Bialecka-Fornal et al., 2012). In this way, the release system adjusts the osmotic gradient, reduces the osmotically driven water influx and curbs membrane tension away from the lytic limit. Under extreme shocks, bacteria may jettison up to 15% of their nonaqueous cytoplasmic content within 20 ms (Figure 4.5a) as determined by the stopped-flow light scattering technique (Cetiner et al., 2017). The low-threshold MscS (1 nS conductance in 200 mM KCl) that becomes active at tensions of 6–7 mN/m and the 3 nS high-threshold MscL that activates at 11–14 mN/m are the main constituents of the release system, and the presence of two populations is readily observed in inside-out patches excised from giant spheroplasts of wild-type *E. coli* (Figure 4.5b). While MscL is an emergency valve that opens at near-lytic tensions, MscS opens considerably below the lytic limit and shows a characteristic adaptive behavior resolving in complete inactivation (Akitake et al., 2005). The solved crystal structure of MscL (Chang et al., 1998) and several crystal and cryo-EM structures of MscS (Reddy et al., 2019; Steinbacher et al., 2007; Zhang et al., 2021) have informed computational modeling, simulations and reconstruction of the main functional states based on experimental parameters of transitions (Akitake et al., 2007; Belyy et al., 2010a; Chiang et al., 2004; Corry et al., 2010; Rajeshwar et al., 2021; Vasquez et al., 2008).

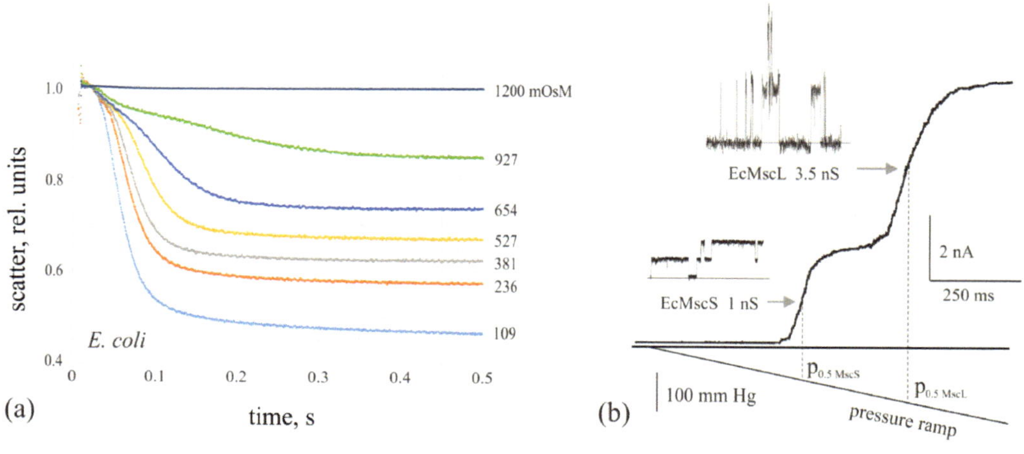

FIGURE 4.5
The macroscopic realizations of the action of the bacterial tension-driven osmolyte release system. (a) Light scattering kinetics recorded in stopped-flow experiments from *E. coli* cell suspension (Frag1 strain) shocked from 1200 mOsm to different end osmolarities (shown by the traces). The short period of swelling is followed by a steeper downward phase of internal osmolyte release. The intensity of scattering reflects the ratio of refractive indexes inside and outside the cell, which drops as a result of osmolyte release. (From Cetiner et al., 2017.) (b) The current trace recorded from an inside-out patch excised from a giant bacterial spheroplast (Frag1 strain). The stimulus is a linear ramp of negative pressure applied to the pipette, which stretches the patch. The two waves of current represent the populations of the low-threshold MscS followed by the high-threshold MscL.

4.5.3 The Large-Conductance Channel MscL

The MscL channel was initially identified and experimentally studied in *E. coli* (Sukharev et al., 1994). The solved crystal structure, however, was of its homolog from *M. tuberculosis* (TbMscL) (Chang et al., 1998), which is 29% identical to EcMscL and shares the same organization. We will first describe the structure of TbMscL and illustrate its iris-like gating mechanism with MD simulation of its transition, and then we will discuss how accurately the simulation predicts the experimentally determined parameters of gating.

Figure 4.6a (left) shows the TbMscL structure and the steps in the simulated opening transition obtained using locally distributed tension molecular dynamics (LDT-MD)

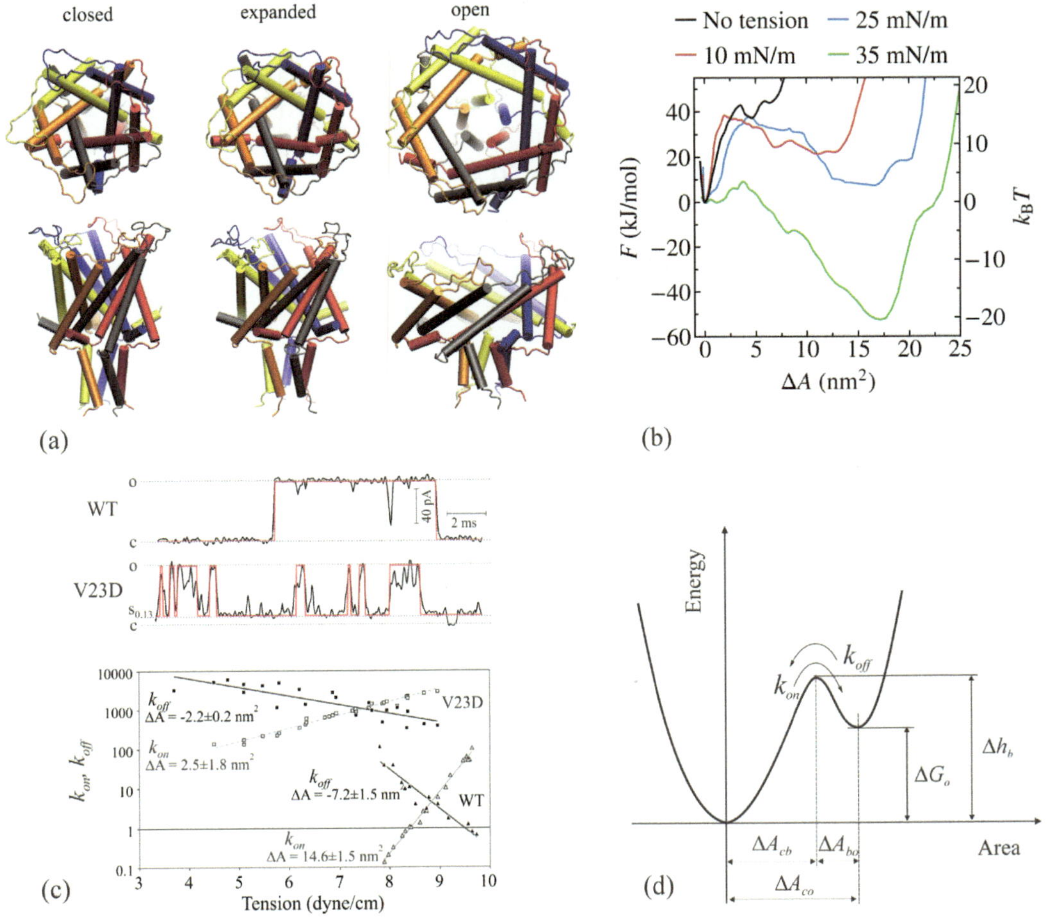

FIGURE 4.6
The gating mechanism of MscL. (a) The gating transition of TbMscL simulated using the LDT-MD technique. The narrow conformation on the left represents the crystal structure (PDB ID 1MSL), which was pulled by radial forces applied to annular lipids. Through an intermediate, the channel arrived to the fully open state characterized with higher tilt of TM2–TM2 helical pairs (right panel). (b) The energy profiles computed at different tensions along the expansion coordinate. (c) The examples of single-channel traces for WT and V23D MscL, their idealization (red trace), and the logarithmic plot of the k_{on} and k_{off} rates on tension. The slopes indicate area changes between the open state and the transition barrier, ΔA_{cb}, and between the barrier and the open state, ΔA_{bo}. (From Anishkin et al., 2005.) (d) The energy profile for EcMscL reconstructed from the kinetic data shown in panel (c). See text.

(Rajeshwar et al., 2021). The crystal structure (2OAR) shows a pentamer of 125 amino acid subunits spanning the membrane twice; the TM1 and TM2 helices are swapped between adjacent subunits such that periplasmic loops physically interconnect all subunits. Both the N- and C-terminal ends are in the cytoplasm where the C-terminal cytoplasmic helices bundle together. Five tightly packed internal TM1 helices form the hydrophobic gate lined by the aliphatic side chains of I14 and V21 in the constriction. In all reported simulations this gate is desolvated (vapor locked). The predicted transition, which is driven by membrane tension transmitted primarily through the boundary regions of the lipid bilayer, results in gradual tilting of the TM1–TM2 helical pairs with simultaneous pore expansion. Proceeding through a low-conducting intermediate expanded state, the channel arrives to an open state characterized by ~30° helical tilt and a 17–20 nm^2 lateral protein expansion. The partially flattened complex now forms a fully solvated pore 1.5 nm in radius with an estimated 2.5 nS conductance (in 200 mM KCl). The C-terminal helical bundle remains associated in the open state, which, together with unstructured linkers, forms a prefilter at the cytoplasmic entrance to the pore. LDT-MD simulations combined with multiple-walker well-tempered metadynamics allowed Rajeshwar et al. (2021) to calculate the free energy profile of the gating transition at different tensions applied to the closed conformation (Figure 4.6b). The barrier separating the closed- and open-state wells is positioned near 4 nm^2, whereas the energy minimum for the open well at high tension was located near 17 nm^2 on the expansion coordinate. Are these positions of the well and the barrier consistent with experiments?

The first step in the experimental description of MscL gating was the determination of the midpoint tension for channel activation, which required protein reconstitution into giant unilamellar liposomes and visualization of patch curvature in the course of patch stimulation. Several studies came to the consensus that the tension of half-activation ($\gamma_{1/2}$) for MscL is near 12 mN/m (Moe and Blount, 2005; Nomura et al., 2012; Sukharev et al., 1999). With this parameter, activation curves were fitted to the Boltzmann equation (Equation 4.1) and revealed that the intrinsic bias stabilizing the closed conformation in the absence of tension (G_o) is ~50 kT (125 kJ/mol) and the expansion area associated with the transition (ΔA) is near 20 nm^2 (Chiang et al., 2004). Therefore, the 17 nm^2 channel expansion computed for TbMscL is in good correspondence with experimental area change determined for homologous EcMscL. Multiple experiments have supported the iris-like character of helical movements characterized by similar barrel expansion. These experiments involved disulfide cross-linking of the open state (Betanzos et al., 2002) and determination of lipid and water accessibilities of different helical domains using electron paramagnetic resonance (EPR) (Perozo et al., 2002a) as well as single-molecule FRET (Corry et al., 2010).

While the in-plane area expansion parameter for MscL was confirmed by experiment, the parameter describing the position of the barrier separating the closed and open states in EcMscL was different. Analysis of tension dependencies for the separate opening (k_{on}) and closing (k_{off}) rates provides additional information about the shape of the energy profile along ΔA chosen as the reaction coordinate (Sukharev et al., 1999). Eyring-type equations for k_{on} and k_{off} take into account the height of the transition barrier as it is seen either from the bottom of the closed well or from the open-state well, respectively (Figure 4.6c).

$$k_{off} = k_o e^{-\frac{\Delta h_b - G_0 - \gamma \Delta A_{bo}}{kT}}$$

$$k_{on} = k_o e^{-(\Delta h_b - \gamma \Delta A_{cb})/kT}$$

Differentiation with respect to γ produces a simple relationship that equates the slope of $kT \bullet \ln(k_{on})$ on tension to ΔA_{cb}, the area change from the bottom of the closed-state well to the top of the transition barrier. By the same token, $kT \bullet d\ln(k_{off})/d\gamma = \Delta A_{bo}$, which is the area change from the center of the open-state well to the top of the barrier. Logarithmic plots of k_{on} and k_{off} on γ for WT and V23D MscL are shown in Figure 4.6d. The slopes assign the transition barrier position at about 0.65 of the total area change between the closed and open states for WT MscL. The location of the barrier closer to the open state makes the parabolic well for the closed state wider. The closed conformation is therefore predicted to be "softer" than the open conformation, and the channel should undergo a substantial "silent" expansion before it overcomes the transition barrier and opens. The computational prediction of the barrier position for TbMscL was 0.24. It would require further studies to explain the discrepancy. The experimental ΔA_{cb} and ΔA_{bo} values for V23D MscL are lower and the entire transition is of smaller scale because the channel is pre-expanded in the closed state due to excessive pore hydration (Anishkin et al., 2005).

This iris-like transition is highly consistent with the computed lateral pressure profile of the lipid bilayer (Gullingsrud and Schulten, 2004), where the peaks of tension correspond to the hydrophobic–hydrophilic boundaries inside the membrane. Tension transmitted through these boundary layers is predicted to act on the ends of TM helices causing tilt. Since the lateral pressure profile is sensitive to the chemistry of both headgroups and aliphatic tails, channel gating is modulated by lipid composition (Moe and Blount, 2005) and the presence of lipid-intercalating amphipathic substances. Asymmetrically inserted lysolipids act as potent activating agents (Perozo et al., 2002b).

The intrinsic energy difference between the states, G_o, appears to have three major components. One is the spring-like action of the periplasmic loops. Another is the elastic energy stored in distorted lipids at the protein–lipid boundary. Helical tilting associated with pore expansion produces a substantial flattening of the channel complex. The thickness mism atch between the protein and the surrounding bilayer due to elastic deformation of the boundary lipids was calculated to provide a substantial component of the returning force contributing to G_o (Ursell et al., 2007). The third contributor is unfavorable hydration of the hydrophobic gate that largely defines G_o and the barrier between the states (Anishkin et al., 2010). The energy and cooperativity of transitions critically depend on hydration properties of the pore as illustrated by the comparison of wild-type MscL with the V23D mutant (Figure 4.6d). This mutation places charged side chains in the hydrophobic gate and produces pre-expanded easily opened (gain-of-function) channels with high occupancy of subconductive states (Anishkin et al., 2005).

4.5.4 MscS Channel and Its Adaptive Gating Mechanism

The mechanosensitive channel of small conductance, MscS, opens a 1 nS pore at tensions of 5–7 mN/m with the slope of $P_o(\gamma)$ curve corresponding to ΔA of 15–18 nm^2 (Akitake et al., 2005). Steps of tension applied to patches excised from giant spheroplasts produce transient slowly decaying current responses, which represent a combination of three processes: closure due to stimulus adaptation within the patch membrane (Belyy et al., 2010), intrinsic adaptation and complete inactivation.

The functional manifestation of this adaptive behavior is the ability of MscS to perceive the rate of tension onset. Figure 4.7a shows a series of traces recorded on a ~100 channel population in an excised patch stimulated by linear pressure ramps applied with different rates (Boer et al., 2011). The end pressure in all ramps is identical. As seen from the currents, the full population readily responds to an abrupt pressure onset, but only a fraction

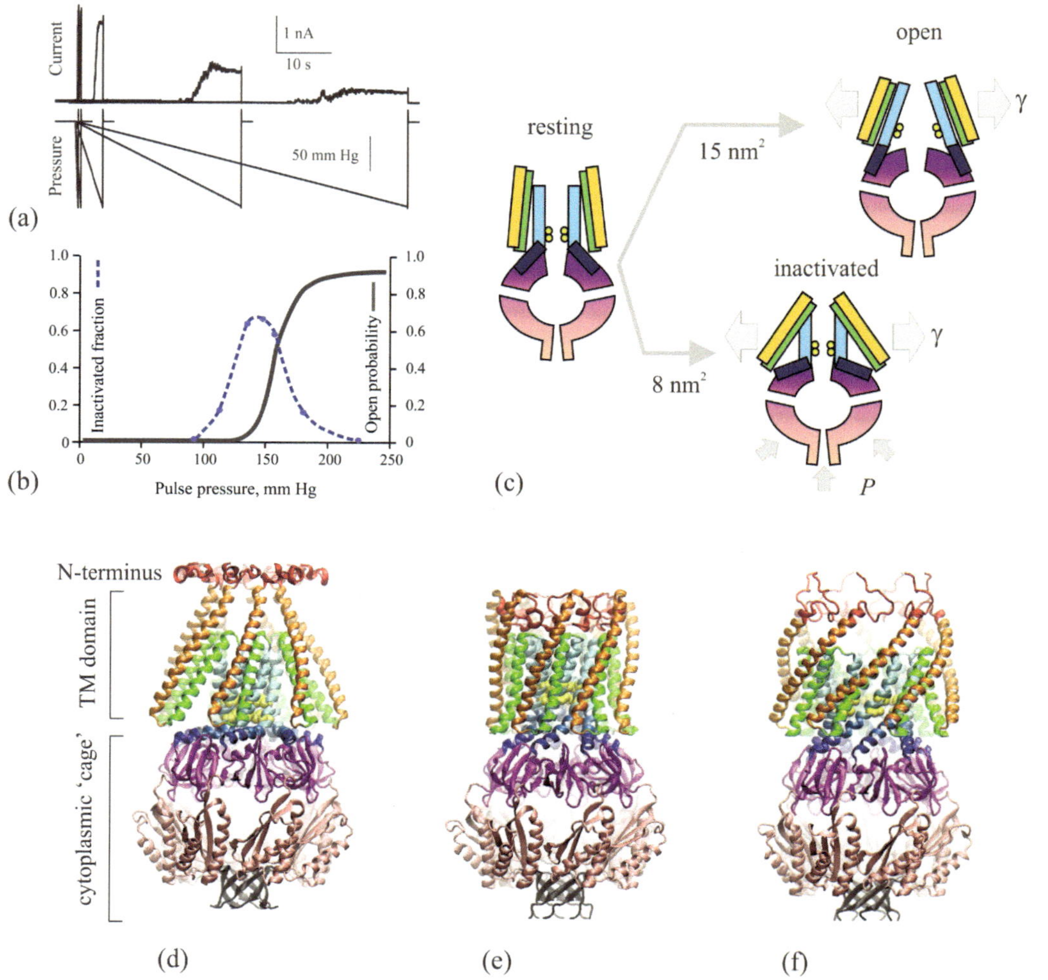

FIGURE 4.7
The adaptive functional cycle of bacterial mechanosensitive channel MscS. (a) Responses of MscS population in a patch stimulated by ramps of pressure applied with different speed. (From Boer et al., 2011.) (b) The probabilities of opening and inactivation transitions as a function of patch pressure. (From Akitake et al., 2005.) (c) A representation of hypothetical MscS transitions to the open and inactivated states with estimated changes of in-plane areas. The inactivated state is stabilized by moderate tension (γ) and crowding pressure (*P*) in the cytoplasm. (d) The complete cryo-EM structure of MscS obtained in nanodiscs (PDB ID 6PWP) characterized by peripheral (TM1–TM2) helices detached from the gate-forming TM3s. (e) The compact model of the resting state. (f) The model of open state generated from the cryo-EM structure obtained in short-chain lipids (PDB ID 6VYL). Modeling and simulations produced the compact resting state by restoring the contact between splayed TM1–TM2 pairs and the central TM3s. Opening involves a ~15 nm^2 expansion of the TM barrel creating a 16Å wide aqueous pore surrounded by kink-free TM3 helices; the gate is colored yellow. Entering the inactivated state involves detachment of TM1–TM2 pairs from TM3s and reformation of the crystallographic kink at G113 associated with a ~8 nm^2 barrel expansion.

of channels responds to a slow pressure ramp while the majority "ignore" it. This behavior is attributed to the ability of MscS to enter the inactivated state, which is nonconductive and tension insensitive.

Analysis has shown that inactivation, like activation, is driven by tension from the closed state (Kamaraju et al., 2011). On the pressure (tension) coordinate, the maximum of inactivation probability is at the foot of the activation curve near the threshold (Figure 4.7b). The channels remain in the inactivated state as long as tension persists. Upon tension release, the inactivated MscSs return back to the resting state within 2 s. Importantly, the rate of inactivation critically depends not only on membrane tension, but also on the presence of macromolecules or polymers (crowders) on the cytoplasmic side of the patch (Rowe et al., 2014). Here we briefly present the hypothesis (Figure 4.7c) and key supporting data on how MscS integrates the two independent mechanical inputs – interfacial tension in the membrane and the bulk parameter of crowding pressure – to disengage its gate as a result of their synergistic action.

The first 3.9 Å crystal structure of *E. coli* MscS (Steinbacher et al., 2007) revealed the channel as a homoheptamer of 3TM subunits (2OAU). A subsequently solved structure of the A106V mutant (2VV5) represented a partially open state (Wang et al., 2008). Today, besides additional crystal structures of *E. coli* MscS solved in detergents, there are several cryo-EM structures solved in membrane-like nanodiscs. Figure 4.7d shows the MscS structure in nanodiscs solved with its periplasmic N-terminal domain (6PWP) (Reddy et al., 2019). This nonconductive conformation shows splayed peripheral pairs of helices (TM1–TM2) that form a ribbed protein–lipid boundary. Inner TM3s line the central pore with a hydrophobic gate at the cytoplasmic end, just above the sharp kink. The horizontal parts of TM3 helices, parallel to the membrane plane, continue into long C-termini from each subunit that combines to form a large hollow cytoplasmic domain with seven side portals, termed the "cage." These side portals define the selectivity of MscS and the repertoire of permeable osmolytes.

Like in the first crystal structure (2OAU), a substantial separation of the peripheral TM1–TM2 helices from the gate-forming TM3s suggests an uncoupled state of the gate, implying that this can be the inactivated conformation in which the link that transmits tension from lipid bilayer to the gate is interrupted. Computational packing of the splayed TM1–TM2 pairs along TM3s was the step that restored the hypothetical physical contact between the gate and the lipid-facing helices. The reconstructed buried hydrophobic TM2–TM3 contact turned out to be the force transmission route from the lipid bilayer to the gate. This "compact" model of the resting state is shown in Figure 4.7e.

The strength of apolar associations between TM2 and TM3 was shown to be sufficient to convey external force to the gate and open the channel. Hydrophilic substitutions at various positions in the buried TM2–TM3 zone led to speedy channel inactivation without opening, indicating that the TM2–TM3 contact is labile and behaves as a dynamic "slip bond" conveying tension to the gate (Belyy et al., 2010a). Disruption of this bond leads to gate uncoupling and formation of the crystallographic kink at glycine 113, which stabilizes the inactivated state. Based on the tension-dependences of the rates of inactivation and recovery (Kamaraju et al., 2011), it has been determined that gate uncoupling is associated with an approximately 8 nm^2 in-plane expansion, which is in good correspondence with the outward splaying motion of the TM1–TM2 pairs as in the crystal-like conformation. Thus, under tension the channel has two alternative pathways, one into the open state and another into inactivation.

The modeled opening transition associated with the experimentally estimated 15 nm^2 expansion of the entire transmembrane domain produces a 16 Å-wide pore that well satisfies a 1.1–1.2 nS unitary conductance. The opening straightens the TM3 helices, which have two conserved hinge points (glycines G113 and G121) that act as "collapsible struts" (Akitake et al., 2007). The model of the fully open state generated after the recent nearly open cryo-EM structure of MscS (Zhang et al., 2021) solved in short-chain lipids is presented in Figure 4.7f.

MscS inactivates very quickly under conditions of high macromolecular content at the cytoplasmic side (Rowe et al., 2014). How does crowding pressure make the inactivation path more preferable? MD simulations of the closed and inactivated models revealed that the positions of the hollow cytoplasmic cage domain in the resting and inactivated states are different. In the latter case, the cage domain has a slightly smaller profile in the cytoplasm, making the inactivated state more compact and therefore more preferable in the presence of crowders (Rowe et al., 2014). In addition, application of axial pressure to the cage domain in molecular dynamics simulations to mimic the effect of crowders produces a bigger splay of TM1–TM2 pairs, which additionally increases the "footprint" of the channel in the membrane. Thermodynamically this means that crowding pressure will assist tension in separating the peripheral helices from the core and promote the uncoupling of the gate. The biological meaning of the observed effect of crowders on MscS inactivation is that in the course of the osmotically induced permeability response the channel receives feedback on the state of the cytoplasm. The increased macromolecular excluded volume, perceived by the cage domain as crowding pressure, disengages the gate and thus limits the period of channel action. Because the fractional volume of free water in the cytoplasm should not decrease below a certain point (~70%), this feedback mechanism prevents cell "overdraining" when excessive cytoplasmic osmolytes are released during the turgor adjustment to a hypotonic medium (Rowe et al., 2014).

This adaptive behavior of MscS has been evolutionary designed to minimize the losses of metabolites during adjustments to a wide range of osmotic shocks. Another important role of MscS inactivation appears to be during the termination of the immense osmotic permeability response. MscS and MscL appear to open and close sequentially. When MscL closes, MscS stays open for some time and reduces tension far below the threshold for MscL. When MscS arrives to its own tension threshold, it inactivates and in doing so completely reseals the membrane. Importantly, both MscS and MscL are characterized by completely dehydrated gates, which allows them to stay completely leak-proof at rest in the inner energy-coupling bacterial membrane typically energized to –180 mV.

4.6 Conclusions and Perspectives

The common principle that the mechanotransducer molecule should comply with the stimulus and change its dimension in the direction of the applied force holds for all mechanically activated channels. The fraction of the external macroscopic stimulus reaching the channel, however, depends on the structural context and is not always known. Although there is a clear evolutionary separation in terms of prevalence of specific types of mechanosensitive channels in different clades of organisms, it seems to follow an ancient separation according to the way the organisms handle ubiquitous osmotic forces.

Organisms surrounded by a rigid cell wall use hydrostatic (turgor) pressure for maintenance of their shape and volume. In these cells, the cytoplasmic membrane, which is the main osmotic barrier, experiences normal pressure against the cell wall. Prokaryotes covered by an elastic peptidoglycan layer evolved MscS to regulate internal turgor and hydration of the cytoplasm and MscL as an emergency valve to prevent osmotic lysis. Note that the crowding pressure acting on MscS's cytoplasmic cage domain are normal to the plane of the membrane, and channel compaction in this direction is facilitated by the presence of the outer wall against which the membrane channel is pressed. Presumably for this reason, MscS and its homologs are present primarily in walled organisms (bacteria, algae, fungi and plants). Plants surrounded by a more rigid cell wall inherited and diversified MscS, which is now present in all plant cells and many symbiotic organelles that deal with turgor pressure.

The opposite trend was taken by the animals who took advantage of cellular contractility, abandoning the external wall and developing a powerful and dynamic internal cytoskeleton that restrains osmotic swelling through attachments to a highly folded membrane. The large membrane area excess in a typical animal cell provides freedom for shape changes and volume adjustments. In the absence of strong perturbations, global tension in the animal cell membrane is presumed to be close to zero, yet cell contractility and attachments to the substrate can generate membrane stresses locally and transiently. The channels from the DEG/ENaC or TRP families, shown or presumed to interact with the cytoskeleton, are strictly animal and not found in higher plants; however, a single Piezo homolog is present in plant genomes. Piezos, which are prevalent in animals, have evolved specifically to function in the pleated and scaffolded plasma membrane to monitor local changes in tension. Piezo channels function specifically in curved areas of membrane and sense minute tension that does not significantly stretch but rather unfolds the membrane. The mechanistic question of whether Piezo channels may have an advantage in sensitivity by forming clusters is currently being addressed (Lewis and Grandl, 2021).

It seems plausible that the earliest design of membrane mechanotransducers utilized the simplest force from lipid (FFL) principle of gating driven by two-dimensional membrane tension. Many eukaryotic channels (TRPs, K2Ps, Piezos) have inherited this universal mechanism, and only highly specialized channels such as NompC, MECs and the auditory transduction channel were made responsive to force in a specific direction and thus were coupled to cytoskeletal/ECM filaments. It is interesting that Gd^{3+} and GsMTx4, the two common blockers/modifiers of mechanosensitive channels, are both working through association with lipids. Gd^{3+} condenses anionic lipids around the channel that exert positive pressure keeping the channel closed (Ermakov et al., 2010). The GsMTx4 peptide binds to the resting membrane in its "shallow" position, but inserts deep into the membrane under tension. By contributing itself into the membrane, GsMTx4 negates part of the perturbing tension and acts as a mobile amphipathic material clamping lateral pressure in the exposed leaflet, thus shifting the Piezo's activation curves to the right (Gnanasambandam et al., 2017).

Recordings made on individual MS channels in reconstituted or in heterologously expressed systems are most informative in terms of a mechanistic understanding of individual components. In vivo experiments, however, document responses of the native system as a whole. It is remarkable that genetic knock-out or knock-down techniques rarely show an all-or-none functional effect attributable to the single protein of interest. Instead, experiments often reveal multiple components with overlapping ranges of sensitivity, partial or complete redundancy, or indicate hybrid assemblies made of different subunits producing new features. This suggests that the principles of evolvability (Kirschner and Gerhart, 1998) fully apply to the evolution of mechanosensory systems, which has picked and perfected several conserved designs of MS channels, diversified them and mixed them in different contexts where a combination works better than any individual type. The clear organizational similarity between lipid scramblases (TMEM16) and mechanosensitive channels (OSCA, TMC) emphasizes a highly opportunistic course of evolution and possible co-evolution of different components. Matching various channels with different types of scaffolding may further change the activation threshold, gain, and adaptive capacity. Comparisons of functionally similar designs (hair cells versus ciliated endothelial cells, sensory flagella in protozoa, or sensory hair in Venus flytrap) may be beneficial in searches for similar components, adaptor domains as well as common physical principles.

Acknowledgments

The authors thank David Corey and Elizabeth Haswell for critical comments.

Suggested Readings

This chapter includes additional bibliographical references hosted only online. Please visit https://www.routledge.com/9780367538156 to access the additional references for this chapter, found under "Support Material" at the bottom of the page.

Akitake, B., A. Anishkin, N. Liu, and S. Sukharev. 2007. "Straightening and sequential buckling of the pore-lining helices define the gating cycle of MscS." *Nat Struct Mol Biol* 14:1141–9.

Anishkin, A., C. S. Chiang, and S. Sukharev. 2005. "Gain-of-function mutations reveal expanded intermediate states and a sequential action of two gates in MscL." *J Gen Physiol* 125:155–70.

Arnadottir, J., and M. Chalfie. 2010. "Eukaryotic mechanosensitive channels." *Annu Rev Biophys* 39:111–37.

Belyy, V., A. Anishkin, K. Kamaraju, N. Liu, and S. Sukharev. 2010a. "The tension-transmitting 'clutch' in the mechanosensitive channel MscS." *Nat Struct Mol Biol* 17:451–8.

Brohawn, S. G., J. del Marmol, and R. MacKinnon. 2012. "Crystal structure of the human K2P TRAAK, a lipid- and mechano-sensitive K+ ion channel." *Science* 335:436–41.

Carattino, M. D., S. Sheng, and T. R. Kleyman. 2004. "Epithelial Na+ channels are activated by laminar shear stress." *J Biol Chem* 279:4120–6.

Cetiner, U., I. Rowe, A. Schams, C. Mayhew, D. Rubin, A. Anishkin, and S. Sukharev. 2017. "Tension-activated channels in the mechanism of osmotic fitness in Pseudomonas aeruginosa." *J Gen Physiol* 149:595–609.

Chang, G., R. H. Spencer, A. T. Lee, M. T. Barclay, and D. C. Rees. 1998. "Structure of the MscL homolog from Mycobacterium tuberculosis: a gated mechanosensitive ion channel." *Science* 282:2220–6.

Cheung, E. L., and D. P. Corey. 2006. "Ca2+ changes the force sensitivity of the hair-cell transduction channel." *Biophys J* 90:124–39.

Chiang, C. S., A. Anishkin, and S. Sukharev. 2004. "Gating of the large mechanosensitive channel in situ: estimation of the spatial scale of the transition from channel population responses." *Biophys J* 86:2846–61.

Corey, D. P., and A. J. Hudspeth. 1979. "Ionic basis of the receptor potential in a vertebrate hair cell." *Nature* 281:675–77.

Coste, B., J. Mathur, M. Schmidt, T. J. Earley, S. Ranade, M. J. Petrus, A. E. Dubin, and A. Patapoutian. 2010. "Piezo1 and Piezo2 are essential components of distinct mechanically activated cation channels." *Science* 330:55–60.

Geffeney, S. L., and M. B. Goodman. 2012. "How we feel: ion channel partnerships that detect mechanical inputs and give rise to touch and pain perception." *Neuron* 74:609–19.

Gnanasambandam, R., C. Ghatak, A. Yasmann, K. Nishizawa, F. Sachs, A. S. Ladokhin, S. I. Sukharev, and T. M. Suchyna. 2017. "GsMTx4: mechanism of inhibiting mechanosensitive ion channels." *Biophys J* 112:31–45.

Guharay, F., and F. Sachs. 1984. "Stretch-activated single ion channel currents in tissue-cultured embryonic chick skeletal muscle." *J Physiol* 352:685–701.

Guo, Y. R., and R. MacKinnon. 2017. "Structure-based membrane dome mechanism for Piezo mechanosensitivity." *Elife* 6:e33660. DOI: https://doi.org/10.7554/eLife.33660.

Hille, B. 2001. "Ion channels of excitable membranes (3d edition)." *Sinauer, Sunderland, Mass* xviii:814.

Honore, E. 2007. "The neuronal background K2P channels: focus on TREK1." *Nat Rev Neurosci* 8:251–61.

Jeong, H., S. Clark, A. Goehring, S. Dehghani-Ghahnaviyeh, A. Rasouli, E. Tajkhorshid, and E. Gouaux. 2022. "Structure of C. elegans TMC-1 complex illuminates auditory mechanosensory transduction." Preprint.

Kunzelmann, K. 2015. "TMEM16, LRRC8A, bestrophin: chloride channels controlled by Ca(2+) and cell volume." *Trends Biochem Sci* 40:535–43.

Levina, N., S. Totemeyer, N. R. Stokes, P. Louis, M. A. Jones, and I. R. Booth. 1999. "Protection of Escherichia coli cells against extreme turgor by activation of MscS and MscL mechanosensitive channels: identification of genes required for MscS activity." *EMBO J* 18:1730–7.

Markin, V. S., and F. Sachs. 2004. "Thermodynamics of mechanosensitivity." *Phys Biol* 1:110–24.

Martinac, B., M. Buechner, A. H. Delcour, J. Adler, and C. Kung. 1987. "Pressure-sensitive ion channel in Escherichia coli." *Proc Natl Acad Sci U S A* 84:2297–301.

Nauli, S. M., F. J. Alenghat, Y. Luo, E. Williams, P. Vassilev, X. Li, A. E. Elia, W. Lu, E. M. Brown, S. J. Quinn, D. E. Ingber, and J. Zhou. 2003. "Polycystins 1 and 2 mediate mechanosensation in the primary cilium of kidney cells." *Nat Genet* 33:129–37.

Rajeshwar, T. R., A. Anishkin, S. Sukharev, and J. M. Vanegas. 2021. "Mechanical activation of MscL revealed by a locally distributed tension molecular dynamics approach." *Biophys J* 120:232–42.

Reddy, B., N. Bavi, A. Lu, Y. Park, and E. Perozo. 2019. "Molecular basis of force-from-lipids gating in the mechanosensitive channel MscS." *Elife* 8:e50486. DOI: https://doi.org/10.7554/eLife.50486.

Ricci, A. J., A. C. Crawford, and R. Fettiplace. 2003. "Tonotopic variation in the conductance of the hair cell mechanotransducer channel." *Neuron* 40:983–90.

Rowe, I., A. Anishkin, K. Kamaraju, K. Yoshimura, and S. Sukharev. 2014. "The cytoplasmic cage domain of the mechanosensitive channel MscS is a sensor of macromolecular crowding." *J Gen Physiol* 143:543–57.

Saotome, K., S. E. Murthy, J. M. Kefauver, T. Whitwam, A. Patapoutian, and A. B. Ward. 2018. "Structure of the mechanically activated ion channel Piezo1." *Nature* 554:481–6.

Steinbacher, S., R. Bass, P. Strop, and D. C. Rees. 2007. "Structures of the prokaryotic mechanosensitive channels MscL and MscS." *Curr Top Membr* 58:1–24.

Sukharev, S. I., P. Blount, B. Martinac, F. R. Blattner, and C. Kung. 1994. "A large-conductance mechanosensitive channel in E. coli encoded by mscL alone." *Nature* 368:265–8.

Zhang, Y., C. Daday, R. X. Gu, C. D. Cox, B. Martinac, B. L. de Groot, and T. Walz. 2021. "Visualization of the mechanosensitive ion channel MscS under membrane tension." *Nature* 590:509–14.

5

Inactivation and Desensitization

William N. Zagotta

CONTENTS

5.1 Introduction

In addition to voltage- and ligand-dependent activation, which cause an increase in ionic current, most ion channels also undergo a subsequent decrease in current in the continued presence of the activating stimulus. This decrease in current following activation frequently results in a transient current time course in response to a maintained stimulus (Figure 5.1). This process has classically been referred to as inactivation in voltage-gated channels and desensitization in ligand-gated channels. Inactivation is not to be confused with deactivation, which is simply the reverse of the process of activation. The terms "adaptation" and "tachyphylaxis" (*tachy*, "rapid"; and *phylaxis*, "to guard or protect") are also used to describe the consequence of channel inactivation or desensitization, with an emphasis on cellular responses.

Inactivation and desensitization play a variety of physiological roles in cells. The inactivation of the voltage-gated Na^+ channels is vital to decrease the Na^+ conductance to allow repolarization of the action potential. Similarly, desensitization of glutamate-activated channels helps to terminate the postsynaptic response to maintained neurotransmitter in the synaptic cleft. For voltage-gated Ca^{2+} channels, inactivation helps to limit the amount of Ca^{2+} entering the cell, which is toxic to the cells when unabated. In sensory signaling, desensitization of channels can provide negative feedback that extends the dynamic range of the sensory stimulus that the sensory receptor can detect.

DOI: 10.1201/9781003096214-6

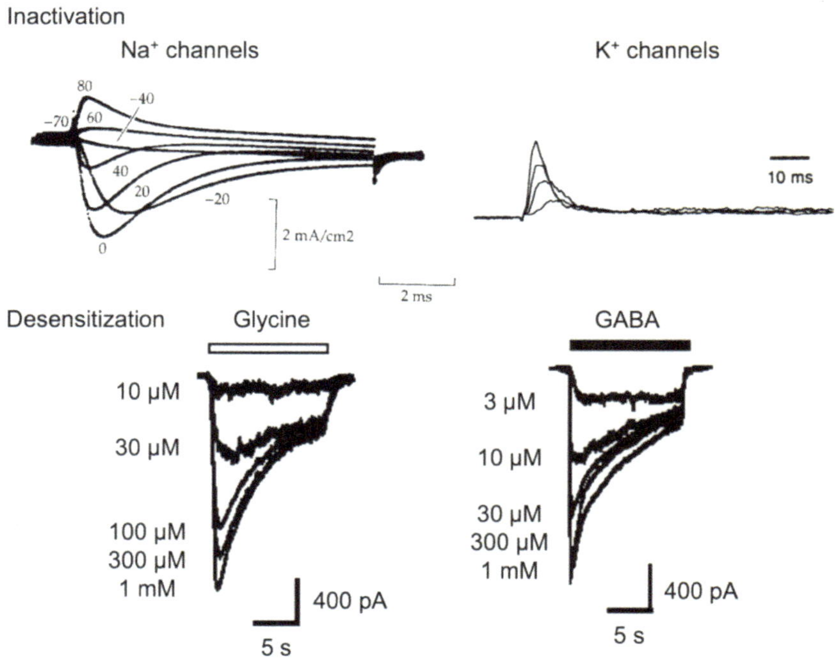

FIGURE 5.1
Examples of inactivation and desensitization. (From Armstrong et al., 1973; Hoshi et al., 1990; Li et al., 2003.)

5.2 Energetics of Inactivation and Desensitization

The molecular mechanisms of inactivation and desensitization are different from channel to channel. However, from an energetic standpoint, there are a number of common themes for most forms of inactivation and desensitization. (1) In all cases, the decrease in ionic current results from a decrease in the open probability of the channel, not a decrease in the conductance of the open channel itself. This indicates that the inactivation or desensitization process involves one or more conformational states of the channel that have a closed pore but are distinct from the resting closed states. (2) In contrast to resting closed states, inactivated or desensitized closed states are very stable in the presence of an activating stimulus so that, at steady state, the pore would prefer to be inactivated or desensitized rather than in a resting closed state or open state. (3) The recovery from inactivation or desensitization almost always requires removing the stimulus for a period of time, either repolarizing the membrane or removing the ligand. (4) Because, in the presence of the stimulus, the inactivated or desensitized state is more stable than the resting or activated state, inactivation or desensitization causes gating charge immobilization (for voltage-gated channels) or increased ligand-binding affinity (for ligand-gated channels). And (5) the inactivation or desensitization process is almost always coupled to activation. Coupling refers to the fact that inactivation or desensitization is much more favorable from activated states of the channel than from resting states. This coupling guarantees that inactivation or desensitization of the channel will only occur after an activating stimulus and not from resting closed states.

The energetics of channel conformational changes can be described by kinetic schemes like those shown in Figure 5.2. In these schemes, resting closed states are indicated by the

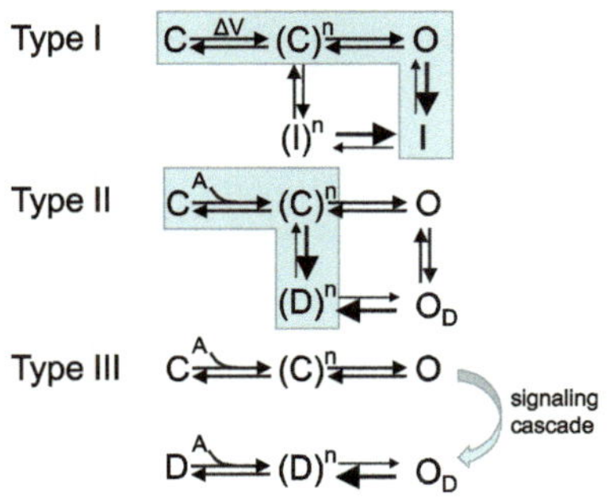

FIGURE 5.2
Mechanisms for inactivation and desensitization.

letter C, open states by the letter O, and inactivated or desensitized states by the letters I or D, respectively. The arrows between the states represent the allowed conformational transitions, with larger arrows indicating more rapid transitions. The activation transitions are represented as horizontal arrows. Activation of most channels requires rearrangements in the sensor domain in multiple subunits, therefore activation requires multiple (n) voltage-sensor rearrangements (ΔV) or agonist-binding events (A), followed by a final opening transition. Inactivation or desensitization transitions are represented as vertical arrows and can, in theory, occur from any of these resting or activated states. The coupling between activation and inactivation/desensitization is reflected in the different sizes of the vertical arrows from different activated states (or the absence of an arrow altogether).

Inactivation and desensitization arise most commonly from one of three different mechanisms with different kinds of gates and coupling (Figure 5.2). In the type I mechanism, inactivation/desensitization involves a gate in the channel separate from the activation gate. The closure of the inactivation gate is coupled to the activation of each subunit or opening of the channel, so that inactivation occurs primarily from the open state. Recovery from this form of inactivation/desensitization can occur either through the open state, producing "hooks" in the tail currents, or through closed states.

In the type II mechanism, inactivation/desensitization involves reclosure of the same gate that opened during activation due to a decrease in the coupling efficiency between the activation conformational change and channel opening (Figure 5.2). This form of desensitization occurs because the activation process of the ligand-binding domain puts a strain on the channel to open. The strain is a thermodynamic consequence of the activated configuration of the sensing domains (voltage-sensor domain or ligand-binding domain) promoting the opening of the pore. This strain can be relieved either by channel opening or desensitizing. Desensitization results from a slippage in the coupling between ligand binding and channel opening, so the channel-activation gate recloses. This mechanism causes desensitization to occur primarily from the closed state immediately before opening and makes subsequent opening very unfavorable (because ligand binding is no longer coupled to channel opening).

The type III mechanism for inactivation/desensitization occurs when the permeant ion initiates a cell signaling cascade that negatively feeds back on the channel (Figure 5.2). This feedback generally makes the opening transition less favorable and causes reclosure of the activation gate similar to the type II mechanism. One common example is Ca^{2+} entering via a Ca^{2+}-permeable channel and binding to calmodulin. If that Ca^{2+}–calmodulin then binds to the channel preferentially to closed states, then it will make opening less favorable and cause reclosure of the channel-activation gate. Ca^{2+}-dependent inactivation of voltage-gated Ca^{2+} channels and desensitization of cyclic nucleotide-gated (CNG) channels, for example, are thought to occur by this mechanism (see Volume II, Chapter 3 and Chapter 11). Similarly, Ca^{2+} entering via a Ca^{2+}-permeable channel can also activate other enzymes, such as protein kinases, lipid kinases or phospholipases. Activation of phospholipase C, for example, causes hydrolysis of PIP_2. If the channel is normally opened by PIP_2, then the channel will reclose as a result of PIP_2 hydrolysis. Transient receptor potential (TRP) channels are thought to desensitize by this mechanism. Since, for the type III mechanism, the feedback varies widely from channel to channel, this form of inactivation/desensitization will be discussed in greater detail in the chapters on the individual channels.

The remainder of this chapter will discuss the molecular basis of the type I and type II inactivation/desensitization mechanisms for three of the most well-understood channels: voltage-gated Na^+ channels, voltage-gated K^+ channels and ionotropic glutamate receptors.

5.3 Na^+ Channel Inactivation

Ion channel inactivation was first described for the voltage-gated Na^+ channel in squid giant axons by Hodgkin and Huxley (HH) (Hodgkin and Huxley, 1952b). They showed with voltage-clamp recordings that depolarizing voltage steps caused an increase, and subsequent decrease in Na^+ current (Figure 5.1). They then went on to model the time course and voltage dependence of the Na^+ and K^+ currents and showed that these currents could account for the shape, voltage threshold, temperature dependance and conduction of the action potential (Hodgkin and Huxley, 1952a). In particular, their model revealed that the inactivation of the Na^+ channels was largely responsible for spontaneous repolarization of the action potential.

The HH model for Na^+ channels (the sodium-carrying system, as Hodgkin and Huxley called them) assumed the independent movement of three activation particles (*m*) to open the channel and one inactivation particle (*h*) to inactivate the channel. Thus, the Na^+ current can be described by the following equation:

$$I_{Na} = m^3 h g_{Na} \left(V - V_{Na}\right)$$

where *m* and *h* are the voltage- and time-dependent probabilities of each gating particle existing in its permissive state (one that allows the channel to be open). This model does not have any coupling between the activation and inactivation particles (Figure 5.3A left; note that the inactivation rates, as depicted by the size of the arrows, are the same for each of the closed and open states). The voltage-dependent decay rate arose from voltage dependence of the inactivation particle h.

Decades later, an alternative model for Na^+ channel inactivation was proposed, a type I mechanism (Aldrich et al., 1983; Armstrong and Bezanilla, 1977; Bezanilla and Armstrong, 1977). In this model, the inactivation transition is voltage-independent and coupled to activation (Figure 5.3B, left). The apparent voltage dependence of the steady-state inactivation ("*h* infinity curve") and decay time course arise entirely from the voltage dependence of activation. In other words, the decay time course of the sodium conductance does not reflect the inactivation transition itself but instead the activation transition. This type I mechanism can explain the time course and voltage-dependent decay of the macroscopic

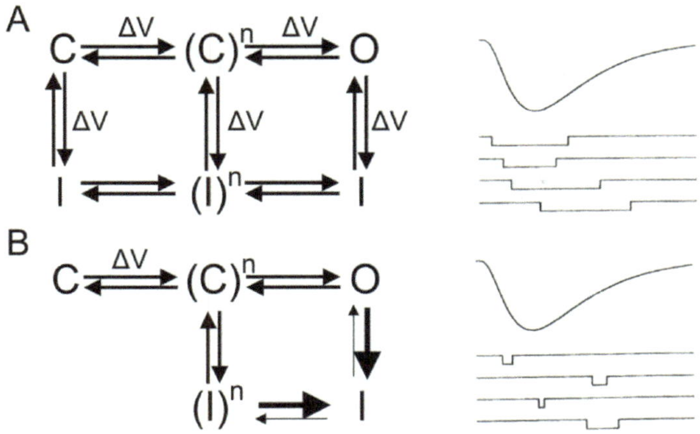

FIGURE 5.3
Possibilities for Na^+ channel gating. (A) Left, HH model. Right, predictions of single-channel currents and macroscopic current. (B) Left, type I model. Right, predictions of single-channel currents and macroscopic current. (From Aldrich, 1986.)

Na^+ conductance as well as the HH model. However, the two mechanisms make different predictions about the behavior of single-channel currents (Figure 5.3, right) (Aldrich, 1986). The HH model predicts long, voltage-dependent open durations, while the type I model predicts short, voltage-independent open durations. In addition, for the type I model, the voltage-dependent first latencies (time to first opening) are long and determine the time course and the voltage dependence of the decay in the macroscopic current. Indeed, single-channel recordings of mammalian voltage-gated Na^+ channels revealed long voltage-dependent first latencies and short, voltage-independent open durations. These experiments established that the inactivation in voltage-gated Na^+ channels is voltage-independent and strongly coupled to activation.

Clues to the molecular basis for inactivation in voltage-gated Na^+ channels also came initially from electrophysiological recordings in squid giant axons. Perfusing the inside of axons with pronase, a proteolytic enzyme, specifically abolished the inactivation process of the channel, while leaving activation intact (Figure 5.4A) (Armstrong et al., 1973). These experiments suggest that the inactivation may arise from an intracellular portion of the Na^+ channel acting as an intracellular gate that plugs the open-channel pore (Armstrong and Bezanilla, 1977). This "ball-and-chain" mechanism could explain why the inactivation is voltage-independent, coupled to activation and susceptible to proteolytic agents.

The identity of a putative intracellular inactivation gate was suggested by mutagenesis experiments. The voltage-gated Na^+ channel consists of four homologous domains, acting as pseudo subunits (Figure 5.4B). Cleavage of the intracellular linker between homologous domains III and IV (III–IV linker) was shown to virtually abolish rapid inactivation (Stühmer et al., 1989). Furthermore, point mutations of just the sequence IFM in the middle of the III–IV linker were sufficient to partially or completely removal rapid inactivation (Figure 5.4B) (West et al., 1992). This led to the proposal that the III–IV linker acts as a "hinged lid" that is coupled to Na^+ channel activation, similar to the previously proposed ball-and-chain mechanism. Furthermore, the inactivation gate seems to be more coupled to the movement of the voltage sensor in homologous domain IV than the other domains.

More recently, structural biology has suggested a modification of this proposal. In multiple cryo-EM structures, the III–IV linker did not directly block the pore, but instead

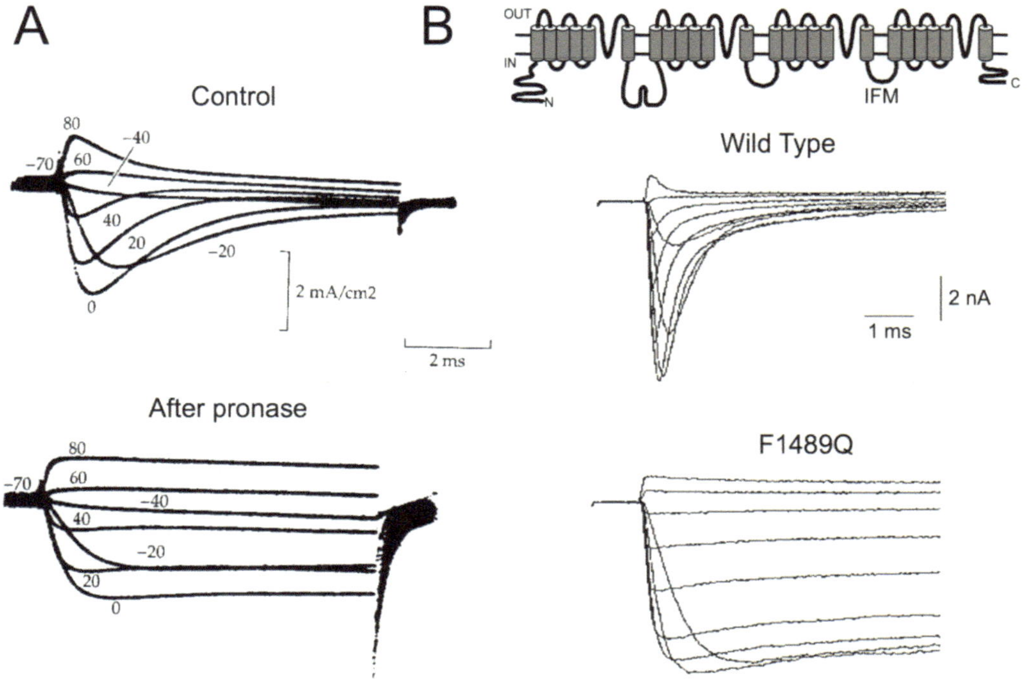

FIGURE 5.4
Inactivation of Na^+ channels is removed by pronase (A) and mutations in the III-IV linker (B). (From Armstrong et al., 1973; West et al., 1992.)

exhibited distinct interactions with the transmembrane core domain under conditions that would favor a closed, open or inactivated state (Figure 5.5) (Yan et al., 2017). This led to the idea that the intracellular linker between homologous domains III and IV might not act as a blocker but, instead, more as an allosteric modulator to promote inactivation, a "allosteric block" mechanism. For a more detailed discussion, readers are referred to Volume II, Chapter 2.

5.4 K^+ Channel N-Type Inactivation

While the HH K^+ conductance did not inactivate, we now know that almost all K^+ channels inactivate on some time scale. One form of K^+ channel inactivation is particularly apparent in the voltage-gated K^+ channel from the Shaker locus in *Drosophila* (see Figure 5.1). Like inactivation in voltage-gated Na^+ channels, this inactivation produces a rapid, voltage-dependent decline in the K^+ current that is abolished by application of intracellular proteolytic enzymes. Also like inactivation in voltage-gated Na^+ channels, single-channel recordings reveal that the inactivation transition is voltage-independent and highly coupled to activation (Zagotta and Aldrich, 1990). These findings suggest that inactivation in K^+ channels also occurs by a type I mechanism.

These similarities to Na^+ channels suggested that this form of inactivation of voltage-gated K^+ channels may occur by the ball-and-chain mechanism. Indeed, the identity of

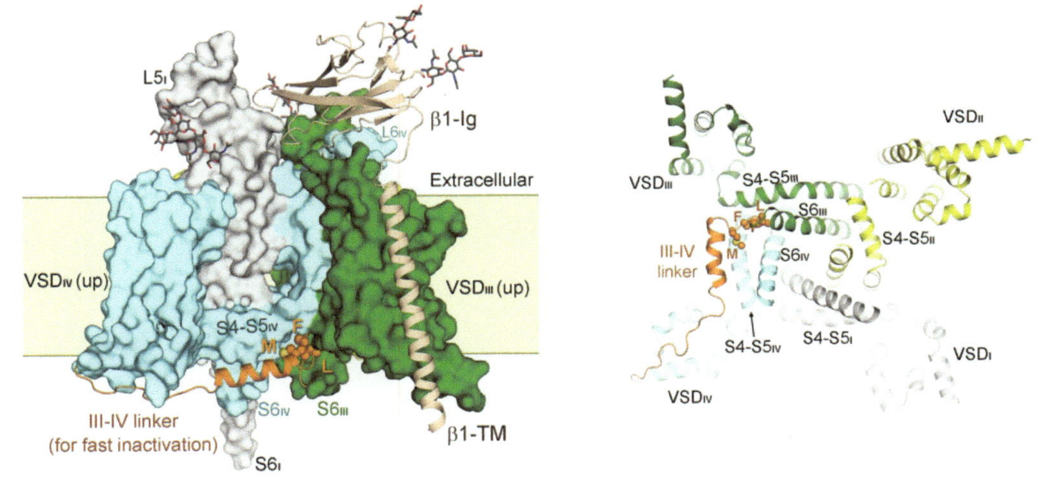

FIGURE 5.5
Cryo-EM structure of the Nav1.4-β1 Na^+ channel from electric eel showing position of the III–IV linker. (From Yan et al., 2017.)

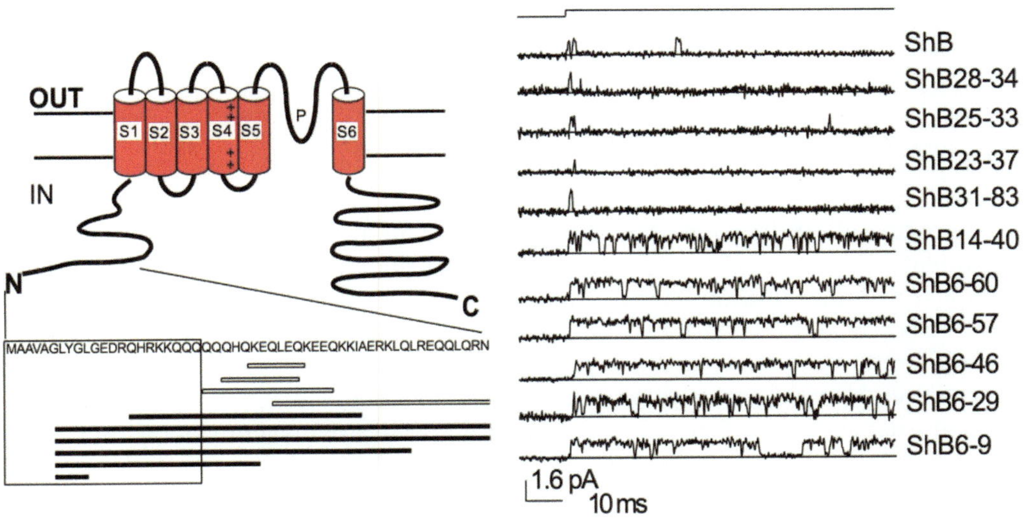

FIGURE 5.6
N-terminal deletions in ShB K^+ channels disrupt inactivation. (From Hoshi et al., 1990.)

an intracellular inactivation gate was suggested by a series of deletion mutants in the N-terminal region of the Shaker K^+ channel (Figure 5.6) (Hoshi et al., 1990). Deletions in the first 20 amino acids of the protein abolished inactivation, whereas deletions in the neighboring region accelerated inactivation. These results suggested that the N-terminal 20 amino acids may constitute an inactivation ball while the neighboring region might form a chain.

To test if the ball was sufficient to produce inactivation, a synthetic peptide corresponding to the first 20 amino acids was synthesized and applied to an amino-terminal-deleted Shaker K^+ channel in inside-out membrane patches (Figure 5.7) (Zagotta et al., 1990). Indeed, application of the peptide to these inactivation deficient channels

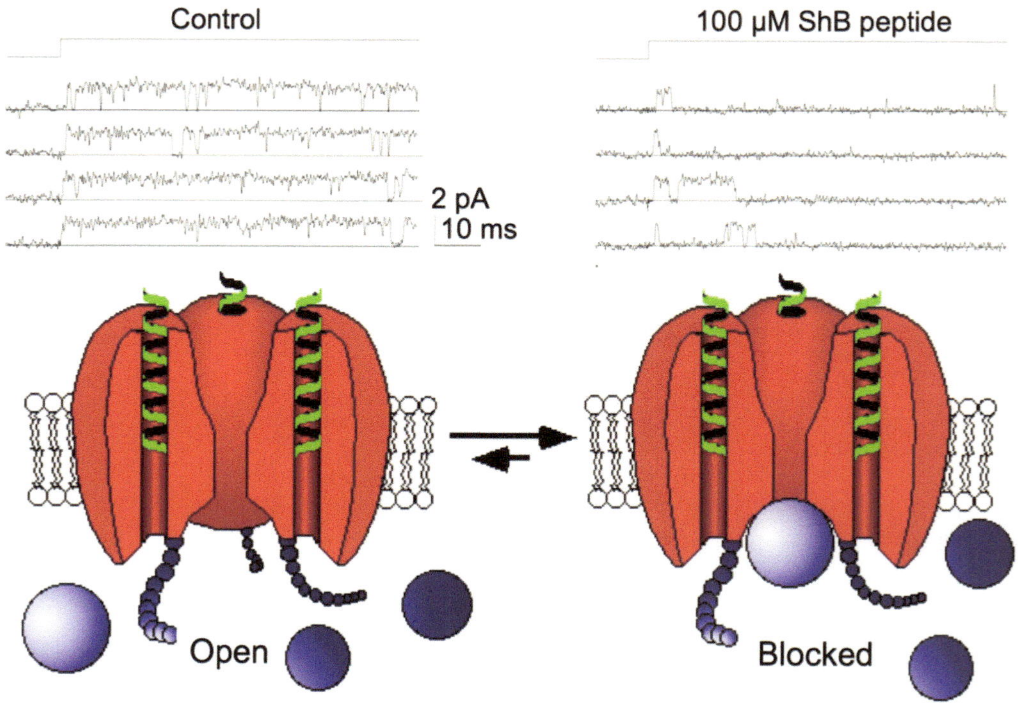

FIGURE 5.7
Inactivation of ShB K^+ channels is restored by ShB(1-20) peptide. (From Zagotta et al., 1990.)

recapitulated all of the properties of native inactivation. Subsequent experiments demonstrated that this peptide was an open-channel pore blocker and that a ball domain from just a single subunit was sufficient to inactivate the channels (MacKinnon et al., 1993). This ball-and-chain form of inactivation was called "N-type inactivation" because it was altered by mutations and alternatively spliced variants in the N-terminal region of the channel. N-type inactivation has been demonstrated in a number of other K^+ channels, and the structure of the N-type inactivated state was recently solved for a bacterial Ca^{2+}-activated K^+ channel (Fan et al., 2020). Also, N-type inactivation is conferred on some channels by the accessory subunit Kvβ1.1, which contributes its own ball-and-chain motif (Rettig et al., 1994).

5.5 K^+ Channel C-Type Inactivation

N-type inactivation is not the only form of inactivation in K^+ channels. This first became apparent when looking at recordings of N-terminal-deleted channels on a much slower time scale (Figure 5.8) (Hoshi et al., 1991). While this type of inactivation is sometimes referred to as slow inactivation, it is not always slow. Recordings from different C-terminal alternatively spliced variants, with deletions in their N-terminal region, revealed that this form of inactivation can be quite fast (Figure 5.8). Because it is altered in C-terminal variants

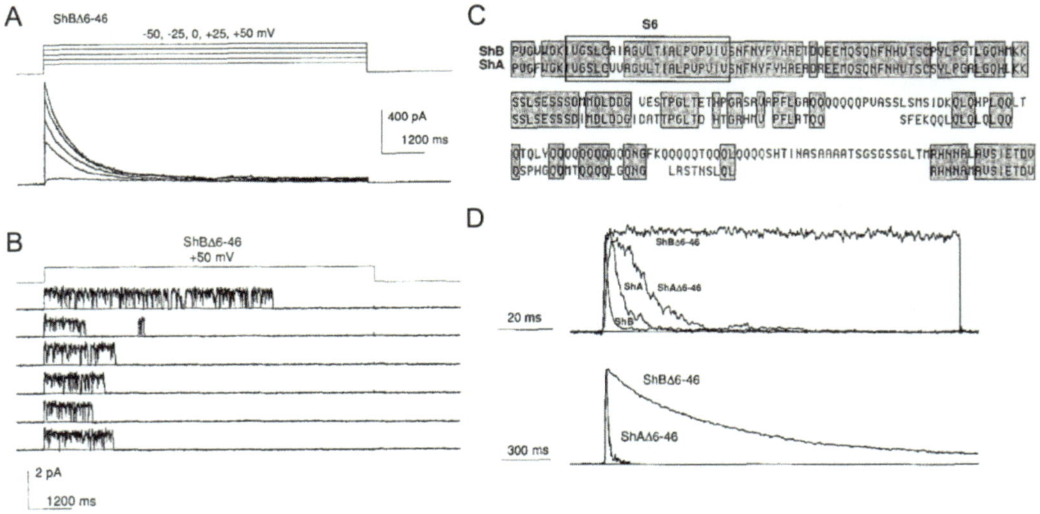

FIGURE 5.8
C-type inactivation of ShB K^+ channels. Macroscopic (A) and single-channel (B) currents from ShBΔ6-46 channels with N-type inactivation removed. (C) Sequence of the C-terminal region of the Shaker variants ShB and ShA. D. Comparison of the inactivation time courses of ShBΔ6-46 and ShAΔ6-46 channels on two different time scales. (From Hoshi et al., 1991.)

of Shaker K^+ channels, this inactivation has been referred to as C-type inactivation. C-type inactivation is similar to other forms of inactivation called P-type or U-type inactivation.

The first clues to the molecular mechanism for C-type inactivation came from examining the rates of C-type inactivation in C-terminal variants of Shaker, ShA and ShB (Figure 5.8) (Hoshi et al., 1991). The rate of C-type inactivation is about 50-fold greater in ShAΔ6-46 channels than in ShBΔ6-46 channels. The sequences of ShA and ShB are identical from the N-terminus to the selectivity filter, then begin to diverge with one difference just before the S6 transmembrane segment, two differences in the S6 and many differences in the intracellular C-terminal region. Surprisingly, the difference in the rate of C-type inactivation could be completely attributed to a single amino-acid difference in the S6…a valine in ShA (V463) and an alanine in ShB (A463). In the structures of K^+ channels, this amino acid lies behind the selectivity filter, suggesting that this C-type inactivation might result from a rearrangement of the selectivity filter, a separate gate from the activation gate located more intracellularly at the S6 bundle crossing. Indeed, other mutations in or near the selectivity filter of K^+ channels have profound effects on the rate of C-type inactivation.

Another clue that C-type inactivation involves the selectivity filter is that permeant ions and pore blockers can alter the rate of C-type inactivation. One particularly insightful study looked at the effects of intracellular and extracellular tetraethylammonium (TEA), a K^+ channel blocker, on the rate of N-type and C-type inactivation (Choi et al., 1991). If the blocker is independent of inactivation (either N-type or C-type), then we would expect that a subsaturating concentration of blocker would not affect the apparent rate of inactivation. However, if the blocker competes with inactivation, interfering with the closing of the inactivation gate, we would expect that the blocker would slow the rate of inactivation, producing a characteristic "cross over" with the current in the absence of blocker.

The result of the experiment on Shaker K^+ channels with primarily N-type or C-type inactivation is that intracellular TEA interferes only with N-type inactivation and

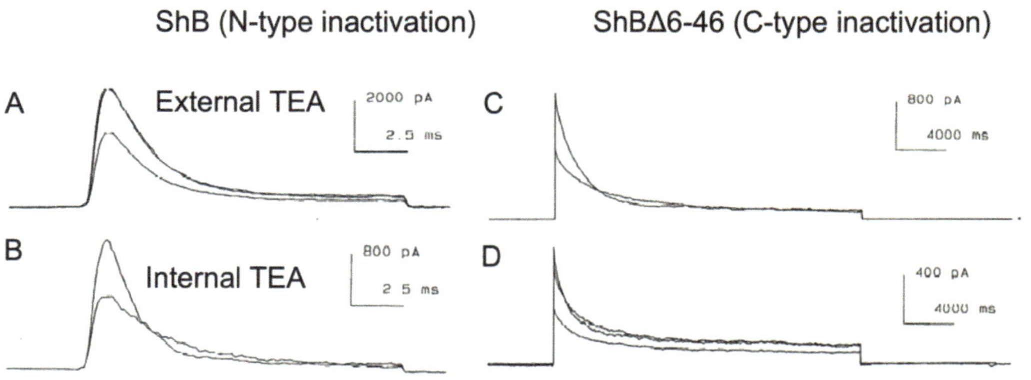

FIGURE 5.9
Block reveals that N-type and C-type inactivation are structurally distinct. Each panel shows the current without and with block by about a 50% saturating concentration of TEA. (A) ShB with external TEA. (B) ShB with internal TEA. (C) ShBΔ6-46 with external TEA. (D) ShBΔ6-46 with internal TEA. (From Choi et al., 1991.)

extracellular TEA interferes only with C-type inactivation (Figure 5.9) (Choi et al., 1991). This experiment reveals that N- and C-type inactivation are structurally distinct. N-type inactivation involves a more intracellular gate, as expected for the ball-and-chain mechanism, and C-type inactivation involves a more extracellular gate. Since the external TEA binding site is known to be at the entrance to the selectivity filter, it has been proposed that the selectivity filter might collapse during C-type inactivation. It has been suggested that C-type inactivation might also involve a widening of the selectivity filter (Hoshi and Armstrong, 2013). Recently, cryo-EM structures have revealed the detailed change in conformation of the selectivity filter during C-type inactivation (Reddi et al., 2022; Tan et al., 2022).

5.6 Ionotropic Glutamate Receptor Desensitization

Desensitization is perhaps best understood in ionotropic glutamate receptors or glutamate-activated channels. In response to a step in glutamate concentration, GluR2 channels open and then rapidly desensitize (Figure5.10A) (Sun et al., 2002). This desensitization occurs preferentially from the closed state immediately before channel opening by a type II mechanism. Two perturbations largely eliminate this desensitization, the L483Y mutation and the drug cyclothiazide (CLZ). These two perturbations are located at the dimer interface between the glutamate-binding domains from neighboring subunits (Figure5.11) and greatly enhance the dimerization affinity of isolated glutamate-binding domains. In fact, there is a very strong correlation between the energy of desensitization(ΔG_{des}) and the energy of dimerization (ΔG_{dd}) (Figure 5.10B) (Sun et al., 2002). These results suggest that the binding of glutamate puts a strain on the intersubunit interaction between the glutamate-binding domains (Figure 5.12). This strain either serves to open the channel gate or breaks the subunit interface causing desensitization. With the subunit interface broken, the conformational change in the ligand-binding domain is no longer coupled to the opening of the pore, and the activation gate recloses.

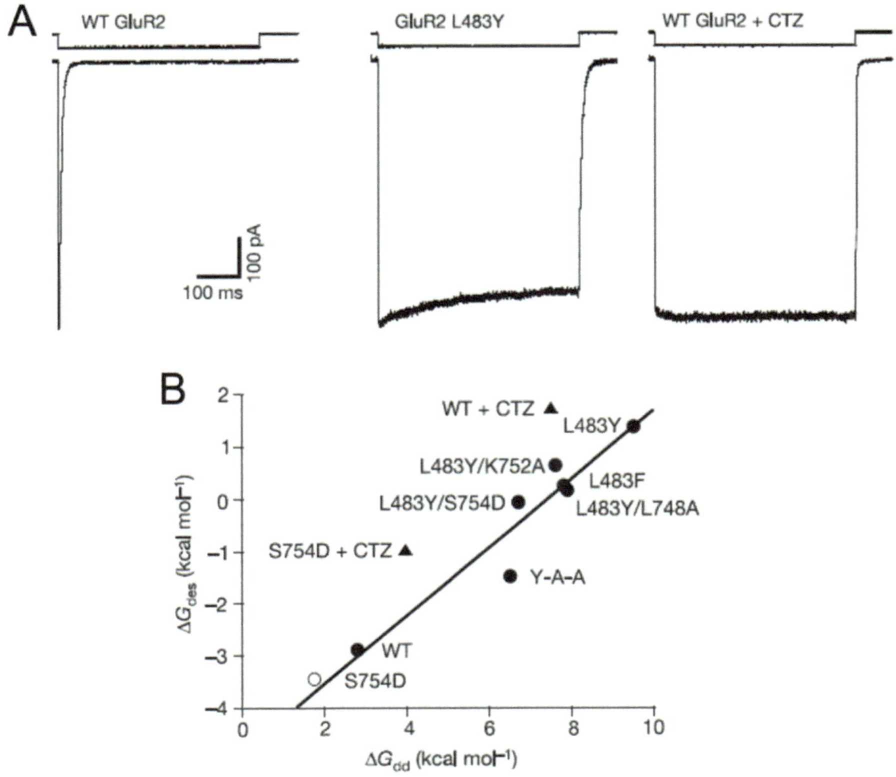

FIGURE 5.10
Desensitization of GluR2. (A) Desensitization of GluR2 and its removal by the L483Y mutation and CLZ. (B) Correlation between the free energy of desensitization and the free energy of dimerization of the ligand-binding domain. (From Sun et al., 2002.)

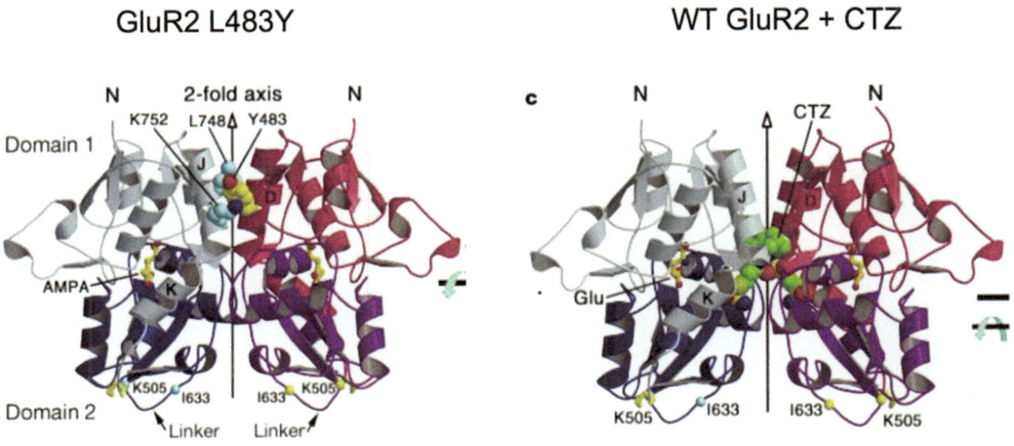

FIGURE 5.11
L483Y mutation and CTZ are located at the dimer interface of the ligand-binding domain. (From Sun et al., 2002.)

5.7 Type II and III Inactivation/Desensitization in Other Channels

It now appears that the type II inactivation/desensitization mechanism is not confined to ligand-gated channels. The inactivation found in a sea urchin hyperpolarization-activated cyclic nucleotide-gated (HCN) channel is thought to involve a "desensitization to voltage" (Shin et al., 2004). It was proposed that activation of the voltage sensor puts strain on parts of the channel involved in coupling the voltage sensor to the pore. This strain can be relieved by either opening the pore or slipping the coupling between the voltage sensor and the gate, producing desensitization (Figure 5.13). A similar type II desensitization-to-voltage mechanism has also been proposed for the closed-state inactivation of the voltage-gated K^+ channel Kv4.2 (Dougherty et al., 2008). This suggests that the mechanism of desensitization may be more widespread than just ligand-gated channels.

Many channels, like HCN channels, are polymodal, activated by multiple stimuli. Because regions of the channel involved in coupling may be different for different stimuli, type II and type III inactivation/desensitization can be stimulus specific. For example, while the sea urchin HCN channel is desensitized by voltage, the ligand cAMP acts to prevent the desensitization, not promote desensitization as seen in other ligand-gated channels. TRPV1 channels are also polymodal, activated by the binding of capsaicin, low pH, PIP_2 and noxious heat. Surprisingly, desensitization of TRPV1 to heat does not prevent activation by capsaicin (Cao et al., 2014). Understanding the mechanisms for the inactivation/desensitization in each channel ultimately has significant bearing on the physiological role for the channel and its inactivation/desensitization.

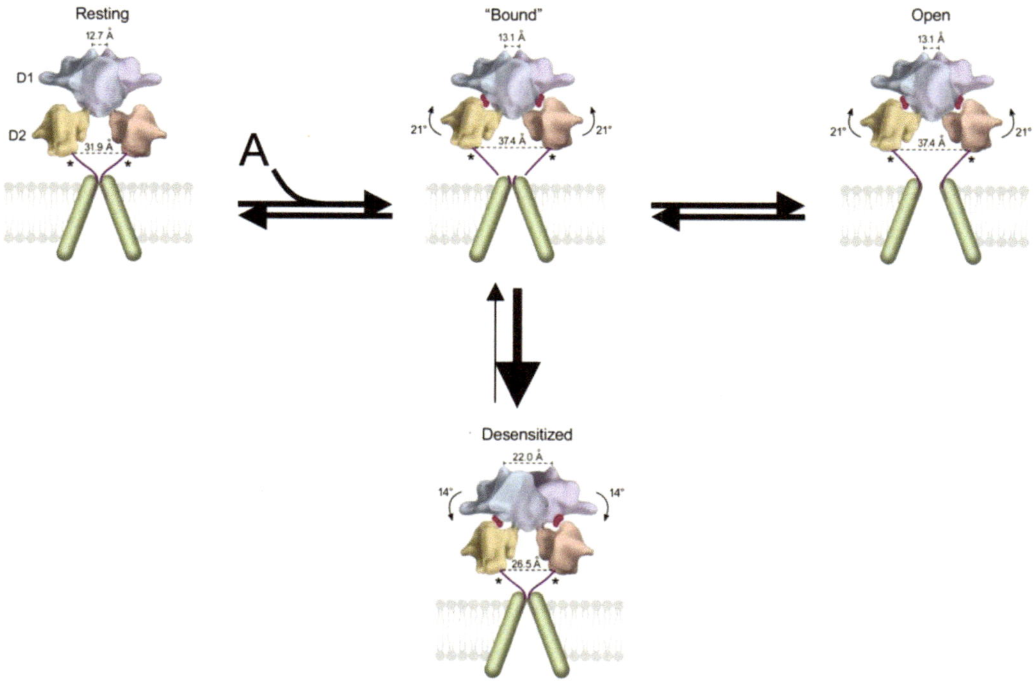

FIGURE 5.12
Mechanism for desensitization of GluR2. (From Armstrong et al., 2006.)

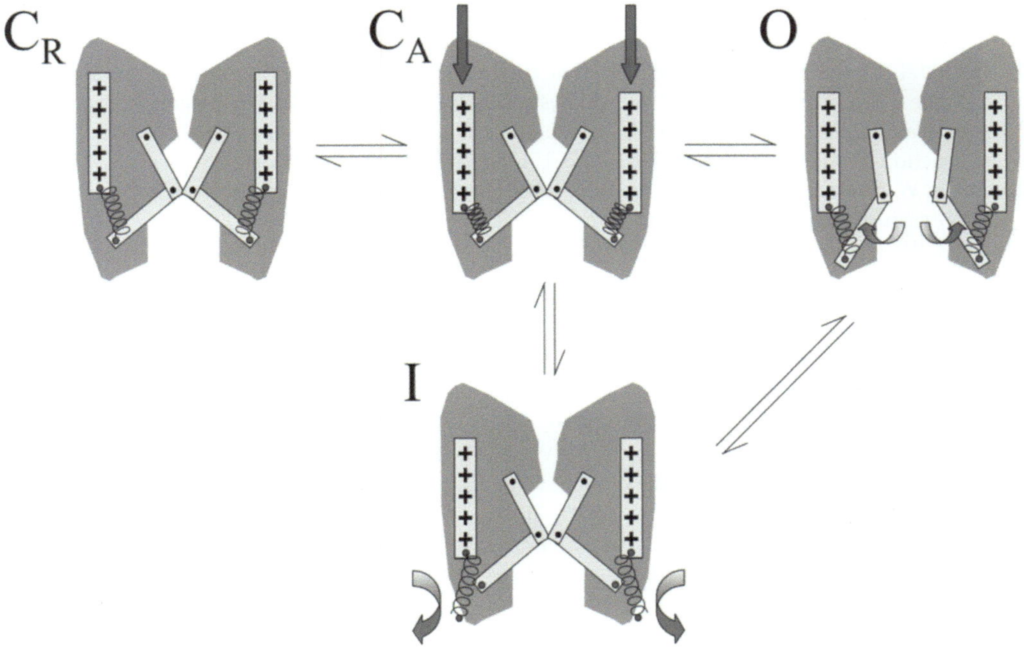

FIGURE 5.13
Mechanism of voltage-dependent desensitization of HCN channels. (From Shin et al., 2004.)

Suggested Readings

Aldrich, R. W. 1986. "Voltage-dependent gating of sodium channels: towards an integrated approach." *Trends Neurosci* 9:82–6. doi:10.1016/0166-2236(86)90028-7.

Aldrich, R. W., D. P. Corey, and C. F. Stevens. 1983. "A reinterpretation of mammalian sodium channel gating based on single channel recording." *Nature* 306:436–41. doi:10.1038/306436a0.

Armstrong, C. M., and F. Bezanilla. 1977. "Inactivation of the sodium channel. II. Gating current experiments." *J Gen Physiol* 70:567–90. doi:10.1085/jgp.70.5.567.

Armstrong, C. M., F. Bezanilla, and E. Rojas. 1973. "Destruction of sodium conductance inactivation in squid axons perfused with pronase." *J Gen Physiol* 62:375–91. doi:10.1085/jgp.62.4.375.

Armstrong, N., J. Jasti, M. Beich-Frandsen, and E. Gouaux. 2006. "Measurement of conformational changes accompanying desensitization in an ionotropic glutamate receptor." *Cell* 127:85–97. doi:10.1016/j.cell.2006.08.037.

Bezanilla, F., and C. M. Armstrong. 1977. "Inactivation of the sodium channel. I. Sodium current experiments." *J Gen Physiol* 70:549–66. doi:10.1085/jgp.70.5.549.

Cao, X., L. Ma, F. Yang, K. Wang, and J. Zheng. 2014. "Divalent cations potentiate TRPV1 channel by lowering the heat activation threshold." *J Gen Physiol* 143:75–90. doi:10.1085/jgp.201311025.

Choi, K. L., R. W. Aldrich, and G. Yellen. 1991. "Tetraethylammonium blockade distinguishes two inactivation mechanisms in voltage-activated K+ channels." *Proc Natl Acad Sci U S A* 88:5092–5. doi:10.1073/pnas.88.12.5092.

Dougherty, K., J. A. De Santiago-Castillo, and M. Covarrubias. 2008. "Gating charge immobilization in Kv4.2 channels: the basis of closed-state inactivation." *J Gen Physiol* 131:257–73. doi:10.1085/jgp.200709938.

Fan, C., N. Sukomon, E. Flood, J. Rheinberger, T. W. Allen, and C. M. Nimigean. 2020. "Ball-and-chain inactivation in a calcium-gated potassium channel." *Nature* 580:288–93. doi:10.1038/s41586-020-2116-0.

Hodgkin, A. L., and A. F. Huxley. 1952a. "A quantitative description of membrane current and its application to conduction and excitation in nerve." *J Physiol* 117:500–44.

Hodgkin, A. L., and A. F. Huxley. 1952b. "The dual effect of membrane potential on sodium conductance in the giant axon of Loligo." *J Physiol* 116:497–506. doi:10.1113/jphysiol.1952.sp004719.

Hoshi, T., and C. M. Armstrong. 2013. "C-type inactivation of voltage-gated K+ channels: pore constriction or dilation?" *J Gen Physiol* 141:151–60. doi:10.1085/jgp.201210888.

Hoshi, T., W. N. Zagotta, and R. W. Aldrich. 1990. "Biophysical and molecular mechanisms of Shaker potassium channel inactivation." *Science* 250:533–8.

Hoshi, T., W. N. Zagotta, and R. W. Aldrich. 1991. "Two types of inactivation in Shaker K+ channels: effects of alterations in the carboxy-terminal region." *Neuron* 7:547–56.

Li, Y., L. J. Wu, P. Legendre, and T. L. Xu. 2003. "Asymmetric cross-inhibition between GABAA and glycine receptors in rat spinal dorsal horn neurons." *J Biol Chem* 278:38637–45. doi:10.1074/jbc.M303735200.

MacKinnon, R., R. W. Aldrich, and A. W. Lee. 1993. "Functional stoichiometry of Shaker potassium channel inactivation." *Science* 262:757–9. doi:10.1126/science.7694359.

Reddi, R., K. Matulef, E. A. Riederer, M. R. Whorton, and F. I. Valiyaveetil. 2022. "Structural basis for C-type inactivation in a Shaker family voltage-gated K." *Sci Adv* 8:eabm8804. doi:10.1126/sciadv.abm8804.

Rettig, J., S. H. Heinemann, F. Wunder, C. Lorra, D. N. Parcej, J. O. Dolly, and O. Pongs. 1994. "Inactivation properties of voltage-gated K+ channels altered by presence of beta-subunit." *Nature* 369:289–94. doi:10.1038/369289a0.

Shin, K. S., C. Maertens, C. Proenza, B. S. Rothberg, and G. Yellen. 2004. "Inactivation in HCN channels results from reclosure of the activation gate: desensitization to voltage." *Neuron* 41:737–44. S0896627304000832 [pii].

Stühmer, W., F. Conti, H. Suzuki, X. D. Wang, M. Noda, N. Yahagi, H. Kubo, and S. Numa. 1989. "Structural parts involved in activation and inactivation of the sodium channel." *Nature* 339:597–603. doi:10.1038/339597a0.

Sun, Y., Olson, R., Horning, M., Armstrong, N., Mayer, M., and Gouaux, E. 2002. "Mechanism of glutamate receptor desensitization." *Nature* 417:245–53. doi:10.1038/417245a.

Tan, X. F., C. Bae, R. Stix, A. I. Fernández-Mariño, K. Huffer, T. H. Chang, J. Jiang, J. D. Faraldo-Gómez, and K. J. Swartz. 2022. "Structure of the Shaker Kv channel and mechanism of slow C-type inactivation." *Sci Adv* 8:eabm7814. doi:10.1126/sciadv.abm7814.

West, J. W., D. E. Patton, T. Scheuer, Y. Wang, A. L. Goldin, and W. A. Catterall. 1992. "A cluster of hydrophobic amino acid residues required for fast Na(+)-channel inactivation." *Proc Natl Acad Sci U S A* 89:10910–4.

Yan, Z., Q. Zhou, L. Wang, J. Wu, Y. Zhao, G. Huang, W. Peng, H. Shen, J. Lei, and N. Yan. 2017. "Structure of the Na v 1.4-β1 complex from electric eel." *Cell* 170:470–82.e411. doi:10.1016/j.cell.2017.06.039.

Zagotta, W. N., and R. W. Aldrich. 1990. "Voltage-dependent gating of Shaker A-type potassium channels in Drosophila muscle." *J Gen Physiol* 95:29–60.

Zagotta, W. N., T. Hoshi, and R. W. Aldrich. 1990. "Restoration of inactivation in mutants of Shaker potassium channels by a peptide derived from ShB." *Science* 250:568–71.

6

Ion Channel Inhibitors

Matthew J. Marquis and Jon T. Sack

CONTENTS

6.1 Mechanisms of Inhibition

Ion channel inhibitors can test a hypothesis or save a life. There is a broad chemical palette of ion channel inhibitors: metal ions, alkaloids, synthetic drugs, venom peptide toxins and endogenous modulatory proteins. These inhibitors act by various mechanisms. Blocker is scientific vernacular for inhibitor. In the context of ion channels, the term blocker can conjure up the imagery of a plugged pore, and many channel inhibitors act in such a fashion. However, not all ion channel inhibitors block the pore, and the term lends itself to imprecise use. To avoid semantic confusion here, the term *pore blocker* connotes a drug that itself occludes the ionic conduction pathway. A mechanistic alternative to pore blockade is allosteric inhibition, where an inhibitor causes a channel to close itself. Allosteric inhibition is also referred to as gating modification or self block. These two mechanisms, depicted in Figure 6.1, describe the fundamental workings of many ion channel inhibitors.

Inhibitors can work by a combination of both pore blockade and allosteric inhibition. An additional inhibitory mechanism involves prevention of a stimulus from acting on a channel, e.g., competitive steric displacement of an agonist by an antagonist. Competitive inhibition is not addressed explicitly in this chapter, as it is comprehensively treated in many other texts (Hilal-Dandan, Brunton, and Goodman 2013; Wyman and Gill 1990). Topics more unique to ion channels are considered here: the physics of pore blockade and gating modulation, with particular focus on how the transmembrane voltage can influence inhibitor efficacy.

DOI: 10.1201/9781003096214-7

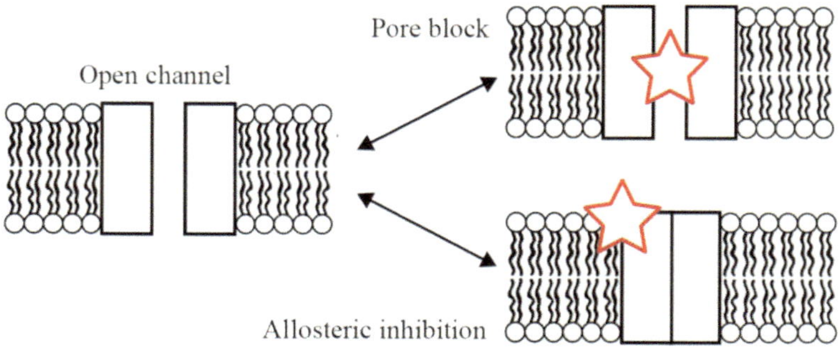

FIGURE 6.1
Types of inhibitors. Pore block results from inhibitors binding in the pore to occlude ion permeation. Allosteric inhibition results from inverse agonists that cause an open channel to close.

6.2 Pore Blockers

The inhibitory mechanism of pore blockers is intuitive: occlusion of the permeant ion path. Ion channel pore blockers play important roles in research and medicine. Tetrodotoxin is produced by symbiotic bacteria in poisonous creatures including pufferfish and newts. Tetrodotoxin blocks the pore of sodium channels and is used in physiology research to suppress sodium currents. Many venomous creatures make pore-blocking peptide toxins that researchers use to identify ion channel types. A peptide toxin from cone snail venom, ziconotide, is a pain therapeutic. The anesthetics phencyclidine and ketamine are NMDA receptor pore blockers. Inadvertent pore block of the hERG potassium channel by drugs such as terfenadine can trigger sudden cardiac death. The expanding pharmacopeia of ion channel pore blockers is far too large to be discussed in a single chapter, so we concern ourselves here with the mechanism of inhibition, which for all pore blockers is essentially the same: binding of the blocker blocks the ion conduction path.

In its simplest form, the concentration–inhibition relation of pore block is described by the physics of classical ligand–receptor interaction. To many physiologists, the relevant term is the fraction of channels that remain conducting at equilibrium, $f_{unblocked}$. For the simple process of an inhibitor blocking the pore of a channel, depicted in Figure 6.2a, this fraction is determined by the inherent dissociation rate, k_{off}, and an association rate constant, k_{on}, that is first-order with respect to the chemical activity of the inhibitor, which we will refer to as the inhibitor concentration, [X]:

$$\frac{f_{unblocked}}{f_{blocked}} = \frac{k_{off}}{k_{on}[X]} \tag{6.1}$$

The rate constants k_{off} and k_{on} differ for each inhibitor–channel pair and in some cases can be measured directly. The ratio of the dissociation and association rate constants yields the dissociation constant, K_D:

$$K_D = \frac{k_{off}}{k_{on}[X]} \tag{6.2}$$

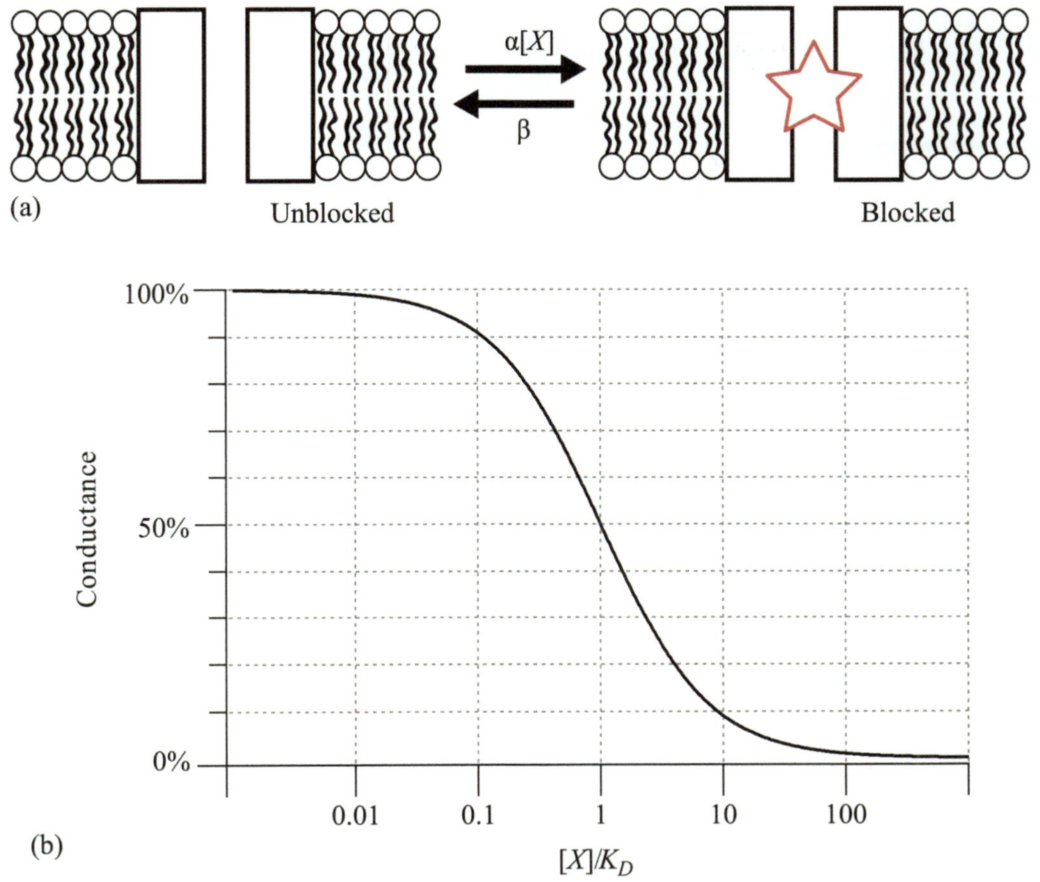

FIGURE 6.2
Pore blocker dose–response. (a) The fraction of channels blocked is determined by an association rate dependent on inhibitor concentration and an intrinsic dissociation rate. (b) When at equilibrium with a single binding site, a pore blocker will inhibit ionic conductance as a function of its concentration, [X]; and dissociation constant, K_D. Curve is plot of Equation 6.3.

The K_D has units of concentration and is a useful value to develop intuition about concentration dependence. In the case of a pore blocker that binds channels in a 1:1 fashion, the K_D is the concentration where 50% of channels will be inhibited, or IC_{50}, making it a straightforward measurable quantity. The K_D is sufficient to describe the concentration–response of the system at equilibrium.

$$\text{Conductance} = \frac{f_{\text{unblocked}}}{f_{\text{unblocked}} + f_{\text{blocked}}} = \frac{1}{1 + ([X]/K_D)} \tag{6.3}$$

A plot of Equation 6.3 demonstrates the dependence of channel block on the concentration of inhibitor (Figure 6.2b). Equation 6.3 is variously referred to as the law of mass action, a Langmuir binding isotherm or Hill logistic with a coefficient of 1. Note that there is a broad concentration range of inhibitor action, with five orders of magnitude increase in inhibitor concentration to span the conductance decrement from 1% inhibition to 99% inhibition (Figure 6.2b). One can never fully inhibit a current, but only approach a saturating value. Thus, when determining what inhibitor concentration to use, consideration needs to be given to how much remaining current is tolerable.

6.3 Perturbation of Pore Block

Structural analyses of blocked ion channels have illuminated atomic-level details of the pore blockade mechanism. These details include which features of the inhibitor form stabilizing interactions with the channel as well as where the inhibitor sits relative to the ions that permeate the channel pore. Structurally, pore blockers are indeed found to be blocking the ion path (Figure 6.3a–c). Consequently, the binding of pore blockers is impacted by permeant ions. The interactions between permeant ions and pore blockers can be quite complex. Importantly, the effects of permeant ions are dynamic. Ion channels open and close, changing the access of permeant ions in solution to blocker sites and hence modulating their effects on inhibitors. Additionally, transmembrane voltage change impacts permeant ions and can exert force on charged blockers directly. The effects of permeant ions and voltage change can dramatically alter the degree of inhibition by an ion channel blocker.

To quantitate the impact of a physical perturbation on the binding of a pore blocker, it is useful to calculate changes in binding energy. Permeant ions and voltage change the energy of the blocked state to make it more or less favorable. The amount of binding energy for a simple pore block is described by Equation 6.4:

$$\Delta G_{blocked} = -k_B T \ln\left(\frac{[X]}{K_D}\right) \tag{6.4}$$

where

$\Delta G_{blocked}$ is the Gibbs free energy for binding
k_B is the Boltzmann constant.
T is the absolute temperature.
ln is a logarithm with base e.

Note that binding energy does not saturate at any concentration. This is due to the binding rates of the blocker continuing to increase as concentration is increased.

When a blocker's binding is altered by a change of some kind, its binding energy will change. We can call ΔG_p a perturbation of inhibitor binding energy, and determine its impact on $\Delta G_{blocked}$:

$$\Delta G_{blocked} = \Delta G^0_{blocked} + \Delta G_p \tag{6.5}$$

$\Delta G^0_{blocked}$ is the energy of inhibitor binding before perturbation. This energetic formulation can describe the effect of many perturbations: voltage changes, ionic concentration, effects of temperature and other factors. Under physiological conditions, voltage and ionic concentration can be relevant perturbations. Effects of the concentration of competitive ligands are treated comprehensively elsewhere (Hilal-Dandan, Brunton, and Goodman 2013; Wyman and Gill 1990). The effect of voltage on block of ion flow is a phenomenon unique to ion channels and is considered further here.

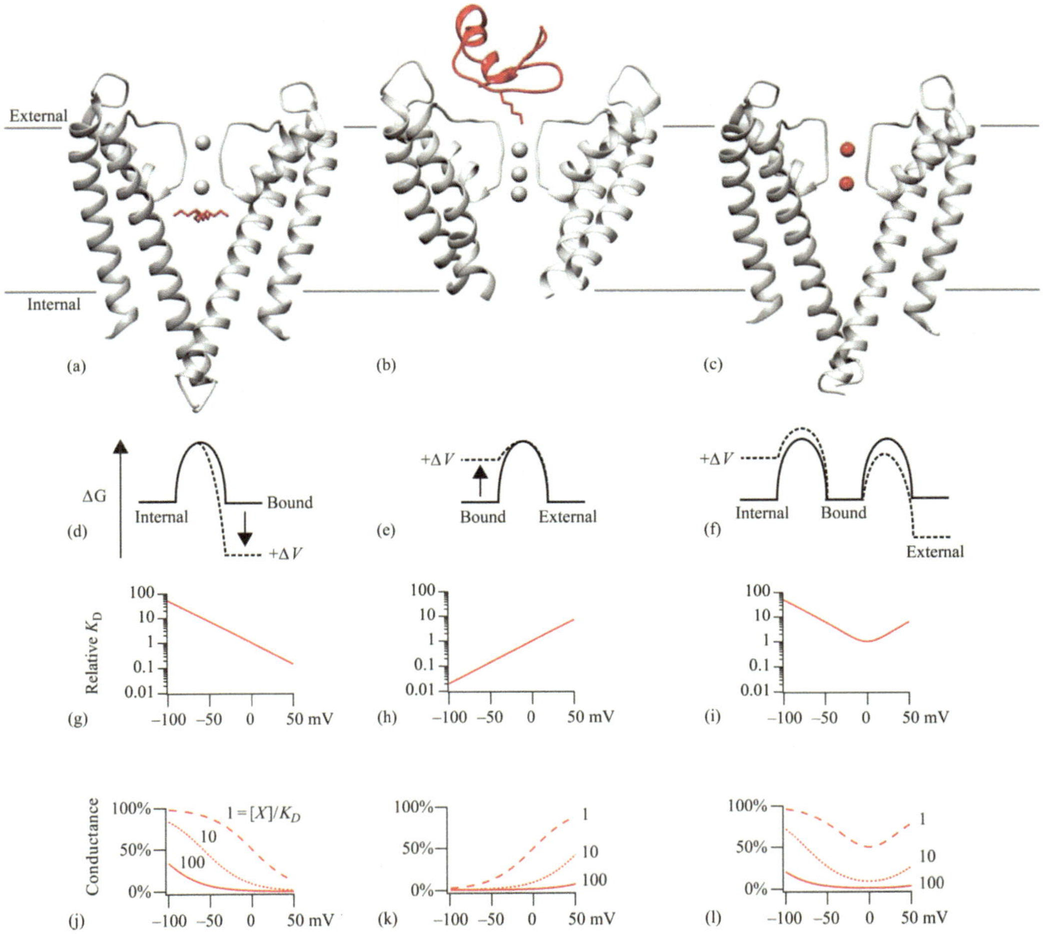

FIGURE 6.3
Pore blockers and voltage-dependent block. (a–c) Crystal structures of K^+ channels (gray) with pore blockers bound (red). Balls indicate permeant ions. (Renderings by Drew Tilley.) The blockers shown in panels (a) and (b) are described in Section 6.4. The blocker shown in panel (c) is described in Section 6.5. (a) Rendering of tetrabutylammonium blocking the pore of the KcsA channel, PDB ID 2HVJ. (b) Charybdotoxin blocking the pore of a Kv channel, PDB ID 4JTA. (c) Ba^{2+} blocking the pore of the KcsA channel, PDB ID 2ITD. (d–f) Reaction coordinate cartoons illustrating the voltage dependence of pore blocker binding. The ordinate represents increasing free energy. Dotted curves indicate energetic changes resulting from an intracellular voltage increase. (g–i) Relative K_D–voltage relation of theoretical pore blockers. Voltage dependence of K_D from Equation 6.8, with $z = 1\ e_0$ (g), $z = -1\ e_0$ (h), or Equation 6.9 (i) with $z_{off,in} = 1\ e_0$, $z_{off,out} = -1\ e_0$, $k_{off,0mV,in} = k_{off,0mV,out}$, $z_{on,in} = 1\ e_0$, $z_{on,out} = 0$, $k_{on,0mV,in} = k_{on,0mV,out}$. (j–l) Conductance–voltage relation of theoretical pore blockers. Curves are plots of Equation 6.3 with voltage dependence of K_D from (g–i). Dashed curve indicates an inhibitor concentration equal to the K_D at 0 mV. Dotted curve is $10 \times K_D$, solid curve $100 \times K_D$.

The K_D of a pore blocker can depend on voltage. The physical origins of voltage-dependent inhibition can be complex, especially when impacted by permeant ions. In the most simple, theoretical case, the voltage dependence of a pore blocker can be conveyed with a single parameter, z, the partial charge value that leads to voltage dependence (Woodhull 1973). Measurements of the degree of inhibition at different voltages can empirically constrain z. From z, how the concentration–response of a pore blocker will be affected by voltage can be predicted. With a simple dependence of pore block on z, the binding energy derived from a voltage change, ΔV, can be easily calculated with Faraday's constant, F:

$$\Delta G_V = \Delta VzF \tag{6.6}$$

Hence,

$$\Delta G_{blocked} = \Delta G^0_{blocked} + \Delta G_V \tag{6.7}$$

and the voltage dependence of the K_D is

$$K_D = K^0_D e^{-\Delta VzF} \tag{6.8}$$

The full mechanism of voltage-dependent pore blockade may not be as simple as the theoretical case discussed here, but the preceding equations can serve as reasonable approximations under many conditions. The exact nature of the voltage dependence of inhibition is determined by the geometry of the blocking interaction. A few specific examples are discussed later to elaborate how voltage-dependent interactions can originate.

6.4 One-Sided Pore Blockers

Many pore blockers can reach their binding site from only one side of the cell membrane. For example, in Figure 6.3a, the inhibitor can reach its binding site only from the internal side of the pore; while in Figure 6.3b, the pore-blocking peptide can bind and dissociate only from the extracellular side. The sidedness of inhibitor binding determines the voltage dependence of its inhibition.

One of the best-studied types of pore blockers are the quaternary ammonium (QA) ions that inhibit K^+ channels from the intracellular side (Armstrong 1969). The QA ion was found to fit nicely into the hydrophobic cavity of a K^+ channel, just internal to the selectivity filter, where a permeant ion normally resides (Zhou et al. 2001) (Figure 6.3a). QA ions are too large to squeeze through the narrow selectivity filter of K^+ channels. To dissociate, they must exit inward, and their voltage dependence arises from this dissociation to the intracellular side of the channel. As the voltage inside a cell is decreased, QA ions, being positively charged, can be directly affected by a transmembrane electric field. Additionally, K^+ ions are electrostatically forced through the channel, displacing the pore blocker from its binding site. The more rapid dissociation leads to a decreased affinity (larger K_D) for the pore blocker at more negative voltages. This effect is schematized in the energy diagram of Figure 6.3d. When voltage increases, an internal blocker of a cation channel is liable to be more stable in the bound configuration. The kinetics of QA ion interactions with K^+ channels can be complex. For the purposes of demonstration, however, it is useful to discuss an idealized simple scenario predicted by Equation 6.8 (Figure 6.3g). Note that within the physiologically relevant voltage range, of –100 to 50 mV, the percent conductance inhibited is altered dramatically. A concentration of pore blocker that inhibits only a small fraction of the conductance at a negative resting potential may inhibit the majority of the current if the cell is brought to a positive potential. The voltage dependence of an internal blocker can be far steeper than that depicted in Figure 6.3g. This type of voltage dependence forms the basis of inward rectification of inward-rectifier K^+ channels, where endogenous polyamines inhibit the outward flow of ions (Lu 2004). Given the dramatic changes in inhibition with cell potential, it is worth carefully considering the voltage changes in any experimental preparation with voltage-dependent inhibitors.

Pore blockers that act from the external side of the pore will have the opposite voltage dependence to internal blockers. Some ion channel inhibitors are pore-blocking toxins that bind the extracellular side of the pore. One of these is the scorpion peptide charybdotoxin, which has been crystallized bound to a K^+ channel (Banerjee et al. 2013). The resultant structure provides an example of how a peptide can physically occlude the conducting pore (Figure 6.3b). Akin to the internal pore blockers discussed earlier, the peptide mimics the chemistry of the channel's permeant ions to bind tightly to the extracellular side of the channel pore. Charybdotoxin displaces a K^+ ion from its binding site with a positively charged lysine residue. Interactions with permeant ions enhance the voltage dependence of toxin dissociation from the channel (Park and Miller 1992). As the voltage inside a cell is increased, the force driving K^+ ions through the channel out of the cell leads to a decreased affinity for the toxin, which is pushed out to the external solution by the flow of K^+ ions (Figure 6.3e). Hence, the voltage dependence is opposite of blockers that exit to the intracellular solution (Figure 6.3g, h). Interestingly, this opposing voltage dependence of extracellular dissociation has been harnessed by NMDA receptors to allow conduction only when block by external Mg^{2+} has been relieved by positive cellular voltage (Mayer, Westbrook, and Guthrie 1984). Akin to inward-rectifier K^+ channels, pore block of NMDA forms a voltage-dependent gate important for normal physiological function.

6.5 Slowly Permeating Blocking Ions

Some inhibitors are ions that more slowly pass through the channel pore. A common class of inhibitors used in electrophysiology experiments is small metal ions that block pores. In experimental preparations, many Ca^{2+} channels can be blocked with Cd^{2+} or Co^{2+}, and K^+ channels with Cs^+ or Ba^{2+}. Ba^{2+} ion block of K^+ channels has been carefully investigated (Neyton and Miller 1988) and serves as an excellent case study of a slowly permeating blocker. Crystal structures have revealed the location of Ba^{2+} in K^+ channels (Jiang and MacKinnon 2000), where it can be seen replacing K^+ in the selectivity filter (Figure 6.3c). The Ba^{2+} ion is nearly the same size as K^+, allowing it to fit snugly into these sites. Yet, due to its greater charge, Ba^{2+} dynamics differ, and it remains in the channel pore for long periods of time, preventing the flux of other ions. Ba^{2+} eventually dissociates, and like K^+, it can exit from its binding site to either the internal or external solution.

The ability of permeant ions to enter and exit the pore from both sides of the channel can lead to biphasic voltage dependence. In addition to interactions with permeant ions, slowly permeating ions have an innate voltage dependence. As voltage is increased inside the cell, cationic blocking ions are driven into the channel from the internal solution and bound ions punch through to exit the external side (Figure 6.3f). Thus, the rate of dissociation of a slowly permeating blocker to opposing sides of the membrane have opposing polarity. For this situation, voltage dependence has multiple components, and different steps in the slow permeation process can dominate at different voltages. A description of such behavior is as follows:

$$K_D = \frac{k^0_{off,internal} \cdot e^{-\Delta V z_{off,internal} F} + k^0_{off,external} \cdot e^{-\Delta V z_{off,external} F}}{k^0_{on,internal} [X] e^{-\Delta V z_{off,internal} F} + k^0_{on,external} [X] e^{-\Delta V z_{on,external} F}} \tag{6.9}$$

The multiple voltage-sensitive components lead to multiphasic voltage dependence. This is depicted in Figure 6.3i, where at negative voltages, inhibition decreases because ions exit to the internal side, and at positive voltages, the inhibition also decreases because ions exit to the external side. Multiphasic voltage dependence is a hallmark of slowly permeating blockers that was identified in careful studies of the proton block of sodium channels (Woodhull 1973).

Voltage-dependent inhibition can be problematic for researchers seeking to pharmacologically eliminate a current, as an inhibitor that produces near-complete inhibition at some voltages can be ineffective at others. This variable inhibition is due to voltage dependence alone, and the efficacy of inhibitors becomes even more complicated when effects of channel conformation are considered.

6.6 Gated Inhibitor Access

The ability of a pore blocker to inhibit can depend on the conformational state of the channel. Block kinetics can exhibit state dependence when a blocker binds in the interior of a channel, to a site that is only accessible when the protein adopts certain conformations. This is often referred to as gated access. A common form of gated access is when inhibitors are only able to enter open channels (Figure 6.4a). Open channel block of K^+ channels was classically described by Clay Armstrong (Armstrong and Hille 1972). A prominent example of open channel block is N-type or ball-and-chain inactivation of K^+ channels, where a protein amino terminus acts as an inhibitor (Zagotta, Hoshi, and Aldrich 1990). A feature of N-termini and some other open channel blockers is that channels cannot close their access gate when the blocker is bound. Other blockers that require channel opening to access an interior blocking site can become trapped inside the channel when it closes. The gate that controls inhibitor access is not necessarily the gate that controls ion permeation. Channels can undergo many conformational changes prior to opening and, when inhibitor access is gated differently than ion permeation, blockers may enter partially activated closed conformations in addition to the open ones (Figure 6.4b). This has been demonstrated for the block of BK channels by intracellular QA compounds (Tang, Zeng, and Lingle 2009). Blockers can also have multiple access pathways that are gated differently. For example, lidocaine can access Na^+ channels through the cytoplasmic Na^+ gate and can also access the same blocking site from the lipid bilayer through a fenestration in the side of the channel (Nguyen et al. 2019; Hille 1977) (Figure 6.4c).

A hallmark of gated inhibitor access is that channel conductance activates, then decays as blockers enter and inhibit channels (Figure 6.4d, e). While gated access necessarily affects the rate of inhibitor binding, it does not necessarily affect binding affinity. However, if the channel does not close normally with the inhibitor bound, the inhibitor will affect gating, and channel gating will affect binding affinity. Such allosteric effects on channel conformation are discussed further next.

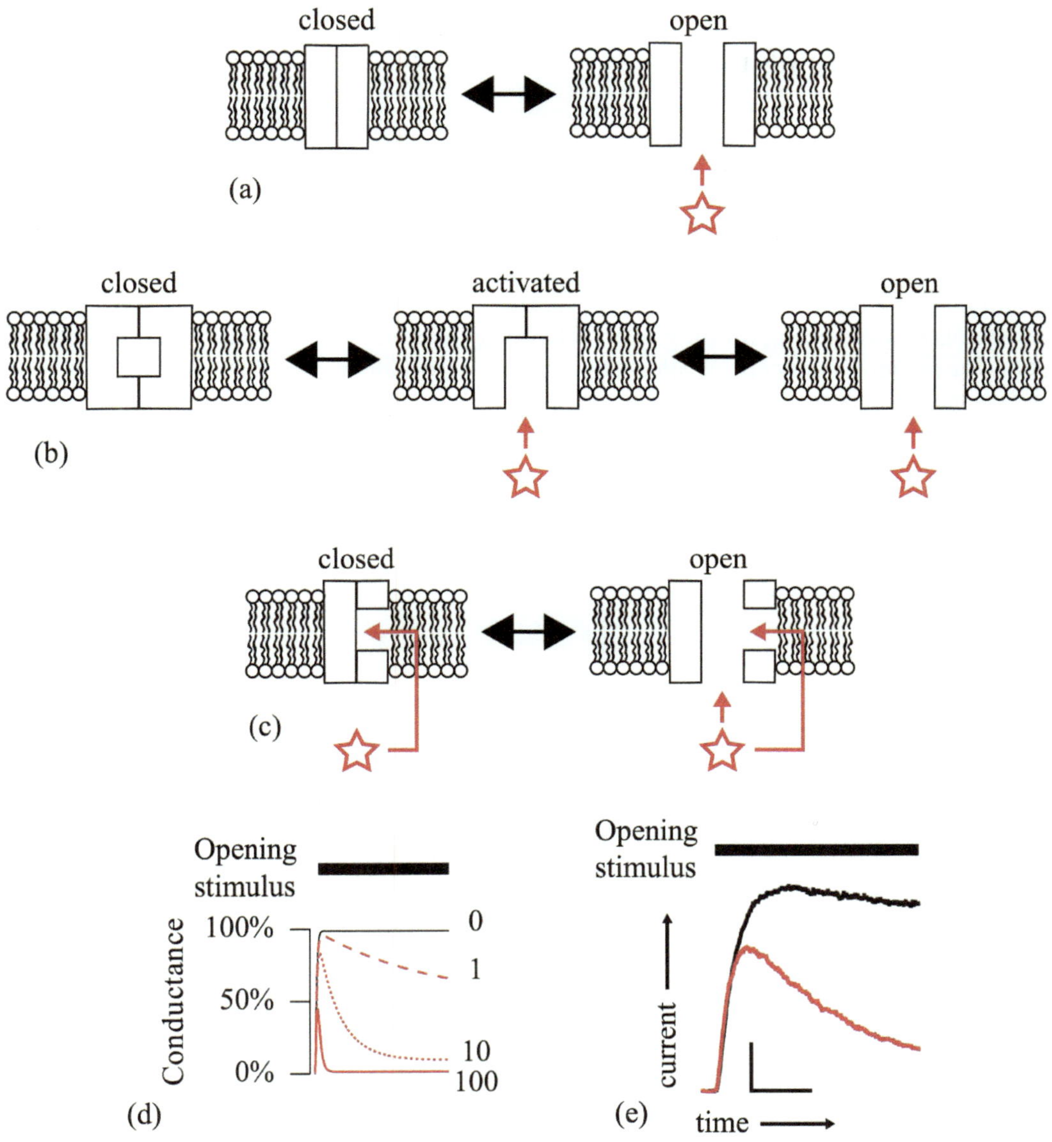

FIGURE 6.4
Gated inhibitor access. (a) An open channel blocker can only occlude the pore when the channel opens. (b) Blocker access may be gated differently than ion conduction. (c) There may be multiple paths, gated differently, to an inhibitor binding site. (d) Time dependence of pore blockade in a kinetic model of an open channel block. Black curve indicates opening in the absence of inhibitor. Dashed red curve indicates an inhibitor concentration equal to the K_D of the open state. Dotted red curve is 10 × K_D, solid red curve 100 × K_D. (e) Block of the voltage-gated K^+ channel Kv2.1 by the blocker RY785. To access its binding site, RY785 requires voltage stimulation but not channel opening.

6.7 Allosteric Inhibition

Not all ion channel inhibitors block the pore. Many act by inducing the channel to close. This mechanism is fundamentally distinct from pore blockade, as the inhibitory action is not due to the inhibitor obstructing the flow of ions, but to the inhibitor stabilizing channels in nonconducting conformations (Figure 6.5). Despite these very different mechanisms, distinguishing pore blockade from allosteric inhibition can be difficult.

Inhibitors can act by a variety of physical means, such as membrane perturbation, surface charge screening or direct binding to channels. Direct binding can competitively inhibit channels by preventing binding of an activating ligand or act by an allosteric mechanism. The fundamental concept of allostery is that a modulatory molecule can selectively stabilize protein conformations, shifting the equilibrium between conformational states. A molecule that stabilizes nonconducting conformations is a negative allosteric modulator.

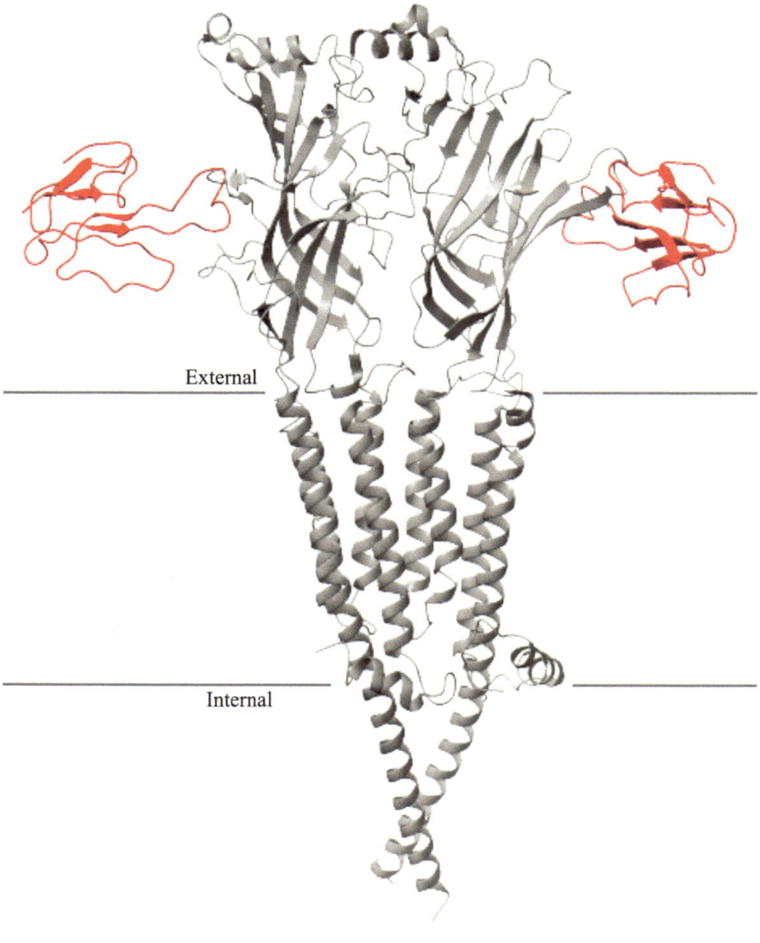

FIGURE 6.5
α-bungarotoxin inhibits acetylcholine receptors by an allosteric mechanism while binding far from the ion conduction pathway. Cryo-EM structure of the nicotinic acetylcholine receptor from *Tetronarce californica* (gray) with bound α-bungarotoxin (red), PDB ID 6UWZ. For clarity, only two of five channel subunits are shown. (Renderings made in ChimeraX; Pettersen et al. 2021.)

By shifting the equilibrium of a channel to nonconducting, the channel will open less often and be inhibited.

Allosteric inhibitors act by energetically stabilizing closed states relative to open ones. In the simplest case, an inhibitor binds to a single site and prevents a channel from opening (Figure 6.6a). This type of allosteric inhibition is technically inverse agonism, as the channel cannot open with the inhibitor bound. For an inverse agonist, the fraction of channel current inhibited is the fraction bound, with the remaining conductance determined by

$$\text{Conductance} = \frac{f_{\text{unbound}}}{f_{\text{unbound}} + f_{\text{bound}}} = \frac{1}{1 + \left([X]/K_D\right)} \tag{6.10}$$

The thermodynamics of this type of inhibition can make the concentration–response indistinguishable from pore block (note that Equation 6.3 is identical). However, other aspects of the response to the inhibitor can distinguish between pore block and allostery. The differences can be subtle or radical. In many cases, the binding of an allosteric inhibitor is sensitive to the conformational changes associated with channel gating. Changing the stimulus that opens the channel will change the binding of the inhibitor. Figure 6.6c depicts alleviation of inverse agonism in response to increasing opening stimulus. The inverse agonist fights the opening stimulus (Figure 6.6b), and the greater the concentration of inhibitor applied, the more stimulus energy is needed to open the channel. The effects of a full inverse agonist saturate only with complete inhibition: otherwise, more inhibitor produces more inhibition. The opening stimuli for many channels are ligands or voltage. As the interactions of inhibitors with ligand-gated proteins are discussed in great detail elsewhere (Wyman and Gill 1990; Hilal-Dandan, Brunton, and Goodman 2013), we here discuss the particular interactions of allosteric inhibitors with voltage-gating processes.

In the case of many voltage-gated channels, voltage increase leads to more channel opening, and inverse agonists modulate these voltage-gating processes. Since voltage alters the probability that a channel will be closed, the fraction of channels with inhibitors bound will change with voltage. The fact that voltage can also affect pore blocker binding can make these two types of inhibition even tougher to distinguish and requires careful interpretation of inhibition data.

An example of an inverse agonist can be found in the defensive mucus of a marine snail. The gastropod *Calliostoma canaliculatum* secretes 6-bromo-2-mercaptotryptamine (BrMT) to deter predators. BrMT is an inverse agonist of voltage-gated K^+ channels (Sack, Aldrich, and Gilly 2004). BrMT acts by selectively binding to closed channels, as in Figure 6.6a, to prevent channels from opening. Its effects are consistent with full inverse agonism where the inhibitor must dissociate from the channel before it can open. In the presence of BrMT, the more positive voltage is required for activation. Increasing concentrations require progressively more positive voltages to open the channels (Figure 6.6d). The major effect of inverse agonists can be due to kinetic rather than equilibrium properties. It can be shown with a reaction coordinate diagram (Figure 6.6b) that a full inverse agonist that requires unbinding before opening will increase the total activation barrier, $\Delta G^{\ddagger}_{opening}$, by an amount equal to the free energy change of its binding to the closed state of the channel, $\Delta G_{closed\ binding}$. Thus, if the inhibitor prevents rate-limiting steps in channel opening, a full inverse agonist will slow the time course of channel opening by a factor of at least

$$\frac{\tau_{inhibited}}{\tau_{control}} = 1 + \frac{[X]}{K_{D\ closed}} \tag{6.11}$$

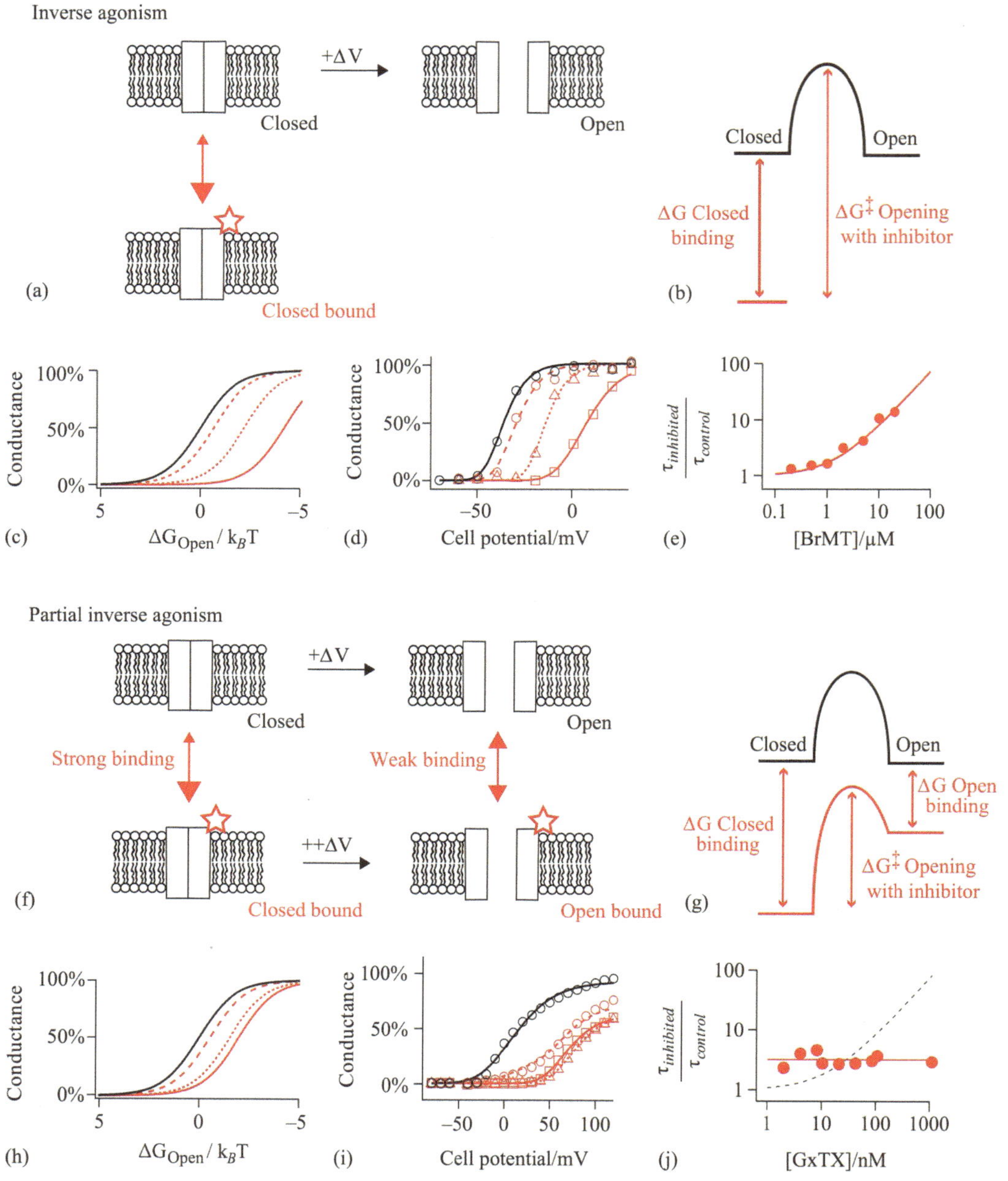
Inverse agonism
+ΔV
Closed
Open
Closed bound
(a)
Closed
Open
ΔG Closed binding
ΔG‡ Opening with inhibitor
(b)
Conductance
100%
50%
0%
5
0
–5
ΔG_Open / k_BT
(c)
Conductance
100%
50%
0%
–50
0
Cell potential/mV
(d)
τ_inhibited / τ_control
100
10
1
0.1
1
10
100
[BrMT]/μM
(e)
Partial inverse agonism
+ΔV
Closed
Open
Strong binding
Weak binding
++ΔV
Closed bound
Open bound
(f)
Closed
Open
ΔG Open binding
ΔG Closed binding
ΔG‡ Opening with inhibitor
(g)
Conductance
100%
50%
0%
5
0
–5
ΔG_Open / k_BT
(h)
Conductance
100%
50%
0%
–50
0
50
100
Cell potential/mV
(i)
τ_inhibited / τ_control
100
10
1
1
10
100
1000
[GxTX]/nM
(j)

where τ represents the time constant of the rate-limiting kinetic process in activation. With increasing concentrations of an inhibitor, channel opening will be progressively slowed (Figure 6.6e). This progressive alteration of kinetics with increasing dose is a hallmark of a full inverse agonist.

6.8 Partial Inverse Agonism

Some allosteric inhibitors will allow channels to open while the inhibitor remains bound. Inhibitors that act in such a way are partial inverse agonists (Figure 6.6f). A partial inverse agonist lowers the probability that a channel will be in a fully open state. The degree to which the binding of a partial inverse agonist leads to channel closing is the coupling between binding and inhibition. This can be quantified as a coupling energy, $\Delta G_{coupling}$, which can be understood as the amount of energy the inhibitor uses to close the channel. But where does $\Delta G_{coupling}$ arise from? The first law of thermodynamics demands that energy cannot just appear. It must be accounted for. It turns out that $\Delta G_{coupling}$ is derived from the binding energy of the inhibitor. When channels open with an inverse agonist bound, it weakens inhibitor binding (Figure 6.6f).

An energy diagram for this type of inhibition is given in Figure 6.6g. It depicts a general model of allosteric interaction where a portion of an inhibitor's binding energy is used to keep the channel from opening. How much coupling energy will arise from a change in binding can be calculated from the binding energies to the different channel states. The coupling energy comes from the difference in binding energy of the open and closed states:

$$\Delta G_{coupling} = \Delta G_{open\,binding} - \Delta G_{closed\,binding} \tag{6.12}$$

FIGURE 6.6 (CONTINUED)
Allosteric inhibition. (a) A full inverse agonist needs to dissociate before the channel can open. Cartoon depicts a voltage activated ion channel that only binds inhibitor in its closed state. (b) Inverse agonist binding energetically stabilizes the closed state. (c) The dose–response of a full inverse agonist does not saturate. Each curve is a Boltzmann distribution of conductance arising from increasing open probability. The black curve is control condition. The dashed curve indicates an inhibitor concentration equal to the K_D for the closed conformation. The dotted curve is 10 × K_D; solid red curve 100 × K_D. (d) Shift of conductance–voltage relation of Shaker K^+ channels does not saturate with increasing concentrations of the inverse agonist BrMT (Sack, Aldrich, and Gilly 2004). The black curve is Boltzmann distribution fit to control data points. The dashed curve is 1 μM BrMT, dotted curve 5 μM, solid curve 20 μM. (e) Increasing inverse agonist concentration progressively slows channel opening. Data points are time constants of channel opening in indicated concentrations of BrMT relative to control. The curve is Equation 6.11 with K_D set to 0.8 μM. (f) A partial inverse agonist can remain bound while the channel opens. The cartoon depicts a voltage-activated ion channel that strongly binds inhibitor in its closed state and more weakly in its open conformation. (g) Open-state binding of partial inverse agonists allow the activation barrier for opening to be reduced. (h) The dose–response of a partial inverse agonist saturates. Each curve is a Boltzmann distribution of conductance arising from increasing open probability. The black curve is the control condition. The dashed curve indicates an inhibitor concentration equal to $K_{D\,closed}$ where $K_{D\,open}$ is 10 × $K_{D\,closed}$. Dotted curve is 10 × $K_{D\,closed}$, solid red curve 100 × $K_{D\,closed}$. (i) Shift of conductance–voltage relation of Kv2.1 channels saturates at high concentrations of the partial inverse agonist GxTX (Tilley et al. 2019). The black curve is the Boltzmann distribution fit to control data points. The dashed curve is 10 μM GxTX, dotted curve 100 μM, solid curve 1 μM. (j) The rate of channel opening with partial inverse agonist is concentration-independent. Data points are time constants of channel opening in indicated concentrations of GxTX relative to control. The solid line is 3.1-fold slowing of opening. The dashed curve is Equation 6.11 with K_D set to 13 nM, $K_{D\,closed}$; poor fit indicates GxTX is not a full inverse agonist.

As binding affinities are related by dissociation constants, it can be seen that

$$\Delta G_{coupling} = -k_B T \ln\left(\frac{1+([X]/K_{D\,open})}{1+([X]/K_{D\,closed})}\right) \tag{6.13}$$

and at high concentrations of inhibitor, the effect of the inhibitor saturates:

$$\lim_{[X]\to\infty} \Delta G_{coupling} = -k_B T \ln\left(\frac{K_{D\,closed}}{K_{D\,open}}\right) \tag{6.14}$$

Therefore, the ratio of the dissociation constants for the open and closed conformations limits the potential of a partial inverse agonist to close a channel. This is demonstrated by the simulation in Figure 6.6h; the effects of partial inverse agonists saturate with increasing inhibitor concentration.

An example of a partial inverse agonist is the tarantula toxin guangxitoxin-1E (GxTX). When bound by GxTX, K^+ channels require more stimulus voltage to open, but increasing the toxin concentration fails to completely inhibit the channels (Figure 6.6i). This is due to channels opening with the partial inverse agonist bound. As depicted in Figure 6.6g, a channel opening with an inhibitor bound can have a $\Delta G^{\ddagger}_{opening}$ that is less than an opening pathway that first requires inhibitor dissociation. Because channels can open in the presence of inhibitors, the opening rate saturates and becomes insensitive to higher concentrations of toxin (Figure 6.6j). Due to limited efficacy, a partial inverse agonist mechanism may limit consequences of drug overdose.

6.9 Use-Dependent Pore Block

Pore blockers can also allosterically modulate channels. An open channel blocker may prevent a channel from closing, like a foot in a door jamb, and could be considered a full agonist if it did not also block the permeation pathway. Other blockers that enter through the channel permeation gate can stabilize closed channels. For example, the "closed channel blocker," 4-aminopyridine, destabilizes the fully activated conformations of K^+ channel voltage sensors while bound in the pore (Armstrong and Loboda 2001). A wide variety of sodium channel inhibitors used to treat pain, arrhythmias and epilepsy are state-dependent pore blockers. In the case of these sodium channel inhibitors, state dependence is referred to as use dependence because the degree of inhibition of the channel increases when the channel is stimulated. Upon repetitive stimulus, such as a train of action potentials, open channel blockers or other use-dependent inhibitors will progressively inhibit their targets (Courtney 1975). This is thought to be an important property of drugs that mitigate excitotoxic pathologies, such as epilepsy (Rogawski and Loscher 2004).

6.10 Inhibition by Lipid Bilayer Effects

The preceding treatments of inhibition mechanism assume that the inhibitor interacts directly with the ion channel. However, inhibitors can act without binding to channels.

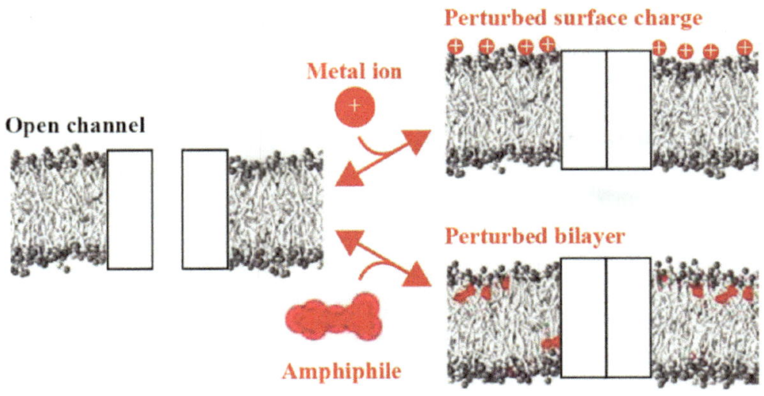

FIGURE 6.7
Membrane perturbation can alter channel function. Illustrations represent allosteric inhibition of channels by perturbing the surrounding membrane. Membrane images are molecular dynamics simulation snapshots of a phospholipid bilayer ±10 mol% resveratrol, an amphiphilic phytochemical. (Simulation and rendering by Helgi I. Ingólfsson.)

All ion channels are suspended in a sheet of lipid bilayer that forms the cell membrane. The bilayer is intimately involved in channel function. Some inhibitors do not appear to bind the ion channels they affect at all, but rather act by perturbing the membrane that surrounds them (Figure 6.7).

A classically studied mechanism of inhibitor action is the surface charge effect. This is where ions adsorb to the surface of membranes, perturbing the electric field near the membrane, and hence the activity of channels in the membrane (Frankenhaeuser and Hodgkin 1957). Multivalent metal ions such as Mg^{2+} can induce classical surface charge effects. The degree of surface charge effect varies from channel to channel and is dependent on the precise conformation of each voltage sensor. Molecules with surface charge effects can also modulate channels by mechanisms distinct from surface charge. For example, Mg^{2+} can also be a pore blocker.

Inhibitors that are hydrophobic or amphipathic in their physical chemistry can partition into the cell membrane. Membrane-partitioning molecules are everywhere and include detergents, lipids, a wide variety of phytochemicals and most clinically used small molecule drugs. By changing the physical properties of the membrane itself, bilayer-perturbing molecules can change equilibria between open and closed channels (Andersen and Koeppe 2007). These changes in equilibria can resemble the effects of directly bound inverse agonists, but may have unusual concentration-response profiles. Membrane perturbing inhibitors have promiscuous effects on many different membrane proteins (Ingolfsson et al. 2014). Whenever working with molecules that are soluble in organic solvents or poorly soluble in water, it is wise to consider their potential to alter ion channel gating by membrane perturbation.

6.11 Concluding Remarks

We have discussed mechanisms by which inhibitors decrease the conductance of ion channels. Inhibitors can bind in the pore and block ion flow, or act allosterically by closing the channel either by direct binding or perturbing the surrounding membrane. Different types of inhibitors are well suited for specific tasks. In attempting to dissect the role of a channel in a complex physiological situation, a selective and complete inhibitor of a specific channel type is called for. In this case, a pore blocker may be preferable. A partial inverse agonist can partially inhibit a channel with a therapeutic window spanning a wide range of concentrations. To inhibit channels only under hyperexcitable pathophysiological conditions, a use-dependent inhibitor may be most appropriate. In any case, understanding how inhibitor efficacy is affected by voltage changes, channel activity and the membrane bilayer allows one to choose ion channel inhibitor doses and interpret results of experiments more wisely.

Acknowledgments

The authors' research on inhibitors has been supported by National Institutes of Health (NIH) grants NS096317, NS114956 and HL128537.

Suggested Readings

Andersen, O. S., and R. E. Koeppe, 2nd. 2007. "Bilayer thickness and membrane protein function: an energetic perspective." *Annu Rev Biophys Biomol Struct* 36:107–30. doi:10.1146/annurev.biophys.36.040306.132643.

Armstrong, C. M. 1969. "Inactivation of the potassium conductance and related phenomena caused by quaternary ammonium ion injection in squid axons." *J Gen Physiol* 54 (5):553–75.

Armstrong, C. M., and B. Hille. 1972. "The inner quaternary ammonium ion receptor in potassium channels of the node of Ranvier." *J Gen Physiol* 59 (4):388–400. doi:10.1085/jgp.59.4.388.

Armstrong, C. M., and A. Loboda. 2001. "A model for 4-aminopyridine action on K channels: similarities to tetraethylammonium ion action." *Biophys J* 81 (2):895–904. doi:10.1016/S0006-3495(01)75749-9.

Banerjee, A., A. Lee, E. Campbell, and R. Mackinnon. 2013. "Structure of a pore-blocking toxin in complex with a eukaryotic voltage-dependent K(+) channel." *Elife* 2:e00594. doi:10.7554/eLife.00594.

Courtney, K. R. 1975. "Mechanism of frequency-dependent inhibition of sodium currents in frog myelinated nerve by the lidocaine derivative GEA." *J Pharmacol Exp Ther* 195 (2):225–36.

Frankenhaeuser, B., and A. L. Hodgkin. 1957. "The action of calcium on the electrical properties of squid axons." *J Physiol* 137 (2):218–44.

Hilal-Dandan, R., L. L. Brunton, and L. S. Goodman. 2013. *Goodman and Gilman's manual of pharmacology and therapeutics*. 2nd ed. New York: McGraw-Hill.

Hille, B. 1977. "Local anesthetics: hydrophilic and hydrophobic pathways for the drug-receptor reaction." *J Gen Physiol* 69 (4):497–515. doi:10.1085/jgp.69.4.497.

Ingolfsson, H. I., P. Thakur, K. F. Herold, E. A. Hobart, N. B. Ramsey, X. Periole, D. H. de Jong, M. Zwama, D. Yilmaz, K. Hall, T. Maretzky, H. C. Hemmings, Jr., C. Blobel, S. J. Marrink, A. Kocer, J. T. Sack, and O. S. Andersen. 2014. "Phytochemicals perturb membranes and promiscuously alter protein function." *ACS Chem Biol* 9 (8):1788–98. doi:10.1021/cb500086e.

Jiang, Y., and R. MacKinnon. 2000. "The barium site in a potassium channel by x-ray crystallography." *J Gen Physiol* 115 (3):269–72.

Lu, Z. 2004. "Mechanism of rectification in inward-rectifier K+ channels." *Annu Rev Physiol* 66:103–29. doi:10.1146/annurev.physiol.66.032102.150822.

Mayer, M. L., G. L. Westbrook, and P. B. Guthrie. 1984. "Voltage-dependent block by Mg2+ of NMDA responses in spinal cord neurones." *Nature* 309 (5965):261–3.

Neyton, J., and C. Miller. 1988. "Discrete Ba2+ block as a probe of ion occupancy and pore structure in the high-conductance Ca2+ -activated K+ channel." *J Gen Physiol* 92 (5):569–86.

Nguyen, P. T., K. R. DeMarco, I. Vorobyov, C. E. Clancy, and V. Yarov-Yarovoy. 2019. "Structural basis for antiarrhythmic drug interactions with the human cardiac sodium channel." *Proc Natl Acad Sci U S A* 116 (8):2945–54. doi:10.1073/pnas.1817446116.

Park, C. S., and C. Miller. 1992. "Interaction of charybdotoxin with permeant ions inside the pore of a K+ channel." *Neuron* 9 (2):307–13.

Pettersen, E. F., T. D. Goddard, C. C. Huang, E. C. Meng, G. S. Couch, T. I. Croll, J. H. Morris, and T. E. Ferrin. 2021. "UCSF ChimeraX: structure visualization for researchers, educators, and developers." *Protein Sci* 30 (1):70–82. doi:10.1002/pro.3943.

Rogawski, M. A., and W. Loscher. 2004. "The neurobiology of antiepileptic drugs." *Nat Rev Neurosci* 5 (7):553–64. doi:10.1038/nrn1430.

Sack, J. T., R. W. Aldrich, and W. F. Gilly. 2004. "A gastropod toxin selectively slows early transitions in the Shaker K channel's activation pathway." *J Gen Physiol* 123 (6):685–96.

Tang, Q. Y., X. H. Zeng, and C. J. Lingle. 2009. "Closed-channel block of BK potassium channels by bbTBA requires partial activation." *J Gen Physiol* 134 (5):409–36. doi:10.1085/jgp.200910251.

Tilley, D. C., J. M. Angueyra, K. S. Eum, H. Kim, L. H. Chao, A. W. Peng, and J. T. Sack. 2019. "The tarantula toxin GxTx detains K(+) channel gating charges in their resting conformation." *J Gen Physiol* 151 (3):292–315. doi:10.1085/jgp.201812213.

Woodhull, A. M. 1973. "Ionic blockage of sodium channels in nerve." *J Gen Physiol* 61 (6):687–708.

Wyman, J., and S. J. Gill. 1990. *Binding and linkage: functional chemistry of biological macromolecules*. Mill Valley, CA: University Science Books.

Zagotta, W. N., T. Hoshi, and R. W. Aldrich. 1990. "Restoration of inactivation in mutants of Shaker potassium channels by a peptide derived from ShB." *Science* 250 (4980):568–71. doi:10.1126/science.2122520.

Zhou, M., J. H. Morais-Cabral, S. Mann, and R. MacKinnon. 2001. "Potassium channel receptor site for the inactivation gate and quaternary amine inhibitors." *Nature* 411 (6838):657–61.

Section 2

Methodologies

7

Expression of Channels in Heterologous Systems and Voltage-Clamp Recordings of Macroscopic Currents

Victor De la Rosa and León D. Islas

CONTENTS

7.1 Introduction

Decoding biophysical mechanisms in ion channels has been achieved since the beginning of the discipline by making use of voltage and current recordings. It was soon recognized by the early pioneers that the most useful recording mode to understand the action potential and the underlying molecular machinery was to measure the flow of ions as the voltage is maintained constant. Under these conditions, the interpretation of ionic currents in terms of particular physical mechanisms is easier.

The advent of several specialized voltage-clamp techniques suited to particular cell types in conjunction with the development of heterologous expression systems allowed the study of molecularly defined ion channels when these began to be cloned.

DOI: 10.1201/9781003096214-9

7.2 Heterologous Expression of Channels in Cultured Cells and Oocytes

7.2.1 *Xenopus laevis* Oocytes

The oocytes of the *Xenopus laevis* frog have been used for several decades to efficiently express large quantities of heterologous membrane proteins. Considerable advances have been accomplished by studying the physiology, pharmacology and biophysics of ion channels and other membrane proteins expressed in this system. Due to their large size, oocytes can be easily manipulated, facilitating culture and electrophysiological measurements. In this system, it is possible to study single membrane proteins or macromolecular protein complexes. Protein expression is achieved through mRNA injection into the cytoplasm of the cell. DNA injection into the nucleus is also possible but technically more difficult because the nucleus must be located and is more susceptible to damage. On the downside, mRNA transcripts are more unstable, although some manipulations can help improve protein expression (Goldin 1991): a poly adenine (A) tail and 7-methylguanosine cap on the eukaryotic mRNAs help direct translation and prevent degradation.

7.2.2 Advantages and Disadvantages of *Xenopus* Oocytes

One of the primary advantages of *Xenopus* oocyte system is the ease of handling. These cells do express many endogenous channels and receptors, but endogenous responses are small, and consequently the current from exogenous channels is usually larger. However, researchers must consider the endogenous responses depending on the experimental setup. In some cases they can be used as an advantage, as is the case of second messenger signaling, for example, intracellular calcium release is mediated solely by the IP_3 receptor. Another advantage is the possibility to adjust the mRNA ratio for different subunits of a multimeric channel or receptor. Also, it is possible to express multiple proteins in the same cell. With respect to electrophysiological recordings, some techniques are unique to oocytes, as is the case of the cut-open voltage clamp (Taglialatela, Toro, and Stefani 1992).

A disadvantage is that it is not possible to express every channel in oocytes, however, there are some channels that can only be expressed in this system and not in mammalian cells. It is likely that differences in the environmental requirements for correct channel function are more favorable in one system than another. These requirements include post-translational steps required for the correct folding, trafficking to the membrane and specific lipid requirements . It is not possible to predict which channels fit in each category, so screening is advisable. Likewise it is worth co-expressing known channel-specific accessory subunits when possible. The oocyte cell membrane contains several invaginations and a vitelline membrane surrounding the cell that might interfere with the delivery of drugs for pharmacology studies, hence the absolute concentration of the drugs is usually higher than for native tissues (Goldin 2006). Removing of the vitelline membrane is usually accomplished by placing the oocyte in a hypertonic solution and the use of forceps. However, oocytes without the vitelline membrane are less stable. Another disadvantage is the quality variability of the cells, attributable to frog maintaining conditions, such as water quality and feeding regimes, and seasonal variability. The major disadvantage is that oocytes are not mammalian cells, hence the functional properties of the channels may not be identical to the native tissues.

7.2.3 Mammalian Cells

Expression systems utilizing mammalian cells provide a better environment to produce proper protein folding, posttranslational modifications and proper assembly, which are important for appropriate biological activity of ion channels. A wide selection of expression systems is available for large-scale recombinant protein production, therefore several factors must be considered when choosing a particular cell line: posttranslational modifications of the protein of interest, toxicity of the protein, desired yield of protein, purpose for which the protein is required and the cost–benefit relation (Khan 2013). Chinese hamster ovary (CHO) cells, human embryonic kidney (HEK 293) cells, HeLa and COS-7 cells are the most widely used cell lines for ion channel expression either by transient transfection or formation of stable cell lines.

An appropriate expression vector for the target DNA must contain an efficient promoter that allows the expression of the gene of interest suitable for the chosen cell line. Most of the vectors also contain an antibiotic-resistant gene and a poly A sequence for efficient transcription and stabilization of the mRNA. Transfection of these plasmids depends on the cell type and efficiency desired. Methods for DNA incorporation into a cell include calcium phosphate transfection and the use of other agents based on cationic lipids. DNA complexes formed with calcium or lipids enter the cell by endocytosis and reach the nucleus for gene expression. Other methods such as electroporation are used on low-proliferating cells like primary culture cells. This method is based on passing an electric current to the cells and cDNA mixture in order to form temporary pores in the membrane and allowing charged molecules to enter the cells.

7.2.4 Advantages and Disadvantages of Mammalian Cells

One major advantage of mammalian cell culture is posttranslational modifications and glycosylation of the protein of interest. This ensures expression of protein in a system that resembles their natural environment. The physicochemical environment of the culture can be precisely controlled: pH, osmolarity, gases and nutrients. Additionally, continuous production of a specific protein by the generation of stable lines is easily accomplished by combining the coding sequence with a selectable marker such as an antibiotic-resistance gene.

Conversely, mammalian cell expression levels are difficult to control. Expression of different proteins or subunits is set by the cDNA concentration used for transfection. In stable cell lines, where the plasmids are incorporated in the genome of the cell, expression levels are no longer under control, usually leading to high levels of expression. This is an important disadvantage for some experiments such as single-channel recording for which low expression levels are necessary. On the contrary, membrane composition of the host cell might limit expression or proper folding. Some strategies have been developed to account for this problem such as the addition of histone deacetylase to the culture media and engineering of optimized vectors.

Mammalian cells express a wide variety of endogenous channels; usually the expression levels of these native channels are low, but some of them have been found to produce a significant ionic current that can interfere with the recording of the exogenous channels. Additionally, cell strains are derived and maintained by subculturing, which can introduce genomic variability, leading to heterogeneity on the cell culture.

7.3 Voltage Clamp

The invention of the voltage-clamp technique initiated the modern study of ion channels and exploded our understanding of the physiology of excitable cells (Cole and Moore 1960). By allowing recordings of the current that flows through the channels in the membrane, inferences can be made of the mechanisms underlying ion permeation and channel gating and modulation. The basic idea is to set the membrane potential (V_m) of a cell to a desired value (V_t). With membrane potential clamped, the membrane current is measured. The simplest electrical model of the cell is an RC circuit, with R_m the resistance of the membrane, including passive and active (ion channel) components (the membrane conductance, $g_m = 1/R_m$), and C_m the capacitance of the membrane. Figure 7.1a shows a simplified voltage-clamp circuit employing two electrodes. An intracellular electrode reads V_m, and a second electrode injects current, I_{inj}. This current is equal to the total current through

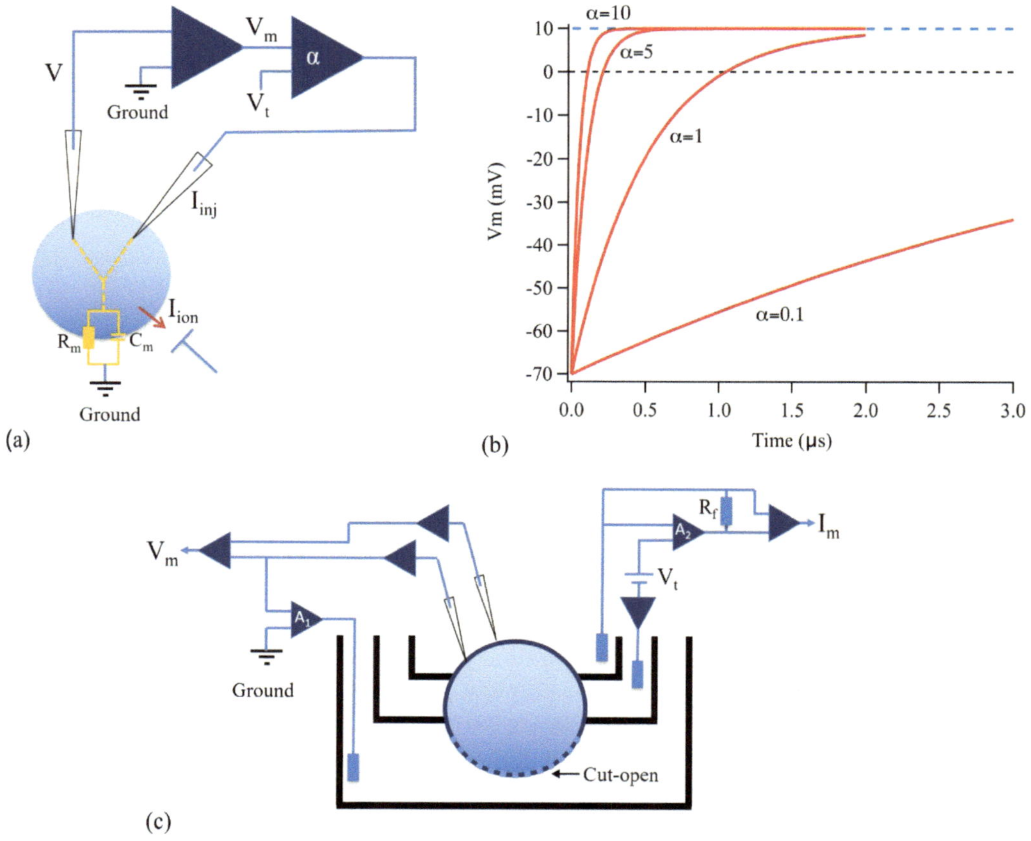

FIGURE 7.1
Simplified schematic of a voltage-clamp circuit. (a) Two microelectrodes are inserted in the cell and interact in a circuit with the membrane equivalent circuit. R_m, membrane resistance; C_m, membrane capacitance. (b) Time course of voltage clamp of a hypothetical cell. The cell parameters used for the simulations of the solution to Equation 7.3 are C_m = 500 nF, R_m = 1.8 MΩ, V_r = –70 mV, V_t = 10 mV. The values of α are indicated in the figure. (c) Simplified scheme of a cut-open voltage clamp. The oocyte sits in a chamber with three compartments. The amplifier A_2 clamps the top and middle section of the oocyte to V_t. The lower section of the oocyte is permeabilized and clamp to ground by amplifier A_1 via a low-resistance electrode inserted on the top membrane.

the circuit, which is the sum of I_{ion}, the ionic current flowing through the membrane and I_C, the capacitive current moving to charge the membrane when there is a time-dependent change in the membrane potential V_m: $C_m(dV_m/dt)$.

If there is a current, I_{inj}, being injected into the cell:

$$I_{ion} + C_m \frac{dV_m}{dt} = I_{inj} \tag{7.1}$$

The basic principle of the voltage-clamp technique is to keep the membrane voltage, V_m, constant and equal to V_t, a desired test voltage, by injecting a current I_{inj} into the cell while measuring I_{ion}, the total ionic current flowing through the membrane and all ionic channels.

In order to apply a clamp potential large enough to compensate for the voltage drop, it is necessary to compare V_m and the I_{inj} and readjust the clamp potential. An operational amplifier serves to amplify the difference between these two inputs by a factor α, the gain. In Figure 7.1b, the amplifier with a gain α produces a current:

$$I_{inj} = \alpha\left(V_t - V_m\right) \tag{7.2}$$

Substituting Equation 7.2 in Equation 7.1 we get

$$I_{ion} + C_m \frac{dV_m}{dt} = \alpha\left(V_t - V_m\right) \tag{7.3}$$

Ohm's law states that a current will flow through a resistance ($R_m = 1/g_m$) proportional to the voltage difference, hence, if I_{ion} is linear with V_m:

$$I_{ion} = g_m\left(V_m - V_{rest}\right) \tag{7.4}$$

Rearranging Equation 7.3:

$$V_m + \frac{C_m}{g_m + G}\frac{dV_m}{dt} = \frac{\alpha V_t + g_m V_{rest}}{g_m + \alpha} \tag{7.5}$$

For a desired change of V_t, the membrane potential $V_m(t)$ will change exponentially from V_{rest} toward V_t with time constant $\frac{C_m}{g_m + \alpha}$. If the gain α is very large, the time constant becomes smaller (the clamp is faster) and the final value of V_m becomes arbitrarily close to V_t. This is illustrated for several values of α, for a hypothetical cell being voltage clamped between –70 and 10 mV (Figure 7.1b).

A current-to-voltage converter is used to measure the ionic current flowing from the cell to ground in response to the voltage-clamp signal. This voltage signal can be displayed in an oscilloscope or digitized and acquired on a computer.

7.3.1 Two-Electrode Voltage Clamp

For recordings in *Xenopus* oocytes and other large cells, two electrodes can be used to control the membrane potential, just as in the configuration discussed in the previous section. The two-electrode voltage clamp (TEVC) in oocytes uses two glass microelectrodes filled with 1–3 M KCl solution to a final resistance of ~1 Mohm. The basic configuration of a TEVC corresponds to the voltage-clamp circuit discussed in Figure 7.1. One microelectrode measures the membrane potential difference with respect to an extracellular reference electrode, while the other injects current to maintain a desired voltage. Membrane currents flowing to ground are measured with an extracellular virtual ground, current-to-voltage converter circuit. Due to the very large membrane capacitance of oocytes, the time constant of voltage changes in the TEVC can be slow (~100 μs), which means that very fast gating events are subject to distortion. This problem can be alleviated by using microelectrodes of less than 1 Mohm resistance or specialized modifications to the basic TEVC circuit (Baumgartner, Islas, and Sigworth 1999). Also, the cut-open voltage-clamp technique can be used for improved recording of fast channel events.

7.3.2 Cut-Open Oocyte Clamp

The large size advantage of the oocytes for TEVC is countered by the large capacitance that limits the voltage-clamp speed. This problem can be alleviated by clamping a smaller membrane area. The cut-open voltage clamp (COVC) (Taglialatela, Toro, and Stefani 1992) combines the oocyte expression system with an improved clamp speed and high signal-to-noise ratio. It allows the perfusion of the oocyte and can be combined with other techniques such as fluorometry. The basic preparation consists of three electrically isolated chambers separated with Vaseline. The oocyte lies in the lower chamber, where the membrane is "cut open" or permeabilized to obtain electrical continuity between the cytoplasm and the internal solution present in this chamber. The middle chamber serves as an electrical guard and the top chamber contains an electrical isolated membrane segment exposed to the external solution. This is the membrane area from which ionic currents are recorded. The electrical arrangement consists of three voltage clamps. The lower chamber is clamped to ground via a low-resistance electrode inserted on the top membrane. The middle chamber is clamped to the same potential as the top to minimize leak currents. The top chamber is clamped to the voltage command. The currents are recorded through a low-resistance electrode in the top chamber.

With the cut-open setup, the voltage clamp can settle in 20–40 μs, and current noise is relatively low, in the order of 1 nA RMS (root-mean-square deviation) at 5 kHz. Recordings are stable for several hours and it allows for internal and external solution control, although internal solution exchange is slow and difficult to accomplish. A perfusion canula or a pipette inserted through the lower chamber can improve this problem.

7.4 Patch Clamp

The patch-clamp technique allows fast measurement of signals with small amplitude, even less than 1 pA. The use of the patch clamp to record single channels, its implementation and analysis procedures are presented in the next chapter. Here we will focus on technical details of the electronics and use of the patch-clamp to measure macroscopic currents.

The simplified electronic design of the patch clamp allows for wider applications, while TEVC or COVC are cell-specific recording techniques. As opposed to TEVC or COVC, a patch clamp makes use of a single electrode to control voltage and record current.

7.4.1 The Electronics

All patch-clamp amplifiers make use of an electronic circuit based on a current-to-voltage converter with very high feedback resistance (Hamill et al. 1981). Some patch clamps use a capacitor instead of a resistance for the feedback loop, with some advantages regarding noise in the electronics, since a capacitor has essentially infinite resistance. The main idea is to measure the very small current that flows into the pipette without changing the voltage at which the pipette is clamped, that is, under voltage-clamp conditions. This is accomplished by having an operational amplifier, which makes the output equal to the difference between the two inputs, connected as an inverting amplifier at the input stage (Figure 7.2a). Several extra electronic stages are needed for signal conditioning and filtering or capacitance transient compensation and series resistance compensation, which can be used in experiments involving changes in voltage. A full account of patch-clamp electronics can be found in Sigworth (1995). It is important to point out that, as with any experimental technique, artifacts can contaminate several aspects of patch-clamp recording.

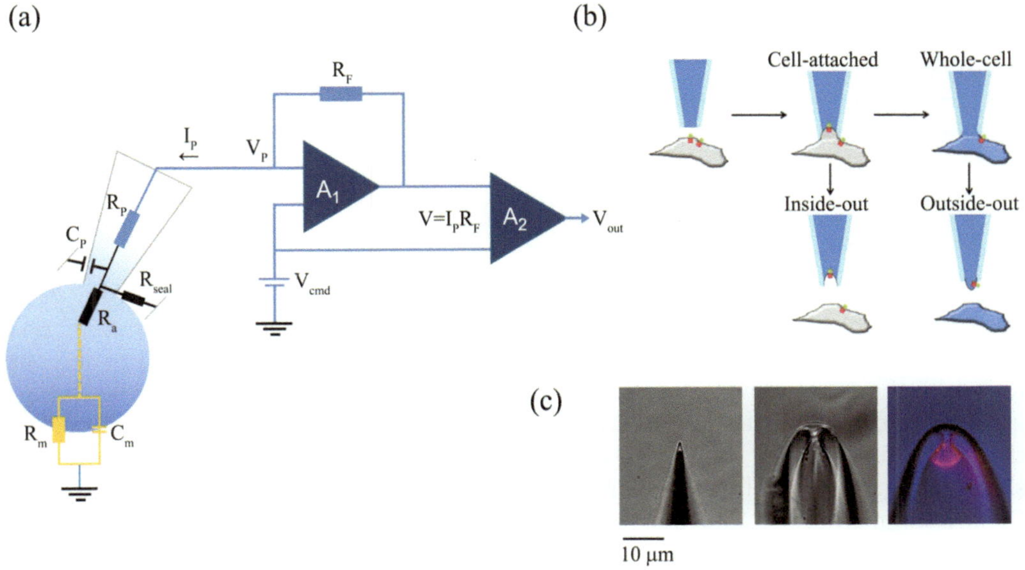

FIGURE 7.2
Simplified schematic of the whole-cell patch-clamp electronics. (a) The microelectrode circuit interacts with the membrane equivalent circuit. R_P, pipette resistance; C_P, pipette capacitance; R_a, access resistance; R_F, feedback resistance; V_{cmd}, command or test voltage. The first operational amplifier A_1 takes the pipette current and converts it to a voltage, which is then compared to a test or command voltage by amplifier A_2. This signal can then be filtered and stored for analysis. The command input voltage V_{cmd} sets the desired clamp voltage at the pipette electrode. (b) Representation of the recording configurations that are attainable in patch-clamp recording. The remarkable mechanical stability of the membrane and glass interaction accounts for the ability to maintain stable recordings in the cell-free configurations. The glass recording pipette is shown in light blue and the filling electrolyte in dark blue. A hypothetical channel is depicted with its extracellular domains represented by the green circles and the intracellular domains by the yellow squares. Notice the opposite orientation of these domains in the inside-out and outside-out configurations. (c) Shape of pipettes and shape of the membrane patch in an inside-out patch. (Left) Small tip pipette appropriate for single-channel recording. Note the very steep approach to the tip. (Center) Large tip pipette that can be used in larger cells such as *Xenopus* oocytes to obtain large membrane patches or macropatches. (Right) A pipette with a membrane patch in the tip (inside-out configuration). The membrane contains fluorescently labeled ion channels, which allows direct visualization of the dome shape of the patch. The size scale is the same for all panels. 60× magnification.

These are especially important in single-channel recording, since we are trying to record picoampere-magnitude signals in a system that may have an RMS current noise value of half a picoampere. Identifying the sources of these artifacts is as important as all other aspects of the experimental technique.

7.4.2 Establishing the Gigaseal

Obtaining a high-resistance seal and recording from a patch-clamped membrane is achieved as follows. The preparation must be as clean as possible, that is, free of extracellular matrix and connective tissue. It is common to use cultured cells, be it in primary culture or cell lines heterologously expressing the channels of interest. Recordings can also be achieved from more complex preparation, such as *Xenopus laevis* oocytes, chronic slices or even from tissues in vivo (Edwards et al. 1989; Kitamura et al. 2008). In these cases, some degree of enzymatic treatment is still required in order to obtain clean membranes (Stuhmer 1992). Cleaning the patch pipette tip, usually by fire-polishing, is helpful. The pipette is then brought into contact with the membrane. It is advisable to allow for steady solution flow out of the pipette by application of gentle positive pressure. By monitoring the electrode resistance, the pressure is released when resistance is approximately twice its initial value.

At this point, gentle suction is applied to the electrode and, in most cases, the resistance (R_s) between the electrode and the cell membrane will increase to gigaohm values within seconds.

7.4.3 Configurations

Once established, a seal with R_s in the gigaohm range is very stable. Recordings can be obtained in this configuration, which is commonly known as cell-attached or on-cell (Figure 7.2b). The versatility of the patch-clamp technique is demonstrated by the ability to detach the piece of membrane attached to the glass electrode without disturbing the gigaseal (Horn and Patlak 1980). These cell-free configurations allow precise control of the composition of the solutions bathing the membrane. If an on-cell patch electrode is slowly retrieved from the cell, the region of membrane under it detaches from the rest of the plasma membrane and remains part of the gigaseal. This configuration is known as inside-out and is useful when access to the intracellular part of the channel is needed, such as when studying intracellular acting blockers or the actions of modulatory substances or activators in the intracellular region of the channels. An example of this is the study of cyclic nucleotide-gated (CNG) channels, in which the ligand is applied to the exposed intracellular face of the channels (Benndorf et al. 1999).

After seal formation, one can apply a pulse of suction or a brief (microsecond duration) high-voltage pulse, both of which will result in the rupture of the patch of membrane without disruption of the gigaseal. In this configuration, the electrode has access to the interior of the cell, making it possible to obtain current or voltage recordings arising from all the plasma membrane. This configuration is termed whole-cell mode. If the pipette is now slowly withdrawn, a membrane patch can be reformed in its tip, but this time with an inverted orientation. In this new configuration, the patch is called an outside-out patch. In this mode, it is possible to very rapidly perfuse the extracellular facing substructures of the channel and study ligand-gated ion channels with extracellular binding sites (Maconochie and Knight 1989; Colquhoun, Jonas, and Sakmann 1992) or the action of peptide blockers in potassium channels (Goldstein and Miller 1993).

Patch-clamp recording is an incredibly versatile technique. For example, it can be used to control the intracellular composition of a cell by perfusing its interior in the whole-cell mode with the solution contained in the pipette (Horn and Marty 1988). Using this configuration it is possible to study the biophysics of macroscopic ion currents (Fox, Nowycky, and Tsien 1987), cell signaling, intercellular communication (Pfaffinger et al. 1988), and several other aspects of cellular dynamics, such as secretion or control of cell volume (Neher and Marty 1982). Macroscopic currents can also be recorded in cell-attached, outside-out and inside-out configurations allowing for fast voltage-clamp recordings and providing some of the best characterizations of the kinetics of channels (Zagotta, Hoshi, and Aldrich 1994; Schoppa and Sigworth 1998). However, the initial aim in the development of the patch clamp was to record the activity of a single ion channel, and it is in this capability that its true uniqueness can be appreciated (see Chapter 8).

7.4.4 Compensation and Voltage Errors; Series Resistance

In whole-cell recordings, the series resistance (R_s) is the sum of all the resistances between the input of the amplifier and the cell membrane. Principally, it is the sum of the pipette resistance (R_P) and the access resistance (R_a) between the tip of the pipette and the interior of the cell.

High R_s can attenuate the voltage command (V_{cmd}; the voltage applied to the cell) by a voltage drop across R_s (Armstrong and Gilly 1992). Therefore, R_s limits the amount of current needed to charge the membrane: $V_m = V_{cmd} - I_m R_s$. If R_s or I_m are small, then V_m approaches V_{cmd}. If the channels are closed, the membrane resistance (R_m) >> R_s and the product $I_m R_s$ is negligible. If R_m decreases and I_m increases due to channel opening, R_m can approximate R_s, and V_m will deviate from V_{cmd}. R_s has a bigger impact when the current is larger, leading to alterations in current kinetics and erroneous characterization of the voltage dependence. Using small ionic currents and selecting a large pipette opening minimize R_s errors. Typically an R_s of twice the pipette resistance is optimal depending on the cell type.

Moreover, a high R_s limits the clamp speed and the resolution of the current recordings. The voltage-clamp time constant is given by $\tau \approx R_s C_m$ when $R_m >> R_s$. Thus, R_s creates a low-pass filter for I_m with a corner frequency of the form $f_c = \dfrac{1}{2\pi R_s C_m}$.

Series resistance is compensated by adding a computed scaled value of the pipette current to the command potential; this is commonly referred as R_s correction (Sherman, Shrier, and Cooper 1999). Because this is a feedback loop it is prone to oscillation. Another compensation method, referred to as R_s prediction or supercharging, adds an overshoot to the voltage applied to the pipette. Consequently, the membrane capacitance charges faster. Since this is not a feedback loop, it does not introduce oscillation nor compensate for the voltage drop. In practice, R_s compensation is accomplished by a combination of both. Most researchers aim for a R_s compensation of ~80 %, although lower values are usually kept.

7.5 Analysis of Macroscopic Ionic Currents

The end result of a voltage-clamp experiment is a measurement of all the components of current flowing through and at the membrane. These include channel-mediated currents, which can be voltage-gated or voltage independent. An important first step in the analysis of membrane currents is the separation of the linear current components, which are usually of little interest, from the component of the current (ionic or capacitive) that is mediated by the channels. This separation can be carried out according to the type of currents that are being recorded. For voltage-dependent channels, a procedure known as *p/n* subtraction is the most reliable. Since voltage-gated channels will be activated over a specific range of voltages, contributing nonlinearly to the total current, the linear current components are recorded over the range of voltages in which channels are not activated. The linear components are recorded in response to a pulse 1/n the size of the pulse p and then appropriately scaled and subtracted (Figure 7.3a). If successful, this procedure allows

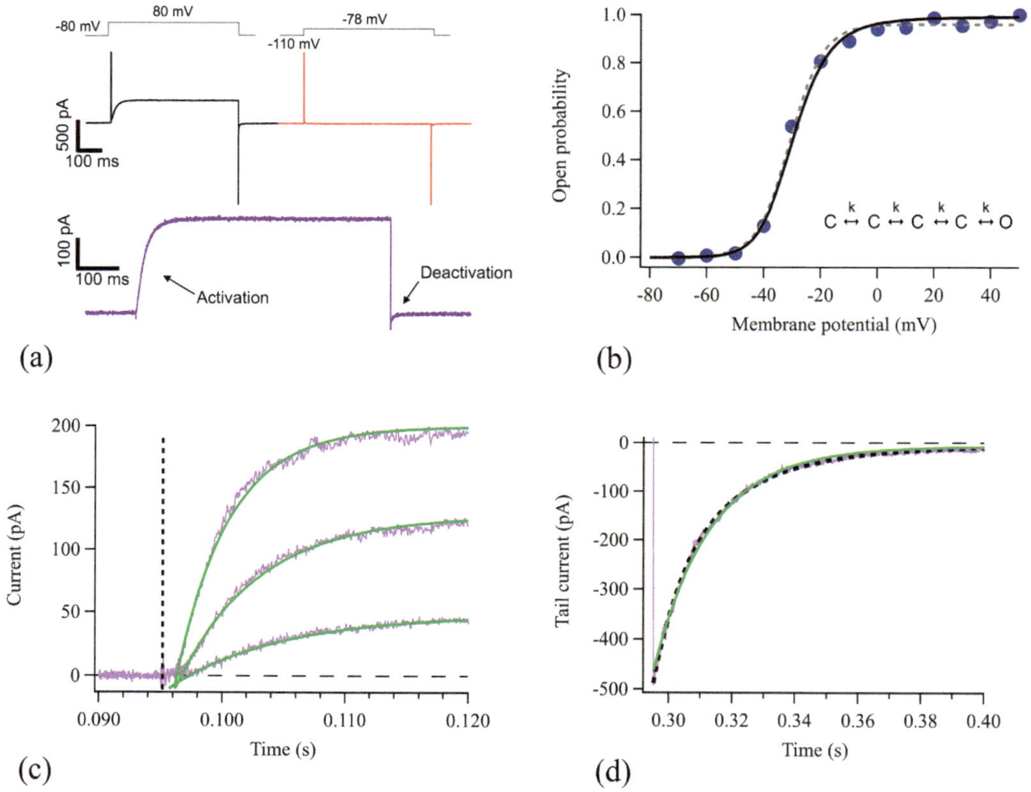

FIGURE 7.3
Analysis of macroscopic currents. (a) Illustration of the p/n subtraction procedure. The black trace is the total current recorded from a patch expressing voltage-gated proton channels. The fast spikes are the capacitive currents at the beginning and end of the 80 mV voltage pulse from a holding potential of –80 mV. The slow current is the proton current through the channels. The red trace is the current obtained at a more negative voltage of –110 mV with an amplitude of 32 mV (160 mv/5; p/5 procedure). This pulse does not elicit channel activation and only contains the linear components, made of capacitive and leak currents.. The pulse is multiplied by 5 and subtracted from the initial current record, yielding the subtracted proton current (purple trace). (b) Fitting an activation function. Normalized open probability of the voltage-gated channel Kv1.2 (blue circles). The dashed line is a least-squares fit to the Boltzmann equation (Equation 7.8). The continuous line is a fit to Equation 7.9 with $I = 5$ states (inset). Notice that this fit is better able to explain the slowly increasing *Po* at positive voltages, while the Boltzmann function is flat. (c) Time course of Kv1.2 channel current activation at three voltages. Notice the delay in the activation after the voltage change (dashed line). The green curves are fits to an exponential function extrapolated to zero current. (d) Current deactivation (tail current) fitted to a single (green) or double (black dashed) exponential.

recording of the time course and magnitude of channel-mediated currents with high fidelity, although it will introduce excess noise if not done properly (see Appendix).

Once the current flowing through the channel of interest is separated from the linear components, information about the underlying gating mechanism(s) can be obtained from these macroscopic (many channels) current recordings. An important initial framework is to consider that the macroscopic current is given by

$$I(a,t) = P_o(a,t) \cdot N \cdot \gamma(V - V_r) \tag{7.6}$$

Here, a refers to the activation factor (voltage, ligand binding, membrane tension, etc.), P_o is the open probability, N is the number of channels, γ is the conductance of a single channel and V_r is the reversal potential. It must be recognized that γ can be voltage-dependent, reflecting the complicated mechanism of channel permeation. In what follows, we will assume that the single-channel current-voltage behavior is linear and thus γ is constant.

7.5.1 Voltage Dependence, Activation, Deactivation

A common workflow to estimate the parameters of Equation 7.6 from macroscopic current recordings is to obtain the activation function and kinetic parameters. Often the activation function is obtained from the stimulus dependence of the macroscopic conductance. In the case of voltage-activated currents and since we have assumed that the microscopic conductance is not voltage-dependent, the macroscopic conductance is obtained as

$$\frac{I(a,t)}{(V - V_r)} = G(V) = P_o(a,t) \cdot N \cdot \gamma \tag{7.7}$$

Since N and γ are unknown most of the time, this function $G(V)$ is normalized to the maximum value (G_{max}) obtained at a very depolarized potential, $G(V)/G_{max}$.

The $G(V)$ is usually fit to one of several activation functions, commonly in the form of a Boltzmann function:

$$\frac{G(V)}{G_{max}} = \frac{1}{1 + e^{-q(V - V_{1/2})/KT}} \tag{7.8}$$

The simple Boltzmann function represents the equilibrium distribution of open channels when the channel has only one closed and one open state. q is the charge moved in going from the closed to the open state in e_o (elementary charges), $V_{1/2}$ is the voltage where 50% of channels are open, K is the Boltzmann constant and T is temperature in Kelvin. . While useful as a first approximation or as a way to parameterize results from voltage-clamp experiments, this simple equation is limited as a representation of channel gating. In general, channel gating involves transitions between multiple voltage-dependent and voltage-independent states. In the case of channels with a single open state and multiple sequential closed states, the equilibrium distribution of the open state can be derived from the general equation:

$$P_o = \frac{\prod_{i=1}^{n} K_i}{\sum_{j=1}^{n} \prod_{i=1}^{j} K_i} \tag{7.9}$$

The K_i's are the equilibrium constants of each transition. These can be voltage-independent or voltage-dependent with the form $K_i = K_i\left(0\ mV\right) \cdot e^{q_i V/KT}$, where K_i(0 mV) is the equilibrium constant at 0 mV and qi is the charge moved in the particular transition. An example of this fitting procedures is given in Figure 7.3b.

The time course of current activation can be complicated and nonmonotonical. In general, activation will occur along a time course that can be approximated by sums of exponential functions of time. Often, one is interested in describing the rate-limiting relaxation within the current time course. In this case, a single exponential function can be fitted to the second half of the activation time course. This fitting procedure also allows estimating the composite time course of non-rate-limiting transitions as the delay obtained by extrapolating the exponential fit to zero current. The time constant of this exponential will give an estimate of the speed of activation, and at extreme positive values it will approach the inverse of the rate-limiting step in activation (Zagotta, Hoshi, and Aldrich 1994). Similarly, channel deactivation time course can be recorded from tail currents at varying negative voltages. It is preferable to use a nonphysiological concentration of permeant ion to increase the magnitude of tail currents, specially at voltages where the channel closes. An example of tail current analysis is shown in Figure 7.3d.

7.6 Estimation of the Number of Channels Using Noise Analysis

Noise or fluctuation analysis was an important tool that provided numerous fundamental insights before the advent of single-channel recording using patch clamp (Stevens 1972). While mostly in disuse, a subsection of it is still very useful, namely, estimating the number of channels in a particular recording. Since many channels transition from nonactive to active states following a specific time course, this time course reflects the change in the number of open channels at a given time, given by the product $P_o\left(a,t\right) \cdot N$. This product is a random variable and each realization of the channel being activated (i.e., by a repeating voltage pulse or rapid change in pressure or ligand concentration) will produce a different fluctuation each time. In general, the mean current is given by

$$I = N \cdot i \cdot P_o\left(a,t\right) \tag{7.10}$$

and the variance

$$\sigma^2 = N \cdot i^2 \cdot P_o\left(a,t\right) \cdot \left(1 - P_o\left(a,t\right)\right) \tag{7.11}$$

Combining these equations, we can obtain

$$\frac{I}{\sigma^2} = i \cdot \left(1 - P_o\left(a,t\right)\right) \tag{7.12}$$

This means that the single-channel current can be estimated from the ratio of mean current to variance if the open probability is small. However, this simplification can be obviated if we rewrite Equation 7.12 in terms of the mean current ⟨I⟩

$$\sigma^2 = N \cdot i^2 \cdot \frac{I}{Ni}\left(1 - \frac{I}{Ni}\right) \tag{7.13}$$

$$\sigma^2 = Ii - \frac{I^2}{N} \tag{7.14}$$

This expression for the variance is valid for every value of $P_o(a,t)$, which means that the variance can be extracted even from time-varying currents (Sigworth, 1980). If an ensemble of current sweeps, $s_i(t)$, is recorded, the variance can be best estimated by subtracting pairwise sweeps (Heinemann and Conti 1992): $\delta\, s_i(t) = s_i(t) - s_{i+1}(t)$. The time-dependent variance is obtained by constructing the ensemble average of squared difference records:

$$\sigma^2(t) = \frac{\delta s(t)^2}{2} \tag{7.15}$$

The ensemble mean current is calculated by the usual

$$I = \frac{1}{N}\sum_{i=1}^{N} s_i(t) \tag{7.16}$$

The mean-variance relationship is then plotted and fitted to Equation 7.14, from which i and N are estimated. Once these numbers are obtained, an estimation of the open probability, including its value at a saturating value of the stimulus, is straightforward (Figure 7.4). Estimation of these parameters from fluctuation analysis gives values that are subject to the bandwidth of the recording and, in the case of whole-cell recordings, are affected by the higher series resistance intrinsic to this configuration. Care should be taken to obtain recordings at the maximum bandwidth possible. Accounting for sources of error is beyond the scope of this chapter, but a thorough discussion can be found in Heinemann and Conti (1992).

7.7 Gating Current Recording

The opening of voltage-dependent ion channels is triggered by a structural motif that detects voltage across the membrane. The voltage sensor can be understood as a charged component that moves in response to a change in membrane potential, favoring the open or closed state of the channel. A charge moving across the electric field is translated as a transient electric current because the charge movement ceases when a new equilibrium is reached. The charge movement is limited to the membrane electric field; consequently, the recorded currents are capacitive, and the total charge moved during a voltage pulse must be equal to the total charge moved when the voltage is returned to the initial value (Armstrong 1981). Unlike macroscopic currents and single-channel currents that only give information about the open state of the channel, the gating current offers information about the events and transitions preceding the open state.

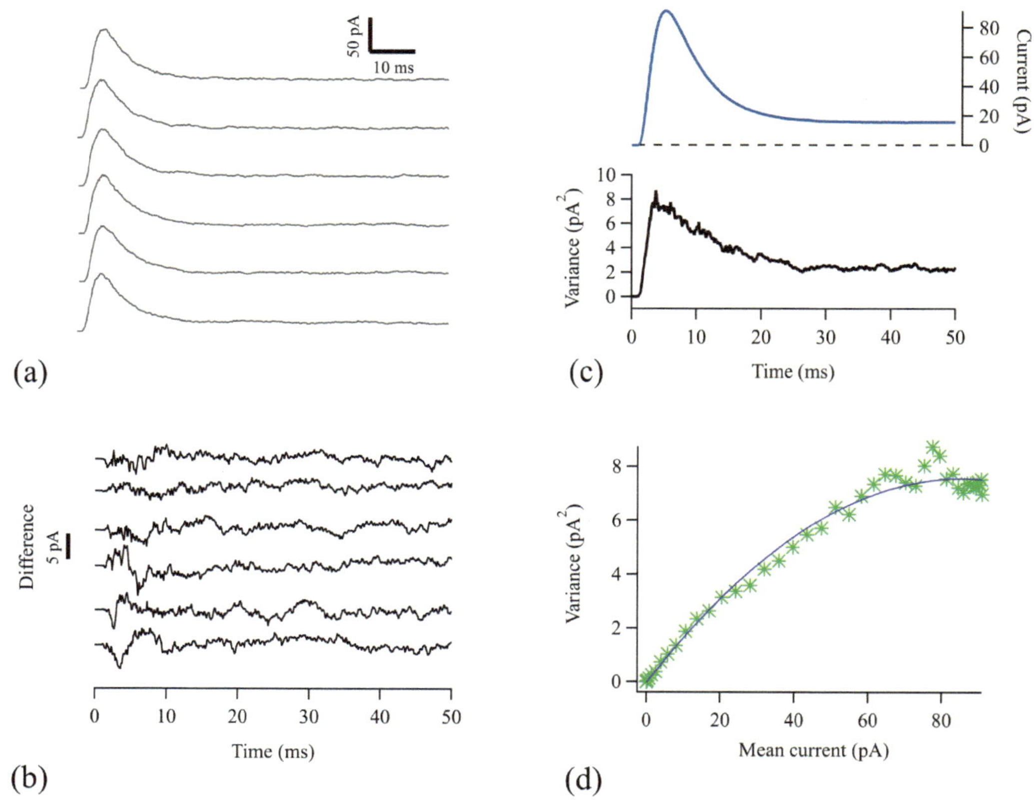

FIGURE 7.4
Extracting channel parameters from nonstationary noise analysis. (a) Simulated macroscopic currents from 1000 channels with a single channel current I = 0.16 pA. Two hundred traces were simulated for this example. (b) Sample pairwise difference traces. Notice how only the fluctuations remain and the current time course has been subtracted. (c) Ensemble mean and variance calculated according to Equations 7.15 and 7.16. (d) The variance versus mean relationship was fitted to Equation 7.14. The parameters estimated from this fit are N = 961, I = 0.17 pA.

The challenge of recording gating currents is the amplitude and temporal resolution. To record gating currents, the ionic current must be eliminated, either by replacing the permeant ion, using specific blockers or by recording nonconducting mutants. In the absence of permeant ions, the membrane behaves as a nonideal dielectric. The current flow continues for a time after a voltage step. The dielectric charge movement can be divided in two parts. First is the rearrangement of the mobile ions in the solution on both sides of the dielectric. This separation of charges is fast and proportional to the voltage change, that is, if the voltage increases, more charges are separated, and it is expected to be linear in the voltage range in which recordings are usually made. Second is the movement of charges confined to the membrane. This reorientation is limited by the membrane thickness and can only occur between certain voltages because the charge movement saturates at extremes voltages (Armstrong 1981). Therefore, it is a nonlinear current. This component is the gating current and can have other components, but the evidence suggest that it mostly reflects the gating process.

Hence, the capacitive current is

$$I_C = I_{C0} + I_{dl} + I_{nl} + I_g \tag{7.17}$$

where I_{C0} is the capacitive current of an ideal capacitor, I_{dl} is the dielectric linear current, I_{nl} is the dielectric nonlinear current not associated with gating and I_g is the gating current. I_{C0} and I_{dl} can be subtracted by a p/n protocol. This procedure allows the isolation of the nonlinear current.

Figure 7.5b shows gating currents recorded during depolarization (on-gating currents) from Kv1.2. The rising phase becomes faster with depolarizing voltages, while the decaying phase is fast at low voltages, become slower at intermediate and turn faster again at more depolarized voltages. A simple interpretation is that the initial transitions carry less charge or move more slowly than subsequent transitions. The decay of the gating current is usually fitted to a single exponential (i.e., Kv1.2) or multiple exponentials (i.e., Shaker K channel).

The integral of gating current (Figure 7.5c) at a given voltage is the gating charge and represents the moving charge times the fraction of the electric field. The QV curve (Figure 7.5e), the voltage dependence of the gating charge, shows sigmoidal behavior due to the saturation of the movement at extreme voltages. If the charge is divided by the number of channels, an approximation of the total gating charge per channel is obtained.

Appendix: Effect of p/n Subtraction on the Current Noise

Let's call a single sweep a current recording obtained at voltage V_T containing a capacitive transient, leak current and channel current, S_T. This sweep has noise with variance var(S_T). A subtraction sweep, without channel openings, S_L, obtained at the same gain but

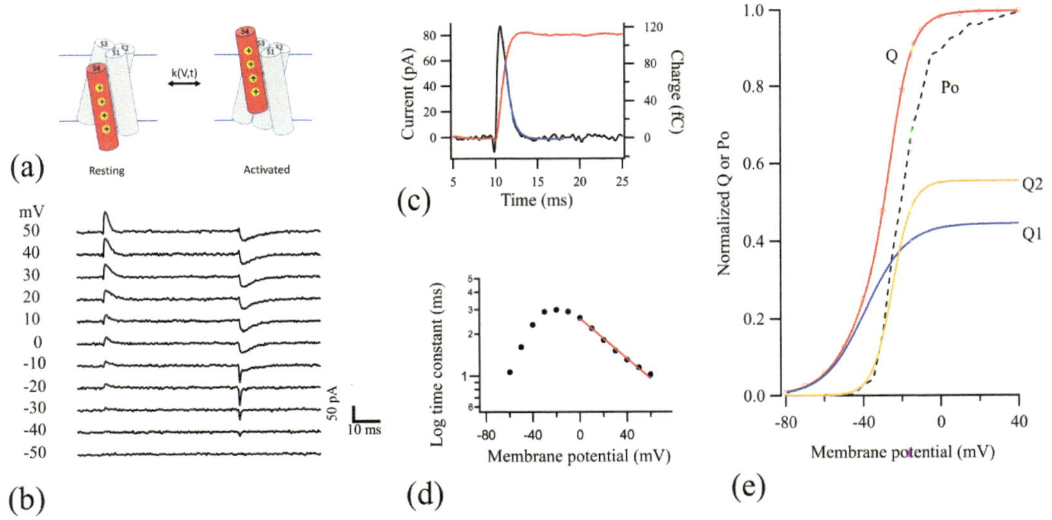

FIGURE 7.5
(a) Scheme of a classic voltage sensor domain in the resting and the activated states, the transition constant is dependent on voltage and time. The movement of the sensing charges in the orange helix gives rise to a nonlinear capacitive current known as the gating current. (b) Gating current recordings elicited by the indicated voltage step from an oocyte in cell-attached patch expressing Kv1.2. (c) Gating current (black) and the corresponding integral (red). The decay of the current is fitted to a single exponential function. (d) The time constant voltage-dependence of the decay fitted to $\tau\ i(V) = \tau\ i(0)\exp(qiV/kBT)$ gives a partial charge estimate of 0.41. (e) Normalized charge versus membrane voltage. The charge (Q) is fitted to a sum of two Boltzmann equations, similar to Equation 7.8, weighted by the charges Q1 and Q2, notice that Q1 is less voltage-dependent. P_o, open probability.

at a more negative voltage V_T/n will have a variance $var(S_L) = var(S_T)$. Before subtraction, the sweep S_L is scaled by n to give a sweep S_{Ln}. This sweep will have the same size of the leak and capacitive transients as S_T, but will have variance $var(S_{Ln}) = var(nS_L) = n^2 var(S_T)$. If S_{Ln} is subtracted from S_T, the variance is increased by n^2, inevitably obscuring any channel openings present in S_T.

A simple property of the variance of uncorrelated variables can be used to see how to reduce the variance of the subtraction sweep. The variance of a sum of variables X_i is

$$\mathrm{var}\left(\sum_{i=1}^{m} X_i\right) = \sum_{i=1}^{m}\left[\mathrm{var}(X_i)\right]$$

If we average m of the scaled sweeps S_{Ln}:

$$\mathrm{var}\left(\frac{1}{m}\sum_{i=1}^{m} S_{Ln}\right) = \sum_{i=1}^{m}\left[\mathrm{var}\left(\frac{1}{m}S_{Ln}\right)\right] = \frac{1}{m^2}\sum_{i=1}^{m}\left[\mathrm{var}(S_{Ln})\right]$$

Here the mean of the m sweeps S_{Ln} is

$$\frac{1}{m}\sum_{i=1}^{m} S_{Ln} = M_L$$

so that

$$\mathrm{var}(M_L) = \frac{1}{m^2}\sum_{i=1}^{m}\left[n^2\,\mathrm{var}(S_T)\right] = \frac{n^2}{m^2}\,m\cdot\mathrm{var}(S_T)$$

This implies that $var(M_L) = var(S_T)$ if $m = n^2$. So, a minimum of n^2 subtraction sweeps must go into the average subtraction sweep M_L before its subtraction from S_T does not significantly increase the noise of the original test pulses.

Acknowledgments

Victor De la Rosa is Cátedra CONACYT supported by FORDECYT-PRONACES/1308052/2020 - CONACyT. León Islas was supported by DGAPA-PAPIIT-UNAM grant No. IN215621.

Suggested Readings

Armstrong, C. M. 1981. "Sodium channels and gating currents." *Physiol Rev* 61 (3):644–83. doi:10.1152/physrev.1981.61.3.644.

Armstrong, C. M., and W. F. Gilly. 1992. "Access resistance and space clamp problems associated with whole-cell patch clamping." *Methods Enzymol* 207:100–22. doi:10.1016/0076-6879(92)07007-b.

Baumgartner, W., L. Islas, and F. J. Sigworth. 1999. "Two-microelectrode voltage clamp of Xenopus oocytes: voltage errors and compensation for local current flow." *Biophys J* 77 (4):1980–91. doi:10.1016/s0006-3495(99)77039-6.

Benndorf, K., R. Koopmann, E. Eismann, and U. B. Kaupp. 1999. "Gating by cyclic GMP and voltage in the alpha subunit of the cyclic GMP-gated channel from rod photoreceptors." *J Gen Physiol* 114 (4):477–90.

Cole, K. S., and J. W. Moore. 1960. "Ionic current measurements in the squid giant axon membrane." *J Gen Physiol* 44 (1):123–67. doi:10.1085/jgp.44.1.123.

Colquhoun, D., P. Jonas, and B. Sakmann. 1992. "Action of brief pulses of glutamate on AMPA/kainate receptors in patches from different neurones of rat hippocampal slices." *J Physiol* 458:261–87.

Edwards, F. A., A. Konnerth, B. Sakmann, and T. Takahashi. 1989. "A thin slice preparation for patch clamp recordings from neurones of the mammalian central nervous system." *Pflugers Arch* 414 (5):600–12.

Fox, A. P., M. C. Nowycky, and R. W. Tsien. 1987. "Kinetic and pharmacological properties distinguishing three types of calcium currents in chick sensory neurones." *J Physiol* 394:149–72.

Goldin, A. L. 1991. "Expression of ion channels by injection of mRNA into Xenopus oocytes." *Methods Cell Biol* 36:487–509. doi:10.1016/s0091-679x(08)60293-9.

Goldin, A. L. 2006. "Expression of ion channels in Xenopus oocytes." In *Expression and analysis of recombinant ion channels*, 1–25. Wiley-VCH, ISBN:9783527312092.

Goldstein, S. A., and C. Miller. 1993. "Mechanism of charybdotoxin block of a voltage-gated K+ channel." *Biophys J* 65 (4):1613–9. doi:10.1016/S0006-3495(93)81200-1.

Hamill, O. P., A. Marty, E. Neher, B. Sakmann, and F. J. Sigworth. 1981. "Improved patch-clamp techniques for high-resolution current recording from cells and cell-free membrane patches." *Pflugers Arch* 391 (2):85–100.

Heinemann, S. H., and F. Conti. 1992. "Nonstationary noise analysis and application to patch clamp recordings." *Methods Enzymol* 207:131–48. doi:10.1016/0076-6879(92)07009-d.

Horn, R., and A. Marty. 1988. "Muscarinic activation of ionic currents measured by a new whole-cell recording method." *J Gen Physiol* 92 (2):145–59.

Horn, R., and J. Patlak. 1980. "Single channel currents from excised patches of muscle membrane." *Proc Natl Acad Sci U S A* 77 (11):6930–4.

Khan, K. H. 2013. "Gene expression in Mammalian cells and its applications." *Adv Pharm Bull* 3 (2):257–63. doi:10.5681/apb.2013.042.

Kitamura, K., B. Judkewitz, M. Kano, W. Denk, and M. Hausser. 2008. "Targeted patch-clamp recordings and single-cell electroporation of unlabeled neurons in vivo." *Nat Methods* 5 (1):61–7. doi:10.1038/nmeth1150.

Maconochie, D. J., and D. E. Knight. 1989. "A method for making solution changes in the sub-millisecond range at the tip of a patch pipette." *Pflugers Arch* 414 (5):589–96.

Neher, E., and A. Marty. 1982. "Discrete changes of cell membrane capacitance observed under conditions of enhanced secretion in bovine adrenal chromaffin cells." *Proc Natl Acad Sci U S A* 79 (21):6712–6.

Pfaffinger, P. J., M. D. Leibowitz, E. M. Subers, N. M. Nathanson, W. Almers, and B. Hille. 1988. "Agonists that suppress M-current elicit phosphoinositide turnover and Ca2+ transients, but these events do not explain M-current suppression." *Neuron* 1 (6):477–84.

Schoppa, N. E., and F. J. Sigworth. 1998. "Activation of Shaker potassium channels. III. An activation gating model for wild-type and V2 mutant channels." *J Gen Physiol* 111 (2):313–42.

Sherman, A. J., A. Shrier, and E. Cooper. 1999. "Series resistance compensation for whole-cell patch-clamp studies using a membrane state estimator." *Biophys J* 77 (5):2590–601. doi:10.1016/s0006-3495(99)77093-1.

Sigworth, F. J. 1980. "The variance of sodium current fluctuations at the node of Ranvier". *J Physiol* 307:97–129. doi:10.1113/jphysiol.1980.sp013426.

Sigworth, F. J. 1995. "Electronic design of the patch clamp." In *Single-channel recording*, edited by B. Sakmann, and E. Neher, 95–198. New York: Plenum.

Stevens, C. F. 1972. "Inferences about membrane properties from electrical noise measurements." *Biophys J* 12 (8):1028–47. doi:10.1016/s0006-3495(72)86141-1.

Stuhmer, Walter. 1992. "Electrophysiological recording from Xenopus oocytes." In *Methods in enzymology: ion channels*, edited by B. Rudy, and L. E. Iverson, 319–39. San Diego, CA: Academic Press.

Taglialatela, M., L. Toro, and E. Stefani. 1992. "Novel voltage clamp to record small, fast currents from ion channels expressed in Xenopus oocytes." *Biophys J* 61 (1):78–82. doi:10.1016/s0006-3495(92)81817-9.

Zagotta, W. N., T. Hoshi, and R. W. Aldrich. 1994. "Shaker potassium channel gating. III: evaluation of kinetic models for activation." *J Gen Physiol* 103 (2):321–62.

8

Patch Clamping and Single-Channel Analysis

León D. Islas

CONTENTS

8.1 Introduction

The patch-clamp recording technique has allowed for the characterization of macroscopic ionic currents, single-channel currents, charge movement and other electrical signals from diverse cell types, and is now a commonplace experimental procedure (Colquhoun, 2007; Sigworth, 1986). One of the main uses of the patch-clamp technique has been the study of ion channels at the single-molecule level. After the first recordings of the activity of single bacterial ion channels in artificial lipid membranes in the late 1960s and early 1970s (Ehrenstein et al., 1974; Mueller and Rudin, 1963), Neher and Sakmann (1976) first reported single-channel patch-clamp recordings in the late 1970s. Since the publication of these and other seminal papers (Hamill et al., 1981; Colquhoun and Hawkes, 1982; Colquhoun and Hawkes, 1981), our understanding of ion channels as molecular machines has increased dramatically and, although alternative patch-clamp methods, such as polymer chips for multiple cell recordings, are now available (Klemic et al., 2002), the essential tools of single-channel recording and analysis have not drastically changed.

As a technique with single-molecule and submillisecond temporal resolutions, the main purpose of a single-channel recording is to measure the average kinetic parameters of a single ion channel, and through these, uncover molecular mechanisms and shed light on the biology of these remarkable proteins. The astonishing versatility and power of patch clamp and single-channel electrophysiology is illustrated by the tens of

DOI: 10.1201/9781003096214-10

thousands of papers published since initial development of the patch-clamp technique in the 1970s and 80s.

8.2 Conditions for Single-Channel Recording

Single ion channel activity can be assayed when one (active) ion channel is confined within the seal established between a patch electrode and a small area of the plasma membrane of a cell. In order for this to happen, the seal must have an extremely high electrical resistance, $R_{seal} >> 1G\Omega$, such that a negligible amount of the current flowing through the channel is lost at the seal resistance and most is collected by the electronics, since the membrane and patch pipette form a voltage divider. Although the exact details are not known, it is thought that strong electrostatic and Van der Waals interactions are responsible for sealing, since the pipette seal is mechanically very stable (Suchyna et al., 2009).

Several factors are responsible for the "sealability" of a particular pipette, and the quality and stability of the seals obtained. The type of glass used to fabricate patch pipettes is very important, with soft borosilicate glasses being the most widely used (Rae and Levis, 1984). Hard glass, although possessing better noise characteristics than borosilicate, tends to be more reticent to seal formation. The shape of the pipette is an important determinant, not only of the seal characteristics, but also of the electrical behavior of the patch. It is preferable to use short stubby pipettes to long pipettes, because the series resistance of the short pipette is smaller. Since suction is applied to the pipette to attain a gigaseal, longer pipettes tend to allow more membrane to be drawn into the tip. This can be problematic since areas of high series resistance and poor membrane attachment can generate membrane blebs with large voltage errors and the patch can get out of voltage clamp (Bae et al., 2011). This is avoided in the short pipettes, where the seal is closer to the tip.

In general, the final shape of the pipette tip is determined by the fire-polishing step. After a pipette is pulled from capillary glass, it is brought into close proximity of a red-hot tungsten filament covered with glass. This process eliminates any roughness of the glass and burns off grease and other contaminants. Pipettes can be obtained with a large range of final tip diameters. For single-channel recordings of small patch sizes, small tips of 1 micron or less should be used. To record larger, macroscopic currents and avoid problems with series resistance, pipettes with a large initial opening (10–20 µm) can be pulled and fire-polished to a desired diameter between 1 and 10 µm. The shape of the patch at the tip of the pipette is generally assumed to be that of a dome or the Greek letter Ω in cross section, as has been visualized in some instances (see Chapter 7 and Suchyna et al., 2009; Hamill et al., 1981).

The capacitance arising from the glass pipette becomes a main noise source and is very relevant in recordings of voltage-gated ion channels, because a change in voltage has to charge the total system capacitance, of which the patch pipette can contribute the largest fraction. To reduce this contribution, patch electrodes are typically covered with a low-dielectric substance, essentially a substance that acts as an electrical insulator and is not easily polarizable. This is achieved by the application to the region close to the pipette tip of polymers such as PDMS (polydimethylsiloxane or Sylgard), dielectric varnishes (Q-dope) or even dental wax or beeswax. These manipulations can and will leave traces of the coating substance at the pipette tip, so, the fire-polishing step is also responsible for the most important characteristic of a new patch pipette: its cleanliness. Any residue accumulated at the tip will be burned off during polishing, with the exception of Sylgard, which will only be cured and hardened by an elevated temperature.

8.3 Analysis of Single-Channel Signals

Single-channel recordings are an electrophysiological report of the activity of the ion channel protein, as it stochastically gates its pore between closed and open conformations. This stochastic gating process is essentially a homogeneous Poisson process, that is, a process producing random events separated by exponentially distributed intervals with mean duration $1/\lambda$ where the rate constant, λ, is time-independent. The formalism used to analyze such processes is that of Markov models, which are discussed in Chapter 10, and assumes memory-less transitions between multiple discrete states joined by time-independent rate constants.

Single-channel analysis has thus two main purposes: to determine the average amplitude of events in the recording, and the duration or lifetimes of these events. The lifetimes and their average values will be modulated by several factors, such as transmembrane voltage, tension, temperature, or the presence of ligands and modulatory substances. It is in the response to these variables where the interesting biology will be found (Hille, 1992).

Recordings can be achieved in two general cases. When the physical variable that controls the activity of the channel is held constant and the recording avoids any transient changes, the measurement is said to be in steady-state or stationary. If the variable is instantaneously (or very rapidly) changed, the experiment is then a nonstationary measurement.

8.3.1 Nonstationary Recordings

Examples of nonstationary single-channel measurements can be found in the study of ligand-activated or voltage-activated ion channels. In these cases, we are interested in the initial response of the channel to a sudden change in ligand concentration or in transmembrane voltage. Suppose we are dealing with a channel that is activated by voltage. In this case, we want to know the voltage dependence of the average time that the channel spends in the closed or open states, also known as dwell times, and the patch-clamp amplifier can change the voltage within a millisecond or less. The changing voltage will charge the capacitance associated with the membrane patch and the patch pipette, producing an initial transient capacitive current. Although in general there will be a delay between the moment of voltage change and the activation of the channel, the capacitance charging process can be slow enough that the first opening of the channel will be distorted by the time course of this capacitive transient. In order to observe clear openings and closings, and to be able to study the time between voltage change and the first opening, also called the first latency or latency to first opening (Figure 8.1b), we must subtract the capacitive transient and the leak current. In principle, the p/n or $-p/n$ subtraction procedure discussed in Chapter 7 could be used, but since the subtraction pulse must be scaled up by n or $-n$, respectively, this operation results in the problem of increased noise in the subtracted pulse, unless a very large number of subtraction pulses can be recorded and averaged. It is thus preferable to find and average several traces without openings or null traces. The null trace average can be used as a leak template to be subtracted from each trace. This procedure should not add significantly to the variance of the noise if enough null traces

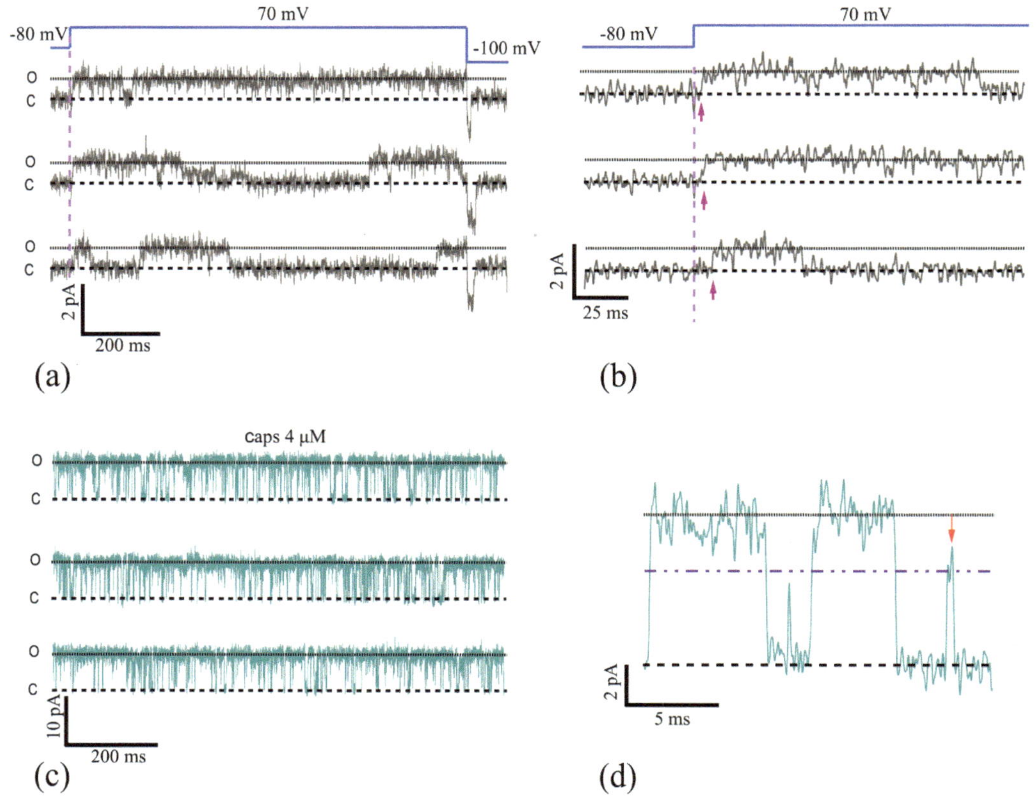

FIGURE 8.1
Nonstationary and stationary recordings. Panel (a) shows recordings from the voltage-gated potassium channel Kv1.2 in response to a voltage pulse of 70 mV. The vertical dashed line indicates the start of the voltage pulse, the horizontal dashed lines indicate the closed state (C) and the dotted line indicates the open state (O). (b) Shows the same recordings in panel (a) plotted in an expanded time scale to indicate the latency to first opening, which is highlighted by the purple arrows. (c) Stationary recordings of a single TRPV1 channel activated by capsaicin. At the indicated capsaicin concentration, channel gating occurs in very long bursts of openings. The dashed line indicates closed state (C) and the dotted line the open state (O). (d) The same channel recording as in (c) is plotted in an expanded time scale to show individual openings and the presence of subconductance state openings, indicated by the red arrow. The blue dotted line indicates the amplitude of the subconductance level.

are accumulated in the average (Sigworth and Zhou, 1992). An example of nonstationary recordings of single voltage-activated Kv1.2 K^+ channels is shown in Figure 8.1a and b.

8.3.2 Stationary Recordings

Another modality of recording single-channel activity is under conditions in which the stimulus has reached a steady and constant value in time, for example, a constant concentration of a ligand in the bath solution. In this case, channel gating has reached equilibrium, and openings and closings should only reflect the constant values of the underlying rate constants. Note that a stationary recording can be applied to any type of ion channel, regardless of its activation modality. In this case, long stretches of channel activity are recorded (several seconds to minutes). Because dwell times are exponentially distributed, longer-lived events are less frequent, and may be entirely missed if only short pulses or short periods of time are recorded. Also, some channels may show different kinetic modes, switching between periods of high and low activity, and these phenomena can be captured only in longer stationary recordings (Rothberg and Magleby, 1998; Auerbach and Lingle, 1986). An example of stationary recordings of a single TRPV1 channel activated by capsaicin is shown of Figure 8.1c and d.

Recordings in steady-state are an important tool for the characterization of ion channel behavior. For example, in the case of a ligand-gated ion channel the probability of finding it in the open state is a function of the ligand concentration. At low agonist concentration the probability might be in the order of less than 1%. In the case of voltage-gated channels, the open probability at low voltages can range from 0.01 to 10^{-8}. This implies that an extremely long time would be necessary to record sufficient events (openings) for statistical calculation of the probability and kinetics of the channel (Islas and Sigworth, 1999; Hirschberg et al., 1995). For example, if the open probability is 10^{-8} and the mean duration of an opening is 10 ms, one expects one opening in approximately 28 hr.

It is often observed that even when recording from a single channel, a few openings do not reach the most common, "full" open state. These openings are called subconductance states and might represent intermediate states or openings arising from different conformations of the pore (Figure 8.1d). Subconductance states call for special care when analyzing channel records and can provide important clues into channel gating mechanisms.

8.3.3 Filtering the Data

Signal conditioning is achieved by low-pass filtering of the current, that is, using a filter that attenuates high-frequency components, usually noise, and thus lets through slower-varying signals with minimal distortion. Filtering is essential to impose the temporal resolution of the current recording and to improve the signal-to-noise ratio by eliminating high-frequency components in the recorded signal. This is most commonly done by passing the current being collected by the patch pipette through an analog filter, usually part of the amplifier's electronics. The best filter for this purpose is the multipole Bessel filter, because it produces less delay of the individual frequency components and thus less distortion of the shape of the signal. The corner frequency, f_c, of the said filter corresponds to a 3 dB attenuation of the original signal or roughly a reduction of 50% amplitude. The f_c is a good parameter to characterize the temporal resolution of the recordings, as shown in Section 8.3.4. The analog current signal, after being filtered, is to be digitally sampled by an analog-to-digital converter. The digital sampling requires an advisable relationship between f_c and how often a digital sample is acquired, the sampling frequency, f_s. This is given by the Nyquist criterion, which states that $f_s \geq 2f_c$. This relationship ensures that aliasing, or the appearance of spurious frequency components in the spectrum of the signal, does not occur. In practice, sampling at $5f_c \leq f_s \leq 10f_c$ is necessary in order to record fast events and improve the detection of events, although at the expense of the signal-to-noise ratio.

Digital filtering prior to analysis is generally advisable. This last filtering step determines the final bandwidth of the recording. Since the analog signal has already passed through several serial filtering stages (the patch-clamp amplifier, any additional analog filter, any additional storage units apart from the computer hard disk), the digital filtering stage will set the bandwidth of the recorded data according to a cutoff frequency, f_c, which can be calculated by the formula

$$\frac{1}{f_c} = \frac{1}{f_1} + \frac{1}{f_2} + \dots \frac{1}{f_n} \tag{8.1}$$

In this equation n is the number of filtering stages, including the Gaussian digital filter.

Digital filtering is generally carried out with a Gaussian filter algorithm, which closely resembles the desirable characteristics of the multipole Bessel filter. Analytical expressions for the Gaussian filter are available, which can be used to obtain expression for other parameters such as the threshold for event detection (Colquhoun and Sigworth, 1995).

8.3.4 Resolution

As mentioned earlier, filtering the data imposes a temporal resolution. This can be calculated for a Gaussian filter as follows. If an instantaneous step input is passed through a Gaussian filter, the output is distorted in several ways. The output can be calculated as a function of time, *t*, with the step response *H(t)* of the filter, which is given by the equation

$$H(t) = \frac{1}{2}\left[1 + erf\left(5.336 f_c t\right)\right] \tag{8.2}$$

In this equation, *erf* is the error function. Figure 8.2a illustrates the response of such a filter to several input pulses of equal amplitude but different durations. As can be gleaned from the figure, shorter pulses are reduced in amplitude by the filter until some become smaller than 50% of the amplitude of the original ones. These pulses will not cross the half-amplitude threshold and thus represent missed events (see Section 8.35). The rise time of the filter can be defined as the time it takes for the output to change from 10% to 90% of its maximum value, and it is given by

$$T_r = \frac{0.332}{f_c} \tag{8.3}$$

For this type of filter, events that are roughly half the duration of T_r will be missed when using the 50% threshold crossing method discussed in Section 8.3.5. The minimum duration of an event that we can expect to detect can be characterized by another metric called the filter death time T_d. The death time can be calculated as $T_d = 0.538T_r$ or $T_d = 0.179/f_c$.

Another consequence of the reduced amplitude of events caused by filtering is that two closely spaced events will be distorted to look like a single one (Figure 8.2b). This will have the effect of missing short-lived events that will be counted as longer ones. Several procedures have been developed to cope with the distortions introduced by this and other types of missed events but are too specialized to be discussed here. The interested reader can consult Qin et al. (1996), Blatz and Magleby (1986) or Crouzy and Sigworth (1990).

8.3.5 Detection of Events

Data obtained from single-channel recordings can be stored directly into the computer memory. Once there, the next step is to analyze the recording and characterize the channel gating events. Two main methods of identifying events in a recording are currently in general use: direct fitting of the time course of the recording (Gibb and Colquhoun, 1991) and the threshold crossing method. This chapter will discuss the latter method. Direct fitting is a more specialized technique, and its advantages and drawbacks are discussed in Colquhoun and Sigworth (1995). The threshold crossing method is also applicable to both stationary and nonstationary recordings.

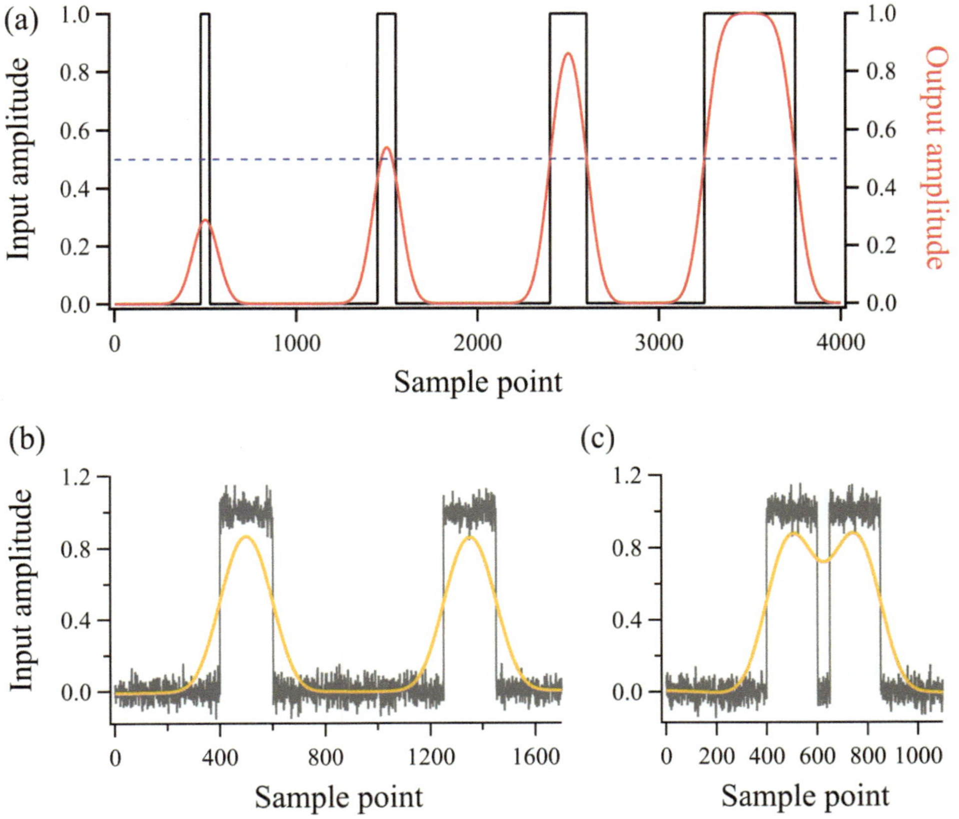

FIGURE 8.2
Response of the Gaussian filter to step inputs of varying duration. (a) Four square pulses of 25, 50, 100 and 500 sample points (black trace) were filtered by a Gaussian filter and the output is superimposed (red trace). The rising and falling time courses of the filtered output response are distorted according to Equation 8.4, which describes the step response of the filter. Note that the shortest step does not reach the 50% value of the amplitude of the longer step and, as a consequence, will be undetected by the 50% threshold crossing technique (this threshold is indicated by the blue horizontal dashed line). Also, the second step barely crosses the 50% amplitude line, and the measured width of the filtered pulse underestimates the real duration of the input step. Two closely spaced openings can be confounded as a single opening due to filtering. (b) Two equal-duration events (gray traces) separated by a longer time than the filter settling time appear as separate events (orange curves). (c) The same duration events separated by a shorter time and filtered by the same filter as in (b), will be counted as a single, longer duration event and the closed dwell time will be missed.

The first step is to find the average amplitude of the events in the recording. This can be done by first compiling an all-points histogram of several events and determining the average amplitude, $\langle I \rangle$ (Figure 8.3). If the channel has a single, well-defined level of conductance, the all-points histogram is approximately distributed according to a Gaussian distribution, and a Gaussian fit can be used to determine the average amplitude, and thus the threshold, as $\langle I \rangle$ θ, where θ is a number between 0 and 1. This value is then fed back into the event detection algorithm. The all-points histogram is a useful tool, since it also permits one to detect the presence or absence of subconductance levels, which can introduce major errors in the analysis by the threshold method and that have to be dealt with in a special way.

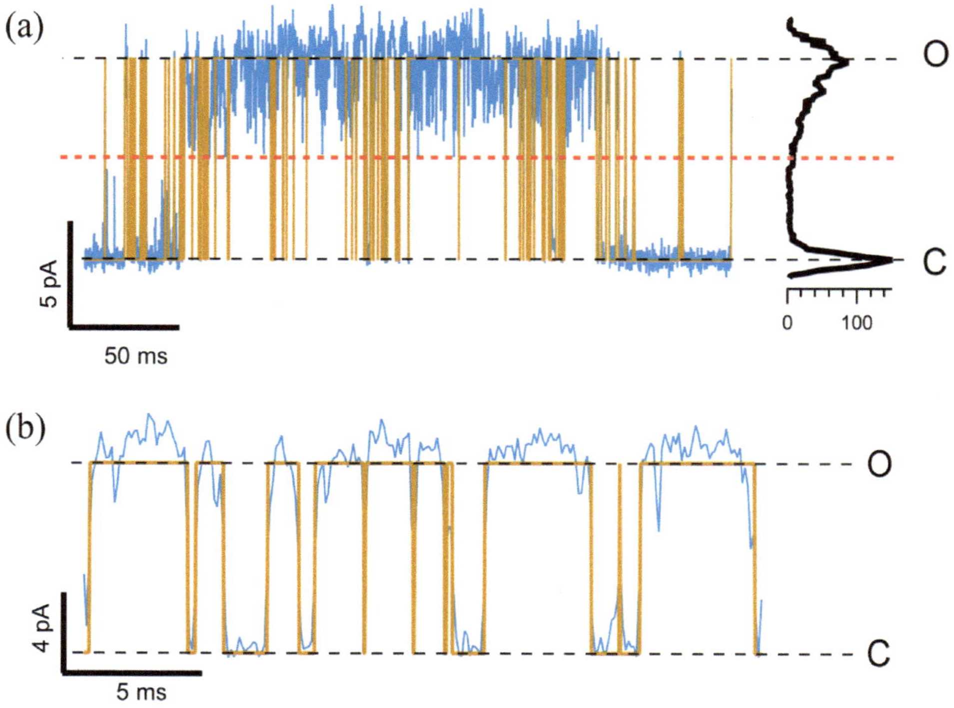

FIGURE 8.3
Detection of events in single-channel recordings. (a) Burst of openings of a single TRPV1 channel (blue trace). An all-points amplitude histogram is shown to the right in black. This histogram indicates a mean open amplitude of ~9.4 pA. The threshold for event detection was set to 5 pA (red dashed line), and the detected channel openings and closings are shown by the yellow trace, which is superimposed on the current trace. Notice the existence of very fast closing events near the open state that do not cross the threshold and thus are not counted. These events constitute what is often termed "open channel noise" and represents very fast transitions to unresolved states. (b) A section of the burst is plotted in an expanded time scale to better show individual open and closed dwell times as well as the identified events. Colors as in (a).

Before events can be detected, the whole record must be interpolated to avoid distortion of the short-lived events as much as possible, and reduce the uncertainty associated with the determination of when exactly the threshold is crossed. A cubic spline is an effective interpolation because it can be easily implemented in software and is not computationally costly. The cubic spline is a polynomial of order 3 that is used to model points between analog samples, in this case the recorded channel current.

Defining the threshold, θ, is an important and complicated issue. The value of θ is critically related to the probability that random, Gaussian-distributed noise events in the recording will cross an arbitrary value defined by $\theta \langle I \rangle$. We want to avoid the detection of too many of these spurious events. If the value of θ is chosen to be too small, many false events will be detected. If it is too close to 1, too many short-lived real events will be missed and fluctuation events present in the noise, when the channel is open, will be counted as closing events. If the channels are being filtered with a Gaussian filter, there is a relatively simple rule of thumb that can be used to choose the value of θ. For small amplitude channel openings, having an amplitude two times the standard deviation of the closed channel noise, heavy filtering has to be employed, perhaps less than 1 kHz and in this case $\theta = 0.7\langle I \rangle$ can be used. In the case of openings with larger amplitude the threshold can be chosen as $\theta = 0.5\langle I \rangle$. This last case is often referred to as the 50% threshold crossing technique and

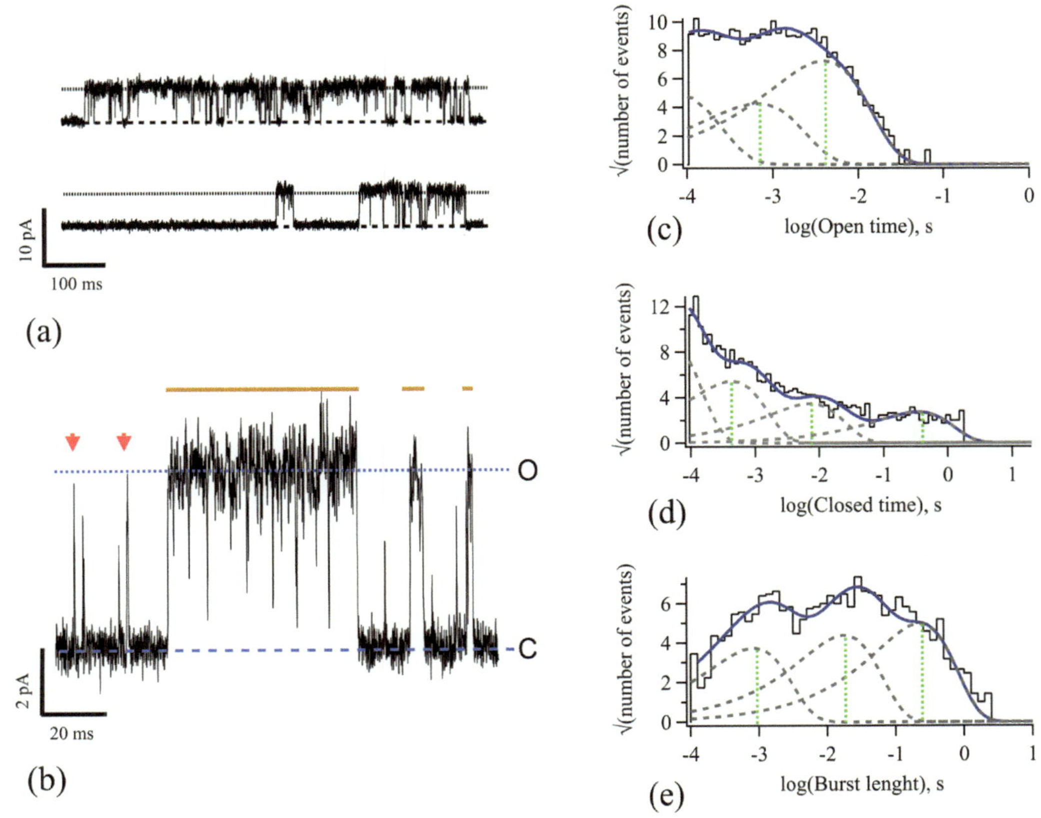

FIGURE 8.4
Distribution of closed, open and burst dwell times. (a) Data were obtained from a single-channel recording of a TRPV1 channel activated by a subsaturating capsaicin concentration. Openings are clustered in bursts separated by closed periods. (b) A few bursts of openings. The arrows indicate single openings and the bars indicate bursts belonging to the long and intermediate burst durations (see panel e). (c and d) The open and closed times were measured with the 50% threshold crossing technique and are displayed according to the Sigworth–Sine transform. Note the presence of several bumps in the distributions, which for the open duration distribution correspond to three exponential components with different mean times, which are individually plotted as dotted curves. The vertical green dotted lines coincide with the peak of each component and represent the mean time of each exponential. The smooth curve over the histogram is the sum of these exponentials estimated by maximum likelihood analysis of the closed times. The closed event distribution is fitted by the sum of four exponentials (d), which are indicated as in (c). (e) The burst duration distribution was calculated as indicated in the main text, from the distributions in (c) and (d). Three exponentials are needed to fit this distribution.

is widely used as the standard method for event detection (Sachs et al., 1982). Once the threshold is selected, the duration of an opening is defined as the sum of all the points above the threshold times the sampling interval. Likewise, closings are recorded as all the events that remain below the threshold (Figure 8.3).

Nonstationary recordings present the opportunity to measure another important lifetime. This is the time between the change in voltage or agonist concentration and the first opening event, which is called the latency to first opening or first latency (Figure 8.1b and Figure 8.5f). This is a special parameter since it cannot be obtained from steady-state recordings. If the channel has only one open state, the first latency represents transitions between all the closed states that the channel has to occupy before opening. Measuring

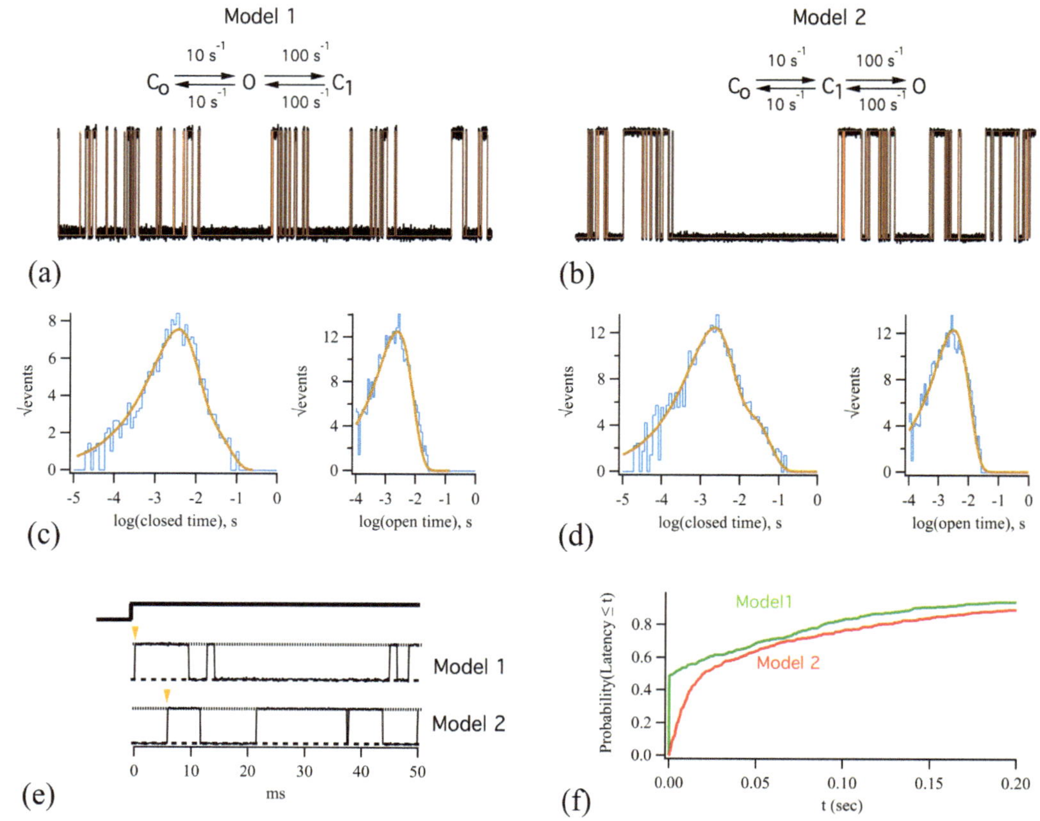

FIGURE 8.5
Different gating mechanisms can be distinguished via single-channel analysis. (a) Linear model of two closed states and one open state. The value of the rate constants is indicated above each arrow signifying the direction of a transition. Lower panel, the black trace is 2 s of simulated openings and closings. The idealized record is superimposed in orange. Notice gating tends to occur in bursts of openings. (b) A different gating model with two closed and one open state. The lower panel also shows 2 s simulation of channel gating as in (a). (c) Histograms of the log-transformed closed and open events for model 1. Notice that the closed-time distribution is fit to a sum of two exponentials, while the open time is fit to a single exponential. The data produced by the two models is visually very similar. (d) Histograms of dwell times as in (c) for model 2. The closed-times histogram is also fit to two exponentials and the open-time histogram to one. (e) Representative simulated traces in response to a voltage pulse. The arrows show the first opening in each trace. (f) Cumulative distribution of first latencies obtained from 400 traces simulated with each model as in (e). Note that this distribution clearly allows to discriminate between the two models, which is not possible from the open- and closed-time distributions alone.

first latencies poses the problem of deciding when the actual change in stimulus starts at the patch-clamped membrane. In the case of a voltage pulse, this is a parameter that depends on the particular instrumentation (amplifier and analog-to-digital/digital-to-analog [AD/DA] converter) and filter settings used, and should be measured in each case. An adequate way to estimate it is to measure an instantaneous change in current, for example, measuring the deactivation of a macroscopic tail current from a voltage-gated channel. The delay is then estimated as the time between the DA pulse and the midpoint of the current change. For voltage pulses, these delays (at the AD/DA converter) are generally less than 1 μs and are not important except at extreme voltages where the intrinsic delay of channel activation might approach these values.

On the other hand, even the fastest changes in agonist concentration carried out in outside-out patches will incur in considerably longer delays, in the order of milliseconds.

The distribution of first latencies is not exponential when there is more than one gating transition. It is common to display it as a cumulative distribution and to compare it with the probability that the channel is open after a time *t* after the change in voltage or ligand concentration. In this representation it can be compared with a model like

$$C_0 \underset{k_{-1}}{\overset{k_1}{\rightleftarrows}} C_1 \cdots C_n \xrightarrow{\alpha} O$$

In this model the channel transitions between *n* closed states, each one denoted by C_i (*i* goes from *0* to *n*) with transition rates k_i and once it reaches the open state *O*, with rate α , it does not close, which is an analogous representation of the measurement of the cumulative distribution of first latencies (Sigworth and Zhou, 1992).

The end result of the event detection step is the generation of lists of event durations. Generally one would obtain a list of open and closed state lifetimes (of which the first latency is the first closed time measured in each sweep), which will be used later for further analysis.

Even though programs of varying complexity can be written to carry out these analyses, it is crucial that the investigator visually inspects each detected event and validates its correct detection. It is important to understand that the selection of the threshold value is valid only for Gaussian-distributed (noise) events. If the recording contains fluctuations other than baseline noise and channel gating, such as changes of the average value of the baseline, noise can be "pushed" to cross the threshold, contaminating the events list with spurious events. If line-frequency (60 Hz) interference is present in the recording, there is a real chance of periodically detecting spurious threshold crossings. Also, the seal resistance of the patch over long periods of recording can change, altering the baseline and the noise characteristics. Analysis under these circumstances can lead to detection of false events if the investigator does not validate each of them. Thus, the programs used for analysis should have a provision to remove false events from the lists of events (Figure 8.2).

8.3.6 Dwell-Time Histograms and Fitting of Distributions

In general, for a Poisson-like random process, such as channel gating, the dwell time in any given state visited by the process is an exponentially distributed random variable (Gardiner, 1994). The events lists are thus distributed as exponentials or sums of exponentials. The data contained in the events list can be compiled in histograms and these can be plotted in a linear scale to graphically display the distribution of the data. This procedure has the drawback that, in the case of a channel with several states, the exponential components can be separated by several orders of magnitude. In a linear representation of such a complex histogram it can be extremely difficult to make out the distinct components. A better visual representation is obtained with the Sigworth–Sine transform (Sigworth and Sine, 1987). For a single exponential, if the intervals or bins of the histogram, T_i's, are transformed logarithmically according to

$$x_i = \ln(T_i) \tag{8.4a}$$

it can be shown that the probability density function (*pdf*) is given by

$$f(x) = a\tau^{-1}e^{\left(x-\tau^{-1}e^{x}\right)} \tag{8.4b}$$

In this equation τ is the mean dwell time, and a is the amplitude of the exponential. This equation can be generalized as a sum of components for distributions with more than one exponential component.

In this display method, the individual dwell times t_i are transformed logarithmically (Equation 8.4a) and the ordinate is displayed as the square root of the number of events, because this transformation evens out the error across all the bin widths. The advantage of this transformation is that it very naturally provides a visual guide to the distribution of the dwell times, because as was shown for the *pdf*, an exponential appears as an approximately bell-shaped curve with its peak value occurring at a time equal to the mean of the exponential (Figure 8.4c–e).

A crucial step in the analysis of single-channel records is the fitting of a theoretical distribution to the actual histograms of dwell times. This has the purpose of finding the parameters that describe the data, specially, the value of mean dwell time. In general, if the channel dwells in a single, discrete state, the distribution of lifetimes in that state will be described by a single exponential whose mean time is the inverse of the sum of all rate constants for leaving that state. Since the distribution of the data is known (it is exponential), the preferred method for fitting a distribution of single-channel dwell times is the use of a maximum likelihood algorithm, as opposed to least squares fitting or other methods. In this method we are not interested in fitting exponential functions to the histogram of dwell times, but in actually using all the dwell times, t_i, in the events list having n entries, to calculate a likelihood function, which, for a single exponential distribution with mean value τ, is defined as

$$Lik(\tau) = \prod^{n} \frac{1}{\tau} e^{(-t_i/\tau)} \tag{8.5}$$

For actual computation of the likelihood, it is more useful to calculate the natural log of the function, also known as the log-likelihood, $L(\tau)$:

$$L(\tau) = \ln Lik(\tau) \tag{8.6}$$

A program can be written to maximize the value of L, which also maximizes Lik, by varying τ, which is the parameter that we are interested in finding out. It can be shown that the value of τ that maximizes L characterizes the exponential distribution that is more probable to describe the data (Horn and Lange, 1983; Colquhoun and Sigworth, 1995). Once τ has been found, an exponential distribution can be generated and superimposed in the actual histogram of the data (Figure 8.4c and d).

For very rapid or very noisy, small-amplitude data, there are methods to obtain the parameters describing the data that involve the use of hidden Markov models (HMMs) (Venkataramanan and Sigworth, 2002).

8.3.7 Burst Analysis

Openings and closings often appear to be clustered in groups separated by long visits to a closed state (Figure 8.4a and b). These groups of openings and closings are referred to as bursts. Important parameters for the establishment and discrimination of kinetic mechanisms can be extracted by analyzing the burst characteristics, such as its duration, number of openings and closings, and the mean duration of these events within the bursts (Colquhoun and Hawkes, 1982). Generally, burst duration is also a random variable with exponential distribution (Figure 8.4e), and this distribution, along with other parameters of the burst such as the total open time, are important in the discrimination of kinetic mechanisms. A classic example of this can be found in the demonstration that blockers that only bind to the open state of a channel can be identified because they produce an increase of the duration of bursts of openings (Neher and Steinbach, 1978).

To carry out burst analysis, one has first to identify a burst. Notice first that channel openings will be clustered in bursts when there are two or more closed states with widely varying mean dwell times. The channel openings will tend to be separated by visits to the short-lived closed state(s), and the bursts will be separated by sojourns to the long-lived closed states. In order to define the burst, one has to predetermine a criterion to set a maximal duration of the closed intervals within the burst, t_c. Any closed interval $t > t_c$ will be considered to be outside the burst and will signal the end of the identified burst. Several criteria have been proposed to define t_c. All of them try to define t_c as a value of t that lies between the fast and slow components of the closed-time distribution. This implies that before burst analysis can be performed, the closed-time distribution must be obtained as explained in the previous section. The method of Colquhoun and Sakmann (1985) is appropriate when the number of long and short closed events is very different. In this method t_c is found by solving the equation

$$\exp\left(-t_c/\tau_f\right) = 1 - \exp\left(-t_c/\tau_s\right) \tag{8.7}$$

In this equation, τ_f and τ_s are the mean durations of the fast and slow components of the closed-time distribution, respectively. Once t_c is defined, the program can use this value to identify bursts and determine their duration and other statistics.

8.4 Inferring a Mechanism

Analysis of single-channel recordings is a powerful method to study ion channel gating. In particular it allows to infer a mechanism of transitions between the closed states connected to the open or multiple open states. As an example, Figure 8.5 shows two different models, each with two closed and one open state but connected in different ways. Simulations of these mechanisms with the indicated rate constants predict steady-state gating at the single-channel level in the form of bursts of openings separated by relatively long shut periods for both models (Figure 8.5a and b). The closed and open dwell-time histograms obtained after 50% threshold analysis do not provide an obvious form to distinguish between the two mechanisms. As expected, the closed-time histogram shows two exponential components and the open-time histogram has a single exponential. The mean times of these exponentials are very similar for both simulated channels (Figure 8.5c and d). However, if a stimulus jump (nonstationary) experiment is performed, as shown by the simulated traces in Figure 8.5e, the first latency can readily distinguish between models. The channels that open after two closed states display a markedly longer latency, while the channels that gate according to model 1 show a much-shortened instantaneous component in the latency. This numerical experiment illustrates that different mechanisms can give rise to very similar single-channel behaviors, but multiple types of experiments can help in distinguishing specific mechanisms of gating, illustrating the power of single-channel analysis.

8.5 Conclusions

Studying ion channels at the single-molecule level is one of the great powers of single-channel recording techniques. Analysis of the activity of channels at this level has produced important insights and continues to push the boundary of knowledge in this particular area. Although in no way comprehensive, this chapter outlines the major procedures and concepts involved in the application of these methods and hopes to be a gateway to the concepts presented in the chapters ahead.

Acknowledgments

This work was supported in part by grants DGAPA-PAPIIT-UNAM No. IN215621 and INV2021 from the School of Medicine, UNAM (National Autonomous University of Mexico).

Suggested Readings

Auerbach, A., and C. J. Lingle. 1986. "Heterogeneous kinetic properties of acetylcholine receptor channels in Xenopus myocytes." *J Physiol* 378:119–40.

Bae, C., V. Markin, T. Suchyna, and F. Sachs. 2011. "Modeling ion channels in the gigaseal." *Biophys J* 101:2645–51.

Blatz, A. L., and K. L. Magleby. 1986. "Correcting single channel data for missed events." *Biophys J* 49:967–80.

Colquhoun, D. 2007. "What have we learned from single ion channels?" *J Physiol* 581:425–7.

Colquhoun, D., and A. G. Hawkes. 1981. "On the stochastic properties of single ion channels." *Proc R Soc Lond B Biol Sci* 211:205–35.

Colquhoun, D., and A. G. Hawkes. 1982. "On the stochastic properties of bursts of single ion channel openings and of clusters of bursts." *Philos Trans R Soc Lond B Biol Sci* 300:1–59.

Colquhoun, D., and B. Sakmann. 1985. "Fast events in single-channel currents activated by acetylcholine and its analogues at the frog muscle end-plate." *J Physiol* 369:501–57.

Colquhoun, D., and F. J. Sigworth. 1995. "Fitting and statistical analysis of single-channel records." In *Single-channel recording*, edited by B. Sakmann, and E. Neher, 2nd ed, 483–588. New York: Plenum.

Crouzy, S. C., and F. J. Sigworth. 1990. "Yet another approach to the dwell-time omission problem of single-channel analysis." *Biophys J* 58:731–43.

Ehrenstein, G., R. Blumenthal, R. Latorre, and H. Lecar. 1974. "Kinetics of the opening and closing of individual excitability-inducing material channels in a lipid bilayer." *J Gen Physiol* 63:707–21.

Gardiner, C. W. 1994. *Handbook of stochastic methods for physics, chemistry, and the natural sciences*. Berlin and New York: Springer-Verlag.

Gibb, A. J., and D. Colquhoun. 1991. "Glutamate activation of a single NMDA receptor-channel produces a cluster of channel openings." *Proc Biol Sci* 243:39–45.

Hamill, O. P., A. Marty, E. Neher, B. Sakmann, and F. J. Sigworth. 1981. "Improved patch-clamp techniques for high-resolution current recording from cells and cell-free membrane patches." *Pflugers Arch* 391:85–100.

Hille, B. 1992. *Ionic channels of excitable membranes*. Sunderland, MA: Sinauer Associates.

Hirschberg, B., A. Rovner, M. Lieberman, and J. Patlak. 1995. "Transfer of twelve charges is needed to open skeletal muscle Na+ channels." *J Gen Physiol* 106:1053–68.

Horn, R., and K. Lange. 1983. "Estimating kinetic constants from single channel data." *Biophys J* 43:207–23.

Islas, L. D., and F. J. Sigworth. 1999. "Voltage sensitivity and gating charge in Shaker and Shab family potassium channels." *J Gen Physiol* 114:723–42.

Klemic, K. G., J. F. Klemic, M. A. Reed, and F. J. Sigworth. 2002. "Micromolded PDMS planar electrode allows patch clamp electrical recordings from cells." *Biosens Bioelectron* 17:597–604.

Mueller, P., and D. O. Rudin. 1963. "Induced excitability in reconstituted cell membrane structure." *J Theor Biol* 4:268–80.

Neher, E., and B. Sakmann. 1976. "Single-channel currents recorded from membrane of denervated frog muscle fibres." *Nature* 260:799–802.

Neher, E., and J. H. Steinbach. 1978. "Local anaesthetics transiently block currents through single acetylcholine-receptor channels." *J Physiol* 277:153–76.

Qin, F., A. Auerbach, and F. Sachs. 1996. "Estimating single-channel kinetic parameters from idealized patch-clamp data containing missed events." *Biophys J* 70:264–80.

Rae, J. L., and R. A. Levis. 1984. "Patch clamp recordings from the epithelium of the lens obtained using glasses selected for low noise and improved sealing properties." *Biophys J* 45:144–6.

Rothberg, B. S., and K. L. Magleby. 1998. "Kinetic structure of large-conductance Ca2+-activated K+ channels suggests that the gating includes transitions through intermediate or secondary states. A mechanism for flickers." *J Gen Physiol* 111:751–80.

Sachs, F., J. Neil, and N. Barkakati. 1982. "The automated analysis of data from single ionic channels." *Pflugers Arch* 395:331–40.

Sigworth, F. J. 1986. "The patch clamp is more useful than anyone had expected." *Fed Proc* 45:2673–7.

Sigworth, F. J., and S. M. Sine. 1987. "Data transformations for improved display and fitting of single-channel dwell time histograms." *Biophys J* 52:1047–54.

Sigworth, F. J., and J. Zhou. 1992. "Analysis of nonstationary single-channel currents." In *Methods in enzymology: ion channels*, edited by B. Rudy, and L. E. Iverson, Vol. 207, 746–762. San Diego, CA: Academic Press.

Suchyna, T. M., V. S. Markin, and F. Sachs. 2009. "Biophysics and structure of the patch and the gigaseal." *Biophys J* 97:738–47.

Venkataramanan, L., and F. J. Sigworth. 2002. "Applying hidden Markov models to the analysis of single ion channel activity." *Biophys J* 82:1930–42.

9

Patch-Clamp Recordings from Native Cells and Isolation of Membrane Currents

Jeanne M. Nerbonne

CONTENTS

9.1 Introduction

Similar to studies of heterologously expressed ion channels discussed in Chapters 7 and 8, the patch-clamp recording technique is a powerful experimental approach to investigate the properties, the functioning, and the regulation of ion channels expressed in native cells at the macroscopic and microscopic levels. The clear advantages of using heterologously expressed channels are that the experimenter controls what is expressed and different types of channels/currents can be studied in isolation. Native cells, however, express a repertoire of different types of ion channels, often with similar time-dependent and/or voltage-dependent properties, complicating the isolation and characterization of the individual currents. This chapter builds on the descriptions of the theory of voltage clamping and the application of patch-clamp recording techniques presented in Chapter 7. The basic equipment, hardware and software requirements for controlling voltage-clamp paradigms, data acquisition and analyses, as well as the design of voltage-clamp protocols for recording ionic currents from native cells are similar to those used in studies of heterologously expressed currents and are not repeated here; readers are referred to Chapter 7. The focus here is on the application of the whole-cell patch-clamp recording technique to facilitate the isolation, thereby allowing the detailed characterization of the ionic currents expressed in native cells. Representative macroscopic whole-cell voltage-clamp recordings from mammalian cardiac myocytes are presented and analyzed for illustration purposes. In addition, the application of specialized voltage-clamp paradigms, and pharmacological and molecular tools to facilitate the isolation of individual current components is illustrated. Additional, more specialized methods, including current-clamp, dynamic-clamp and action potential clamp recordings from native cells are also introduced. Technical and

DOI: 10.1201/9781003096214-11

experimental problems encountered with the analyses and interpretation of whole-cell patch-clamp data and the limitations of the methodologies are also discussed.

9.2 Optimizing Conditions for Patch-Clamp Recordings from Native Cells

A critical factor in whole-cell patch-clamp studies on native cells/tissues is the optimization of the recording conditions to permit the reliable and accurate measurement of the voltage (current-clamp) or current (voltage-clamp) signals of interest. Optimization of the contents of the bath solution to allow membrane potentials (in current-clamp mode) or currents (in voltage-clamp mode) to be recorded reliably is readily accomplished. Using the conventional whole-cell method, it is also possible to control the composition of the intracellular pipette solution (Hamill et al., 1981; Sakmann and Neher, 1984). In combination with well-designed voltage-clamp protocols, the ability to control the composition of the intracellular solution greatly simplifies the isolation of individual currents (see Section 9.3). Bath and pipette solutions should be filtered (at 0.2 μM) to avoid the clogging of the tips of the recording pipettes. The pipette is lowered into the bath with positive pressure applied to help keep the tip clean. With the application of a voltage across the tip, the pipette resistance can be continuously monitored.

Contact between the pipette tip and the cell membrane is signaled by an abrupt increase (two- to tenfold) in the pipette resistance. At this point, the application of gentle suction facilitates the formation of a seal between the pipette and the cell membrane, signaled by a marked increase, from MΩ to GΩ, in pipette resistance as the cell-attached configuration is achieved. After forming a gigaseal, the pipette capacitance should be compensated electronically while still in the cell-attached configuration. The whole-cell configuration is achieved by the application of a brief pulse of suction and is signaled by the appearance of large current transients in response to small (±5–10 mV) voltage steps from the holding potential (HP), reflecting the whole-cell capacitive current. Once the whole-cell configuration is achieved, the cell membrane potential should be clamped to the desired voltage, typically at or near the resting membrane potential, in the range of –50 to –90 mV, depending on the cell type.

In the whole-cell configuration, the access resistance, and the cell input resistance and capacitance can be calculated from the currents recorded during the small (±5–10 mV) voltage steps from the HP and then compensated electronically. Integration of the capacitive transients provides the amount of charge needed to charge the membrane capacitance (Q_C), from which the cell membrane capacitance (C_m) can be calculated. The access or series resistance (R_s) (Figure 9.1A), which reflects the resistances of all components in

FIGURE 9.1 (CONTINUED)
Schematic of the whole-cell recording configuration and exemplar voltage-gated currents recorded from adult mouse ventricular myocytes. (A) Connected to a high-gain, high-resistance amplifier, the pipette voltage (V_p) follows the command voltage (V_{com}). The pipette current (I_p) flows through the feedback resistor (R_f) and is proportional to the voltage output (V_{Rf}) of the amplifier. The (FR) circuit allows for correction of stray capacitance (C_f) and R_f. V_{com} = command voltage; R_s = series resistance; R_m = membrane resistance; and C_m = membrane capacitance. (B) With Cs^+ in the pipette, voltage-gated Na^+ (Nav) currents (I_{Na}) are recorded in response to brief depolarizations from a holding potential (HP) of –90 mV. (C) With tetrodotoxin in the bath to block I_{Na}, voltage-gated Ca^{2+} (Cav) currents (I_{Ca}) are recorded in response depolarizations from a HP of –40 mV. (D) With inward I_{Na} and I_{Ca} blocked and K^+ in the pipette, voltage-gated outward K^+ (Kv) currents are recorded in response to depolarizing voltage steps from a HP of –70 mV. The voltage-clamp paradigms are shown below the records and the current–voltage relations are shown on the right.

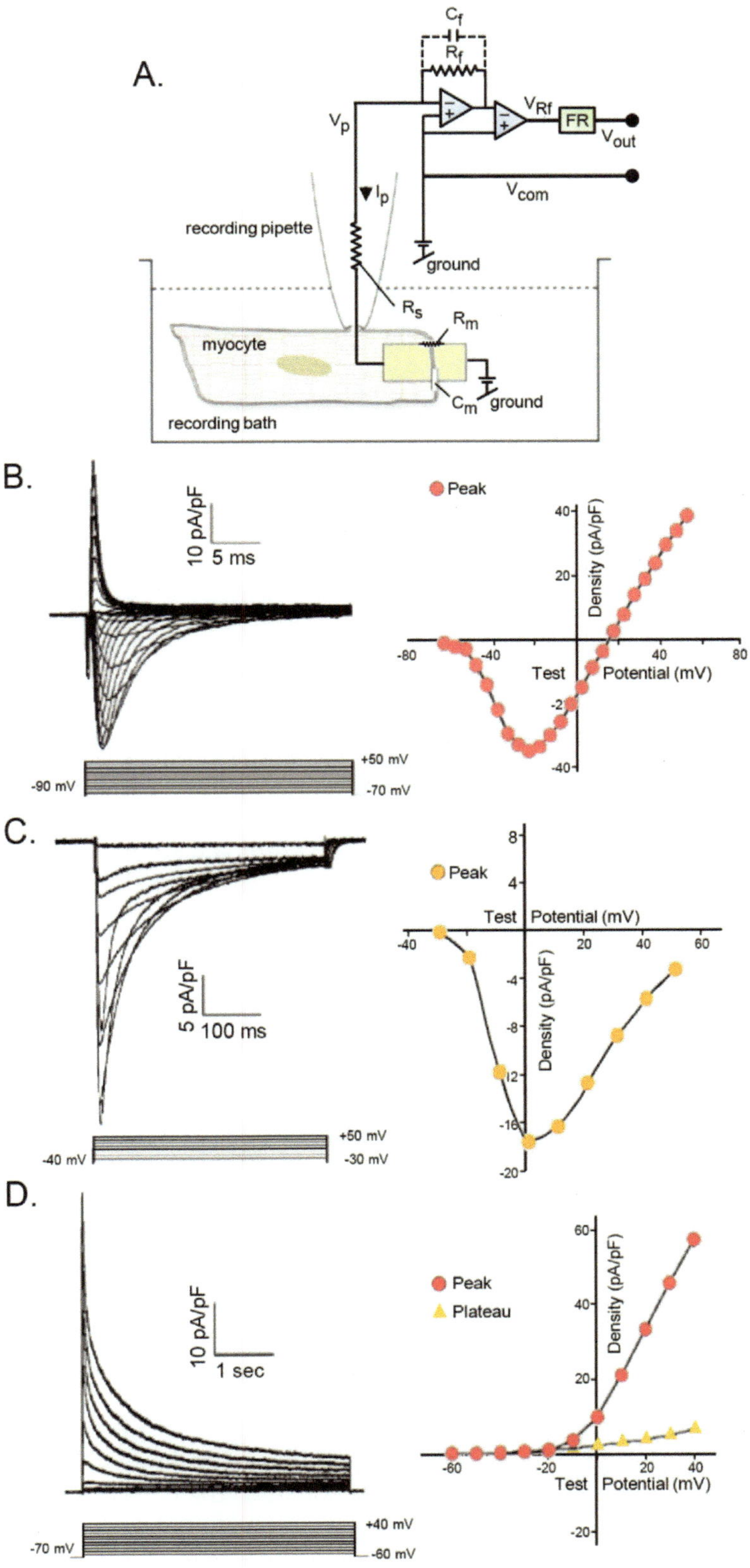
A.
Cf
Rf
Vp
VRf
FR
Vout
Ip
Vcom
recording pipette
ground
Rs
Rm
myocyte
Cm
ground
recording bath
B.
10 pA/pF
5 ms
-90 mV
+50 mV
-70 mV
Peak
Density (pA/pF)
Test Potential (mV)
C.
5 pA/pF
100 ms
-40 mV
+50 mV
-30 mV
Peak
Test Potential (mV)
Density (pA/pF)
D.
10 pA/pF
1 sec
-70 mV
+40 mV
-60 mV
Peak
Plateau
Density (pA/pF)
Test Potential (mV)

series with the cell membrane, should be measured and compensated electronically. It is also important to measure the extent of the compensation achieved, as voltage drops across the uncompensated R_s will result in (voltage) errors that could be substantial, particularly for large amplitude currents. It is also important to report the uncompensated R_s and the resulting voltage errors. Most patch-clamp amplifiers provide a positive feedback R_s (%) compensation circuit used to achieve maximum % compensation.

The ability to control the intracellular medium also enables studies focused on exploring channel modulation by intracellular and extracellular signaling pathways. In addition, combining imaging methods with patch-clamp recordings enables simultaneous analyses of membrane currents and intracellular, e.g., Ca^{2+} and signaling pathways. The elimination of cytosolic components, as occurs in the whole-cell configuration, however, can modify/eliminate intracellular signaling pathways, motivating the development of an alternative strategy, the "perforated patch" (Horn and Marty, 1988). Using a pore-forming hydrophobic agent, such as nystatin, in the method recording pipette, the perforated patch provides electrical continuity between the pipette and the cell. The pore properties of nystatin allow only the movement of monovalent cations and anions, and all other components of the cytosol are maintained.

9.3 Isolation of Voltage-Gated (Inward and Outward) Currents in Cardiac Myocytes

The whole-cell voltage-clamp technique has proven to be very useful in experiments aimed at detailing the properties of the membrane currents expressed in many cell types. To begin, recording conditions are established to eliminate all of the currents other than the current of interest. First, the intracellular (pipette) and extracellular (bath) solutions are optimized. Voltage-clamp protocols, typically involving step depolarizations or hyperpolarizations of fixed durations, are then designed to facilitate the isolation of the current component of interest.

Representative whole-cell voltage-clamp recordings obtained from isolated adult mouse ventricular myocytes are presented (Figure 9.1B–D) to illustrate these points. For recording myocardial voltage-gated inward Na^+ (Nav) currents (I_{Na}), the pipette solution typically contains (135 mM) Cs^+ to eliminate outward K^+ currents. With Na^+ and inorganic (e.g., $CdCl_2$) or organic (e.g., nifedipine) blockers of voltage-gated, high-threshold inward Ca^{2+} (Cav) currents (I_{Ca}) in the bath, I_{Na} is evoked by brief (20 ms) depolarizing voltage steps to potentials between −70 mV and +50 mV (in 5 mV increments at 2 s intervals) from a hyperpolarized, e.g., −90 mV, HP (Figure 9.1B). The recordings shown were obtained with low (20 mM) Na^+ bath to reduce I_{Na} amplitudes to ensure adequate spatial control of the membrane voltage and facilitate the reliable recording of the fast, and typically quite large, myocardial Nav currents. Peak I_{Na} amplitudes at each test potential are measured, and the current–voltage relation is provided by plotting peak I_{Na} as a function of the test potential (Figure 9.1B, right panel).

For recording myocardial I_{Ca}, the Ca^{2+} channel blockers are eliminated from the bath and tetrodotoxin (TTX) is added to reduce I_{Na}. In addition to using a Cs^+-containing pipette solution, tetraethylammonium (TEA) is often added to the bath to ensure the complete block of all K^+ currents and facilitate reliable measurement of I_{Ca}. Under these conditions,

I_{Ca}, evoked in response to (400 ms) depolarizing voltage steps to test potentials ranging from –30 mV to +50 mV (in 10 mV increments at 5 s intervals) from a HP of –40 mV, is recorded (Figure 9.1C). The HP used was –40 mV (Figure 9.1C) to ensure the inactivation of any residual Na^+ currents not blocked by TTX. Peak I_{Ca} amplitudes at each test potential are measured, normalized to the cell membrane capacitance, and the current–voltage relation is provided by plotting peak I_{Ca} as a function of the test potential (Figure 9.1C, right panel).

The experimental conditions are modified for the recording of native myocardial Ca^{2+}-independent Kv currents (Figure 9.1D). With K^+ in the recording pipette and voltage-gated inward Na^+ and Ca^{2+} currents blocked with the inclusion of TTX and Cd^{2+} (or nifedipine), respectively, in the bath solution, Kv currents are routinely recorded from adult mouse ventricular myocytes in response to depolarizing voltage steps to potentials between –60 and +40 mV from an HP of –70 mV (in 10 mV increments at 15 s intervals); the voltage-clamp paradigm is illustrated below the current records. The rates of rise and the amplitudes of the currents increase with depolarization; the largest and most rapidly activating current in Figure 9.1D was evoked at +40 mV. The maximal (peak) and the steady-state (plateau) currents evoked at each test potential are measured and normalized to the whole-cell membrane capacitance, and the current–voltage relations are obtained by plotting the measured peak and plateau current densities as a function of the test potential (Figure 9.1D, right panel).

An interesting alternative to applying fixed voltage steps (Figure 9.1B–D) has been the use of action potential voltage-clamp paradigms, in which the currents evoked in response to voltage deflections that correspond to action potential waveforms are measured (Bányász et al., 2011; Horváth et al., 2021). This approach is discussed and illustrated in Section 9.7.

9.4 Identification of Kinetically Distinct Myocardial Kv Current Components

Although the Kv currents in all adult mouse ventricular myocytes activate rapidly, the amplitudes (densities) and the decay phases of the currents are quite variable (Figure 9.2A). In all myocytes isolated from the right ventricle (RV) or the apex of the left ventricle (LV), for example, there is a rapid component of current decay, consistent with the presence of a transient outward K^+ current (I_{to}), as described in cardiac myocytes in several species (Campbell et al., 1995). In adult mouse ventricular myocytes, however, this current is referred to as the "fast" transient outward current, $I_{to,fast}$ ($I_{to,f}$) to distinguish it from another transient outward current (see later) with slower kinetics (Xu et al., 1999a; Brunet et al., 2004). The decay phases of the Kv currents in adult mouse RV and LV myocytes (Figure 9.2A) are well described by the sum of two exponentials with mean decay time constants (τ_{decay}) of ~70 ms and ~1200 ms, and a non-inactivating, "steady-state" component, I_{ss} (Xu et al., 1999a; Brunet et al., 2004); neither time constant displays any appreciable voltage dependence (Figure 9.2B). The rapidly decaying ($\tau_{decay} \approx 70$ ms) component is $I_{to,f}$ and the slowly inactivating component ($\tau_{decay} \approx 1200$ ms) is referred to as $I_{K,slow}$ (London et al., 1998; Xu et al., 1999a; Brunet et al., 2004).

The peak amplitudes and the waveforms of the Kv currents evoked in cells isolated from the interventricular septum (IVS) are quite different (Figure 9.2A) from those recorded

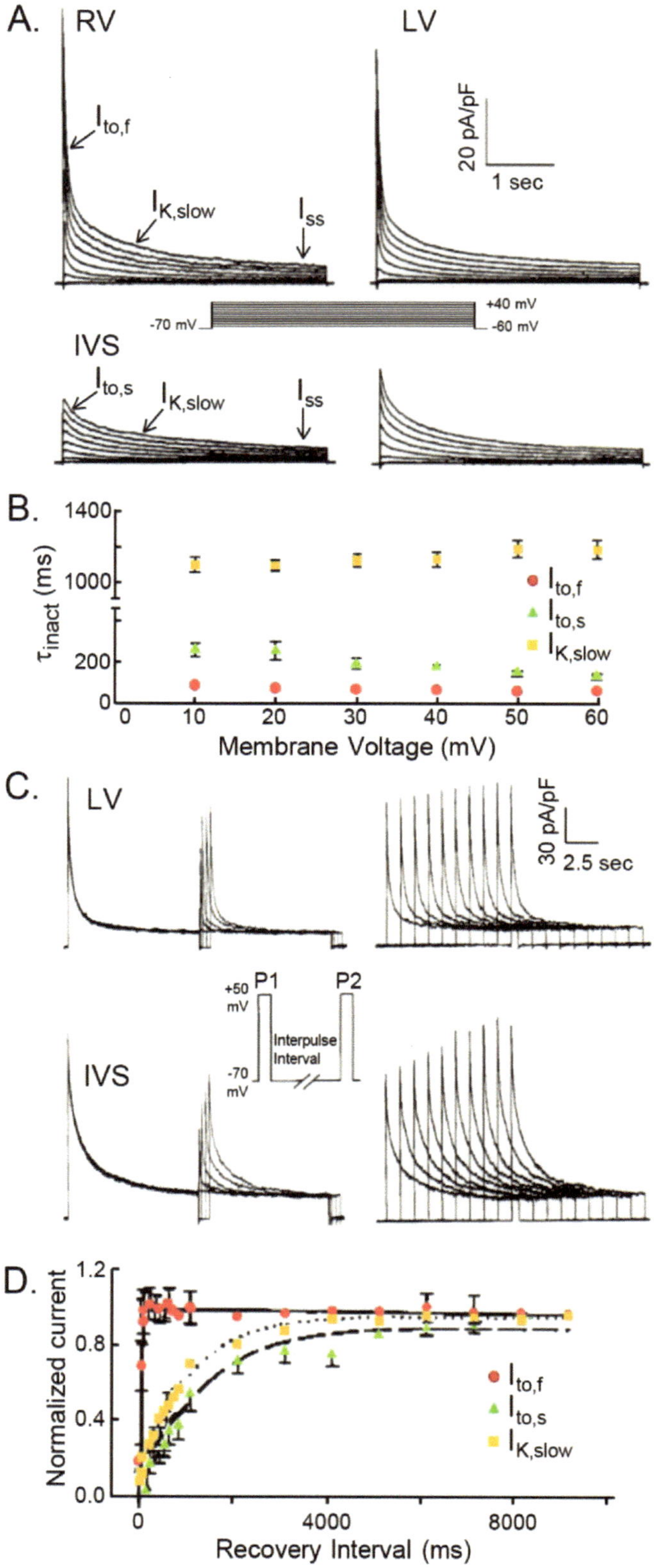
A.
RV
LV
$I_{to,f}$
$I_{K,slow}$
I_{ss}
20 pA/pF
1 sec
-70 mV
+40 mV
-60 mV
IVS
$I_{to,s}$
$I_{K,slow}$
I_{ss}
B.
τ_{inact} (ms)
Membrane Voltage (mV)
$I_{to,f}$
$I_{to,s}$
$I_{K,slow}$
C.
LV
IVS
30 pA/pF
2.5 sec
+50 mV
-70 mV
P1
P2
Interpulse Interval
D.
Normalized current
Recovery Interval (ms)
$I_{to,f}$
$I_{to,s}$
$I_{K,slow}$

from RV and LV cells (Xu et al., 1999a; Brunet et al., 2004). The peak Kv current amplitudes in IVS cells are lower and, in addition, some (~20%) IVS cells clearly lack a rapidly decaying component, i.e., $I_{to,f}$ (Figure 9.2A, lower left). Analyses of the records obtained from these cells, however, also revealed that the decay phases of the currents were best fitted by two exponentials, characterized by mean τ_{decay} values of ~200 ms and ~1200 ms, and a non-inactivating component (Figure 9.2A, lower left); neither time constant displays any appreciable voltage dependence (Figure 9.2B). The τ_{decay} (~200 ms) for the faster component of decay in IVS cells, however, is substantially larger than the τ_{decay} (~70 ms) for $I_{to,f}$ in RV and LV cells and was/is, therefore, referred to as the slow transient current, $I_{to,slow}$ ($I_{to,s}$) (Xu et al., 1999a).

Although not observed in ~20% of IVS cells (Figure 9.2A, lower left), a small rapidly inactivating current is evident in the remaining (~80%) IVS myocytes (Figure 9.2A, lower right), and in these cells three exponentials, with average τ_{decay} values of ~70 ms, ~200 ms and ~1200 ms, are required to describe the decay phases of the currents. The measured time constants of decay for these three components suggest the co-expression of $I_{to,f}$, $I_{to,s}$, and $I_{K,slow}$ (as well as I_{ss}). As is evident in the representative records shown, the density of $I_{to,f}$, when expressed in IVS cells (Figure 9.2A, lower right), is low, averaging ~7 pA/pF (at +40 mV), compared with ~35 pA/pF and ~40 pA/pF (at +40 mV) for $I_{to,f}$ in LV and RV cells (Figure 9.2A, top), respectively (Xu et al., 1999a; Brunet et al., 2004). In addition, $I_{to,f}$ contributes only about 20% to the peak currents in these (IVS) cells, compared with 60%–65% of the peak currents in LV and RV cells. The τ_{decay} for $I_{K,slow}$ (~1200 ms) in IVS cells is indistinguishable from value in RV and LV cells, and the densities of $I_{K,slow}$ (~15 pA/pF) and I_{ss} (~5 pA/pF) in IVS cells are similar to those in RV and LV myocytes (Xu et al., 1999a; Brunet et al., 2004). However, both $I_{K,slow}$ and I_{ss} contribute significantly more to the peak Kv current in IVS compared with LV and RV cells.

Although the voltage dependences of activation and steady-state (i.e., closed state) inactivation are reportedly (Xu et al., 1999a; Brunet et al., 2004; Liu et al., 2011) quite similar (not illustrated), $I_{to,f}$, $I_{to,s}$ and $I_{K,slow}$ display markedly different rates of recovery from inactivation (Figure 9.2C and D). To determine the time dependences of recovery, cells are first depolarized to +50 mV for 9.5 s to inactivate the currents, subsequently hyperpolarized to –70 mV for varying times ranging from 10 ms to 9.5 s (to allow recovery), and then depolarized to +50 mV to activate the currents and assess the extent of recovery

FIGURE 9.2 (CONTINUED)
Multiple Kv currents in mouse ventricular myocytes. (A) Whole-cell Kv currents were recorded as described in the legend to Figure 9.1D from myocytes isolated from the right ventricle (RV), left ventricle (LV) and interventricular septum (IVS) of adult wild-type (WT) mice. The decay phases of the currents in RV and LV cells were well fitted by the sum of two exponentials reflecting the presence of a rapidly inactivating transient component, $I_{to,fast}$ ($I_{to,f}$; ●), and a slowly inactivating component, $I_{K,slow}$ (■). In IVS myocytes, the transient current inactivates slower than $I_{to,fast}$ and is referred to as $I_{to,slow}$ ($I_{to,s}$; ▲). The non-inactivating, "steady-state" Kv current remaining at the end of the 4.5 s depolarizing voltage steps, I_{ss}, is expressed in all (RV, LV and IVS) cells. (B) Mean ± SEM time constants of inactivation of $I_{to,f}$, $I_{to,s}$ and $I_{K,slow}$ are plotted as a function of the test potential. (C) Following a prepulse to +50 mV, cells were hyperpolarized to –70 mV for times ranging from 0 ms to 10 s before a second depolarization to +50 mV to assess recovery; the protocol is illustrated between the records. Representative recordings from LV and IVS myocytes, during the conditioning step and the test steps after varying recovery times, are displayed. The amplitudes of $I_{to,f}$, $I_{to,s}$ and $I_{K,slow}$ at +50 mV after each recovery interval were determined from exponential fits to the decay phases of the currents (see text), and normalized to the amplitudes measured (in the same cell) following the 10 s recovery interval. (D) Mean ± SEM time constants of recovery for $I_{to,f}$ (●), $I_{to,s}$ (▲) and $I_{K,slow}$ (■) are plotted as a function of the recovery interval.

(that occurred at –70 mV). Typical recordings obtained from adult mouse LV and IVS myocytes and the experimental protocol used are presented in Figure 9.2C. To obtain the amplitudes of $I_{K,slow}$, $I_{to,f}$ and/or $I_{to,s}$, the decay phases of the currents at +50 mV after each recovery period are fitted to the sum of two or three exponentials. For the IVS cell in Figure 9.2C, both $I_{to,f}$ and $I_{to,s}$ are present. For each cell, the measured amplitudes of $I_{K,slow}$, $I_{to,f}$ and $I_{to,s}$ after each recovery period are normalized to their respective maximal current amplitudes evoked after 9.5 s. Mean normalized $I_{K,slow}$, $I_{to,f}$ and $I_{to,s}$ amplitudes are plotted as a function of recovery time in Figure 9.2D; the continuous lines represent the best single exponential fits to the average data. The mean normalized recovery data for $I_{to,f}$ are well described by a single exponential with a time constant ($\tau_{recovery}$) of ~30 ms (Figure 9.2D). The mean normalized recovery data for both $I_{to,s}$ and $I_{K,slow}$ also follow monoexponential time courses (Figure 9.2D), with $\tau_{recovery}$ values of ~1300 ms ($I_{to,s}$) and ~1100 ms ($I_{K,slow}$). Similar to inactivation, therefore, $I_{to,f}$ and $I_{to,s}$ also recover from steady-state inactivation at markedly different rates (Figure 9.2D).

9.5 Pharmacological Separation of Co-Expressed Kv Current Components

The identification of multiple, kinetically distinct components of the macroscopic Kv (or other) currents in native cells is often interpreted (as inferred in the presentation earlier) as reflecting the co-expression of distinct types of Kv (or other) channels. It is important to be clear, however, that there are other interpretations of the presence of Kv (or other) current components distinguished based on differences in rates of inactivation and/or recovery. Multiple components of current decay, for example, could reflect the presence of different open and/or inactivated states of the same channel. It is desirable, therefore, to use additional experimental approaches to further characterize the properties of the various macroscopic current components identified based on differences in time- and/or voltage-dependent properties, and pharmacological tools have proven to be very useful for this purpose.

The application of selective and, in some cases specific, K^+ channel blockers has been used often and effectively in studies of native K^+ currents in cardiac (and other) cells. The Kv currents in adult mouse ventricular myocytes, for example, are differentially affected by varying concentrations (10 μM to 5 mM) of 4-aminopyrdine (4-AP) (Xu et al., 1999a). To examine the effects of 4-AP, control Kv currents are recorded prior to the superfusion of 4-AP-containing bath solutions and, when the effect of 4-AP reaches a steady state, Kv currents are again recorded. To obtain the current(s) blocked by 4-AP, records obtained in the presence of 4-AP are digitally subtracted from the controls. As illustrated in Figure 9.3A-B, the currents sensitive to low (10 μM) 4-AP activate rapidly and inactivate slowly, similar to $I_{K,slow}$, suggesting that $I_{K,slow}$ is selectively attenuated by 10 μM 4-AP (London et al., 1998; Xu et al., 1999a). On exposure to higher (0.5 mM) concentrations of 4-AP, the peak Kv current is also affected (Figure 9.3D), and the waveforms of currents sensitive to 0.5 mM 4-AP (Figure 9.3C-D) reveal that both $I_{to,f}$ and $I_{K,slow}$, but not I_{ss}, are affected by 0.5 mM 4-AP. When the 4-AP concentration is increased to 5 mM, further reductions in the currents are observed (Figure 9.3F). Analysis of the currents remaining in 5 mM 4-AP (Figure 9.3F) reveals that both $I_{K,slow}$ and $I_{to,f}$ are blocked completely; only I_{ss} is recorded in the presence of 5 mM 4-AP, although at this (5 mM) concentration, I_{ss} is also attenuated (by ~50%). Moderate concentrations of 4-AP (100-500 μM) also affect $I_{to,s}$ (Xu et al., 1999a).

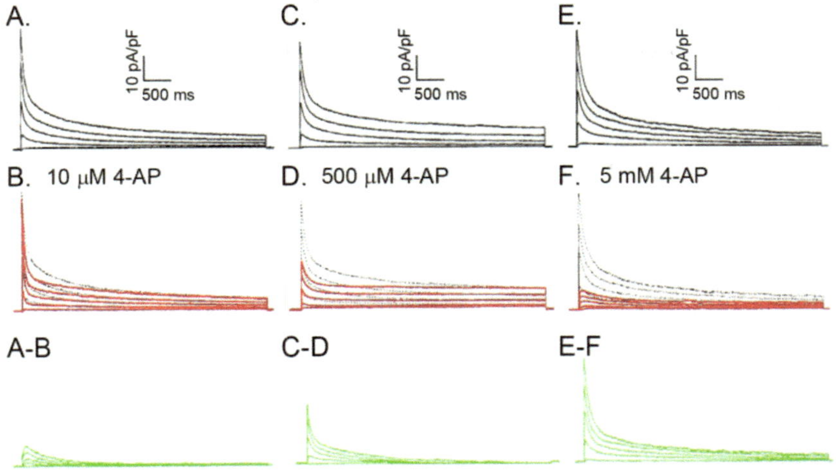

FIGURE 9.3
Differential effects of 4-aminopyridine (4-AP) on mouse ventricular Kv currents. Whole-cell Kv currents, evoked as described in the legend to Figure 9.1D, were recorded from adult mouse LV myocytes under control conditions (A, C, E) and following exposure to (B) 10 μM, (D) 500 μM or (F) 5 mM 4-AP; the currents in the presence of 4-AP are in red (in B, D, F). The 10 μM 4-AP sensitive (A-B) currents (green) were obtained by offline digital subtraction of the currents in the presence of 10 μM 4-AP (B) from the controls (A). The 500 μM (C-D) and 5 mM (E-F) 4-AP-sensitive currents were also obtained by subtraction of the currents in the presence of 4-AP (D, F) from the controls (C, E). Low 4-AP selectively attenuates $I_{K,slow}$ (A-B), whereas at higher 4-AP (C-D and E-F), $I_{to,f}$ is also blocked (see text).

Varying concentrations of another K^+ channel blocker, tetraethylammonium (TEA), also differentially affect mouse ventricular Kv currents (Xu et al., 1999a). $I_{K,slow}$ and I_{ss}, for example, are attenuated by low concentrations (10–25 mM) of TEA, whereas both $I_{to,f}$ and $I_{to,s}$ are relatively insensitive to TEA. $I_{to,f}$, however, is selectively blocked by several spider toxins, including SNX-482 heteropodatoxin-2 and -3 (Xu et al., 1999a; Johnson et al., 2018).

9.6 Molecular Dissection of Native Kv Currents: Kv Pore-Forming (α) Subunits

The finding that varying concentrations of the K^+ channel blocker 4-AP reveals Kv currents (Figure 9.3 A-B, C-D, E-F) that resemble the kinetically distinct Kv current components (Figure 9.2) supports the hypothesis that these components reflect the expression of distinct types of K^+ channels. The identification of the single channel correlates of the macroscopic current components would lend further support to this hypothesis. Nevertheless, similar to the description of the current components based on kinetic differences, there are caveats that remain with using pharmacological and/or single-channel data to argue that the currents identified reflect the expression of molecularly distinct types of channels. With the development and application of molecular genetic strategies in vivo and in vitro, however, it is possible to directly identify the pore-forming subunits and, in some cases, the accessory subunits that contribute to the generation of native K^+ (and other types of) channels.

Like Nav and Cav channel α subunits, Kv α subunits belong to the "S4" superfamily of voltage-gated channels. There are multiple Kv α subunit subfamilies (Kv1–Kv12), each with two to nine members, grouped based on sequence homology. Functional Kv channels comprise four α subunits and only Kv α subunits in the same subfamily co-assemble. In addition, heterologous expression of individual Kv α subunits or of multiple Kv α subunits in the same subfamily gives rise to Kv currents with distinct time- and voltage-dependent properties. In mouse myocardium, for example, multiple Kv α subunits from different subfamilies are expressed (Barry et al., 1995), and heterologous expression of each of these gives rise to K^+ currents with distinct properties. Interest in identifying the molecular determinants of native myocardial Kv currents led to the development of dominant negative strategies that could be applied in vitro or in vivo to probe the roles of the different Kv α subunit subfamilies in the generation of these currents. This powerful approach involves the generation of a mutant Kv α subunit that does not form functional Kv channels when expressed alone, but that co-assembles with wild-type (WT) Kv α subunits (in the same subfamily) and renders the assembled channels nonconducting, resulting in the functional "knockout" of the current(s) produced by the targeted Kv α subunit subfamily.

Illustrating this point, a dominant negative strategy was used to directly test the hypothesis that Kv α subunits of the Kv4 subfamily underlie the generation of mouse ventricular $I_{to,f}$. A point mutation (W to F) was introduced at position 362 in the pore region of Kv4.2 to produce a nonconducting mutant (Kv4.2W362F) subunit (Barry et al., 1998). Co-expression of Kv4.2W362F with Kv4.2 (or the other members of the Kv4 subfamily, Kv4.1 or Kv4.3) in heterologous cells attenuates Kv4-encoded currents, but not the currents produced on expression of Kv subunits in other subfamilies. The mutant Kv4.2W362F α subunit, therefore, functions as a subfamily-specific dominant negative, Kv4.2DN (Barry et al., 1998). Cardiac-specific expression of Kv4.2DN, eliminates the rapidly inactivating (Figure 9.4B) and recovering (Figure 9.4C) Kv current, $I_{to,f}$; $I_{K,slow}$ and I_{ss} are unaffected (Barry et al., 1998). Analysis of the decay phases of the Kv currents in Kv4.2DN LV myocytes, however, revealed that two exponentials, characterized by average τ_{decay} values of ~200 ms and ~1200 ms, were required to fit the data (Barry et al., 1998). The time constant of the faster component of decay in Kv4.2DN LV and RV myocytes is similar to the τ_{decay} (of ~200 ms) for $I_{to,s}$ in WT mouse IVS cells (Figure 9.2B). In addition, this current recovers slowly from inactivation (Figure 9.4D). Together, these findings suggest that $I_{to,s}$ is expressed in Kv4.2DN LV and RV myocytes lacking $I_{to,f}$ (see later). Interestingly, however, $I_{to,s}$ densities in Kv4.2DN IVS myocytes were not significantly different from those measured in WT mouse IVS cells (Barry et al., 1998).

The two members of the Kv4 subfamily, Kv4.2 and Kv4.3, expressed in adult mouse ventricles (Barry et al., 1995) also co-immunoprecipitate, observations interpreted as suggesting that native mouse ventricular $I_{to,f}$ channels are heteromeric (Guo et al., 2002). Voltage-clamp recordings from ventricular myocytes isolated from mice with a targeted disruption of the *Kcnd2* locus (Kv4.2$^{-/-}$), however, revealed that $I_{to,f}$ is eliminated (Guo et al., 2005). In addition, Kv4.3 message/protein expression is not affected by the loss of Kv4.2 (Guo et al., 2005), suggesting that Kv4.3 cannot form functional $I_{to,f}$ channels in mouse ventricular myocytes in the absence of Kv4.2. Similar to the findings in Kv4.2DN ventricular myocytes (Barry et al., 1998), $I_{t,os}$ is also upregulated in Kv4.2$^{-/-}$ cells (Guo et al., 2005). In contrast, the densities and properties of $I_{to,f}$ in ventricular myocytes isolated from mice with a targeted disruption of the *Kcnd3* locus (Kv4.3$^{-/-}$) are indistinguishable from WT cells (Niwa et al., 2008). Using targeted deletion strategies, therefore, it is possible to determine the functional roles of individual α subunits when multiple members of the same α subunit subfamily are co-expressed. Interestingly, $I_{to,f}$ is also expressed in human

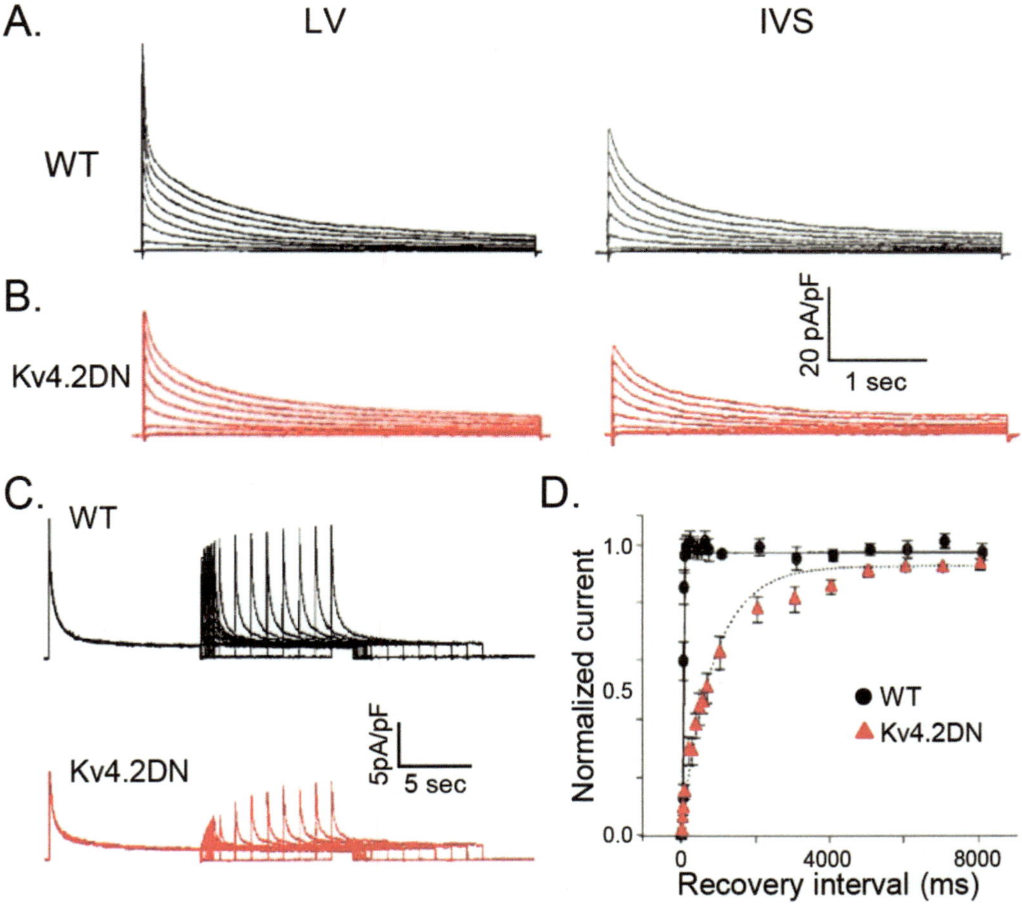

FIGURE 9.4
$I_{to,f}$ is selectively eliminated in adult mouse ventricular myocytes expressing a pore mutant of Kv4.2 (Kv4.2DN) that functions as a dominant negative. (A–C) Whole-cell Kv currents were recorded from myocytes isolated from WT mice and from transgenic mice with cardiac-specific expression of Kv4.2DN. Representative recordings from WT (A) and Kv4.2DN (B) LV and IVS myocytes, evoked as in Figure 9.3, are shown. (C) Representative recordings obtained from WT and Kv4.2DN LV myocytes using the recovery protocol illustrated in Figure 9.3. The rapidly inactivating and recovering (C) current, $I_{to,f}$, that is prominent in WT LV cells, is eliminated in Kv4.2DN cells (see text). (D) Mean ± SEM time constants of recovery of the peak current in WT (●) and Kv4.2DN LV (▲) myocytes plotted as a function of the recovery interval.

ventricular myocytes (Johnson et al., 2018) and, in this case, it seems certain that *KCND3* (Kv4.3) is the critical α subunit encoding $I_{to,f}$, as *KCND2* (Kv4.2) is not expressed.

The finding that the kinetic properties of myocardial $I_{to,s}$ are distinct from $I_{to,f}$ (Figure 9.2), suggests that different Kv α subunits encode these currents. The observation that $I_{to,s}$ is increased in Kv4.2DN (Barry et al., 1998) and Kv4.2$^{-/-}$ (Guo et al., 2005) ventricular cells, in which $I_{to,f}$ is eliminated, is consistent with this hypothesis. Direct experimental support, however, was provided with the demonstration that $I_{to,s}$ is eliminated in IVS myocytes (Figure 9.5) isolated from mice harboring a targeted disruption of the *Kcna4* locus (Kv1.4$^{-/-}$). The properties and the densities of $I_{to,f}$, $I_{K,slow}$ and I_{ss} in interventricular myocytes, as well as in LV and RV myocytes, in contrast, are unaffected by the loss of Kv1.4

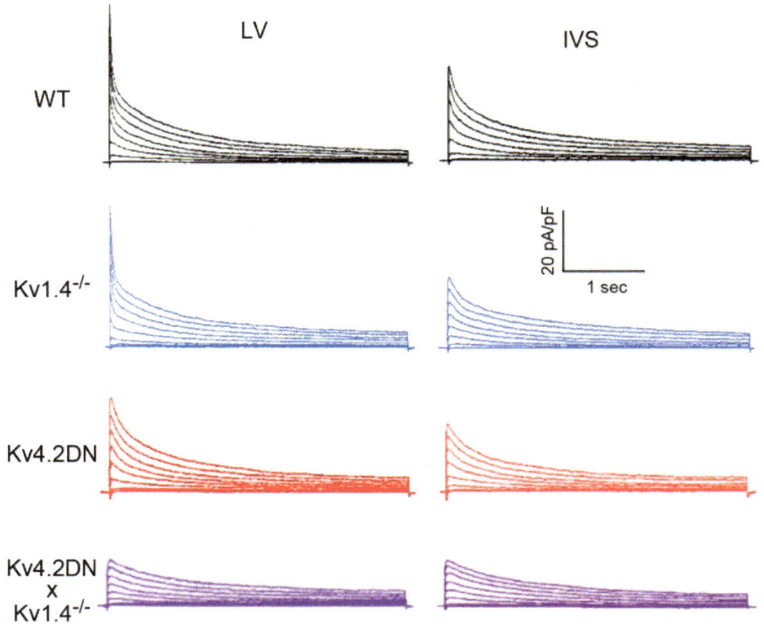

FIGURE 9.5
Molecular determinants of $I_{to,f}$ and $I_{to,s}$ are distinct. Representative whole-cell Kv currents, evoked as described in the legend to Figure 9.1D, from WT, Kv1.4$^{-/-}$, Kv4.2DN, and Kv4.2DN × Kv1.4$^{-/-}$ mouse LV and IVS myocytes, are shown. Note that, in marked contrast with WT, Kv1.4$^{-/-}$ and Kv4.2DN cells, the waveforms of the Kv currents in Kv4.2DN × Kv1.4$^{-/-}$ LV and IVS myocytes are indistinguishable, consistent with the elimination of both $I_{to,f}$ and $I_{to,s}$ (see text).

(Guo et al., 1999). In addition, both $I_{to,s}$ and $I_{to,f}$ are undetectable in myocytes isolated from (Kv4.2DN × Kv1.4$^{-/-}$) animals (Guo et al., 2000) lacking Kv1.4^{-} and expressing Kv4.2DN (Figure 9.5), revealing that the slow transient current that is identified in Kv4.2DN LV cells (Figure 9.4) is, indeed, $I_{to,s}$ and is encoded by Kv1.4. It is also interesting to note that the waveforms of the Kv currents recorded from Kv4.2DN × Kv1.4$^{-/-}$ LV and IVS myocytes Figure 9.4), which lack both $I_{to,f}$ and $I_{to,s}$, are indistinguishable (Guo et al., 2000).

Similar to the transient outward Kv currents, molecular genetic methods have led to the identification of the molecular determinants of functional delayed rectifier Kv channel diversity in the murine myocardium. A role for Kv1 α subunits in the generation of mouse ventricular $I_{K,slow}$, for example, was revealed with the demonstration that $I_{K,slow}$ is selectively attenuated in ventricular myocytes isolated from transgenic mice expressing a truncated Kv1.1 α subunit that functions as a dominant negative, Kv1.1DN (London et al., 1998). It was subsequently shown, however, that $I_{K,slow}$ is also reduced in ventricular myocytes expressing a mutant Kv2.1 α subunit that also functions as a dominant negative, Kv2.1DN (Xu et al., 1999b), revealing that there are actually two, molecularly distinct components of mouse ventricular $I_{K,slow}$: $I_{K,slow1}$, which is sensitive to low (μM) concentrations of 4-AP and encoded by Kv1 α subunits (London et al., 1998); and $I_{K,slow2}$, which is sensitive to low millimolar TEA and encoded by Kv2 α subunits (Xu et al., 1999b). Interestingly, these two Kv current components were first distinguished using molecular genetic methods (London et al., 1998; Xu et al., 1999b).

Molecular genetic strategies can also be developed/applied to probe directly the functional roles of putative channel accessory subunits in determining the expression and properties of native channels. A number of Kv channel accessory subunits, including Kv channel interacting protein 2 (KChIP2) and Kvβ1, for example, co-immunoprecipitate with Kv4 α subunits from adult mouse ventricles and modify the properties of heterologously expressed Kv4-encoded channels (Guo et al., 2002). Targeted disruption of the *Kcnip2* (KChIP2) locus results in the selective elimination of $I_{to,f}$ (Kuo et al., 2001; Foeger et al., 2013), as well as the loss of the Kv4.2 protein (Foeger et al., 2013), demonstrating that the "accessory" KChIP2 protein is actually *required* for the generation of functional mouse ventricular $I_{to,f}$ channels. Interestingly, similar to Kv4.2DN and Kv4.2$^{-/-}$ myocytes, upregulation of $I_{to,s}$ is also evident in KChIP2$^{-/-}$ ventricular myocytes (Foeger et al., 2013). In contrast, loss of Kvβ1 results in reduced $I_{to,f}$ densities, as well as increased $I_{K,slow2}$ (but not $I_{K,slow1}$) densities (Aimond et al., 2005).

9.7 Probing the Functional Roles of Kv (and Other) Channels in Native Cells

Current-clamp recordings of myocyte membrane potentials and action potential (AP) waveforms are also readily accomplished using the whole-cell configuration of the patch-clamp recording technique (Figure 9.6A). Voltage recordings can be obtained from the resting membrane potential, i.e., without any added depolarizing or hyperpolarizing currents injected, or with current added to manipulate the membrane potential and determine the effects of these manipulations on AP thresholds, amplitudes, waveforms and durations. Another important step forward has been the application of the dynamic-clamp technique, which combines mathematical simulations of ion channel currents with patch-clamp recordings and allows the introduction of simulated conductances into cells during current-clamp recordings (Sun and Wang, 2005; Berecki et al., 2014; Dong et al., 2006). Although originally developed and exploited to study synaptic currents (Goaillard and Marder, 2006), several investigators have used dynamic clamping, in combination with conventional whole-cell and perforated-patch (Horn and Marty, 1988) recordings, to directly probe the physiological roles of specific voltage-dependent conductance pathways in regulating the resting and active membrane properties of mammalian, including human, cardiac myocytes (Sun and Wang, 2005; Berecki et al., 2014; Dong et al., 2006; Johnson et al., 2018).

AP waveforms in human ventricular myocytes isolated from the subendocardial and subepicardial surfaces of the LV free wall are distinct (Figure 9.6A). The main difference is the presence of the early rapid phase of repolarization, giving rise to the "notch" in the voltage trajectory (Figure 9.6A), which is correlated with the higher density of $I_{to,f}$ in subepicardial LV myocytes (Johnson et al., 2018). To directly explore the impact of manipulating $I_{to,f}$ density on human LV myocyte AP waveforms, a model of $I_{to,f}$ was generated from whole-cell voltage-clamp data and applied to isolated LV myocytes in real time during current-clamp recordings (Johnson et al., 2018). The most prominent effect of increasing $I_{to,f}$ on AP waveforms in human LV epicardial myocytes is hyperpolarization of the notch potential and an increase in the duration of the notch, with only modest effects on the plateau and the AP duration (Figure 9.6B). Slowing the time constant of inactivation of $I_{to,f}$, however, results in dramatic shortening of the AP duration and the collapse of the plateau (Figure 9.6B).

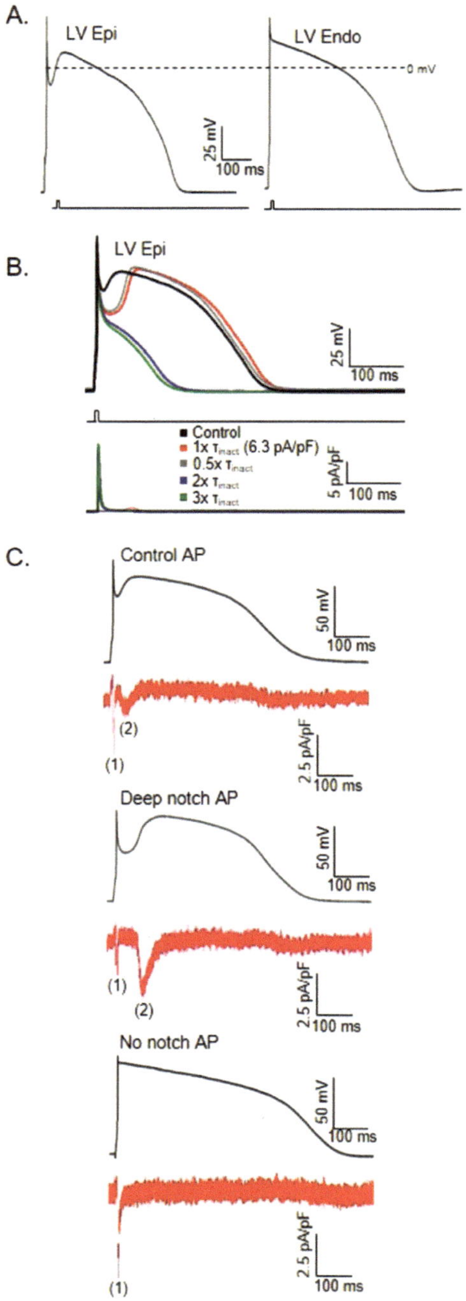

FIGURE 9.6

Dynamic clamp-mediated manipulation of $I_{to,f}$ and action potential clamp recordings of I_{Ca} in human LV myocytes. (A) Representative recordings of action potentials (APs) evoked in human LV myocytes isolated from the subepicardial (LV epi) and subendocardial (LV endo) layers of the LV freewall. (B) Dynamic-clamp-mediated addition of $I_{to,f}$ in a LV epi myocyte affects the notch potential and slowing $I_{to,f}$ inactivation results in collapse of the AP. (C) Modulation of the notch potential by $I_{to,f}$ affects I_{Ca}. Representative recordings of whole-cell I_{Ca} (red traces) in a human LV epi myocyte evoked during action potential clamp recordings with different AP waveforms (illustrated above the current records). With the control AP and the deep notch AP, two components of I_{Ca} were observed: one coincident with the upstroke of the action potential (1) and the other coincident with the depolarization following the notch (2). In contrast, only the early I_{Ca} was observed with the no-notch AP waveform.

The impact of variations in AP waveforms on the amplitudes and time courses of individual membrane currents can also be explored experimentally using AP voltage-clamp paradigms, in which the currents, evoked in response to voltage deflections that correspond to the AP waveforms recorded from native cells, are measured (Bányász et al., 2011; Horváth et al., 2021). The effects of the presence of a notch and of the notch potential on the time course and amplitude of I_{Ca}, recorded as in Figure 9.1 with other voltage-gated (Na^+ and K^+) blocked, in a human LV epicardial myocyte are illustrated in Figure 9.6C. As is evident, two distinct components of I_{Ca} are observed in response to a control AP waveform: one, labeled I_{Ca1}, that is coincident with the upstroke of the AP, and the other, labeled I_{Ca2}, that is coincident with the depolarization following the notch (Johnson et al., 2018). When a deep notch AP is used as the voltage command (in the same cell), both I_{Ca1} and I_{Ca2} are again observed, although the density of I_{Ca2} is larger and the peak I_{Ca2} occurs later than with the control AP (Figure 9.6C). In marked contrast, when a no-notch AP waveform is used as the voltage command (in the same cell), only I_{Ca1} is observed (Figure 9.6C). These (dynamic clamp and action potential clamp) approaches seem certain to be used increasingly to directly probe the physiological roles of individual (or combinations of) membrane currents and the pathophysiological consequences of alterations in these currents associated with inherited and acquired diseases.

Acknowledgments

The author thanks Richard Wilson for expert technical assistance. The author also acknowledges the financial support provided to support the people and the work in her laboratory from the American Heart Association; the Department of Veterans Affairs; the National Institute of General Medical Sciences; the National Institute of Neurological Disorders and Stroke; and the National Heart, Lung and Blood Institute.

Suggested Readings

Aimond, F., S. P. Kwak, K. J. Rhodes, and J. M. Nerbonne. 2005. "Accessory Kvbeta1 subunits differentially modulate the functional expression of voltage-gated K+ channels in mouse ventricular myocytes." *Circ Res* 96:451–8.

Bányász, T., B. Horvath, Z. Jian, L. T. Izu, and Y. J. Chen-Izu. 2011. "Sequential dissection of multiple ionic currents in single cardiac myocytes under action potential-clamp." *J Mol Cell Cardiol* 50:578–81.

Barry, D. M., J. S. Trimmer, J. P. Merlie, and J. M. Nerbonne. 1995. "Differential expression of voltage-gated K^+ channel subunits in the adult rat heart: relationship to functional K^+ channels?" *Circ Res* 77:361–9.

Barry, D. M., H. Xu, R. Schuessler, and J. M. Nerbonne. 1998. "Functional knockout of the transient outward current, long QT syndrome and cardiac remodeling in mice expressing a dominant negative Kv4 a subunit." *Circ Res* 83:560–7.

Berecki, G., A. O. Verkerk, A. C. van Ginneken, and R. Wilders. 2014. "Dynamic clamp as a tool to study the functional effects of individual membrane currents." *Methods Mol Biol* 1183:309–26.

Brunet, S., F. Aimond, H. Li, et al. 2004. "Heterogeneous expression of repolarizing voltage-gated K^+ currents in adult mouse ventricles." *J Physiol* 559:103–20.

Campbell, D. L., R. L. Rasmusson, M. B. Comer, and H. C. Strauss. 1995. "The cardiac calcium-independent transient outward potassium current: kinetics, molecular properties, and role in ventricular repolarization." In *Cardiac electrophysiology: from cell to bedside*, edited by D. P. Zipes, and J. Jalife, 83–96. Philadelphia, PA: Saunders.

Dong, M., X. Sun, A. A. Prinz, and H. S. Wang. 2006. "Effect of simulated I(to) on guinea pig and canine ventricular action potential morphology." *Am J Physiol Heart Circ Physiol* 291:H631–67.

Foeger, N. C., W. Wang, R. L. Mellor, and J. M. Nerbonne. 2013. "Stabilization of Kv4 protein by the accessory K^+ channel interacting protein 2 (KChIP2) is required for the generation of native myocardial fast transient outward currents." *J Physiol* 591:4149–66.

Goaillard, J. M., and E. Marder. 2006. "Dynamic clamp analyses of cardiac, endocrine, and neural function." *Physiol* 21:197–207.

Guo, W., H. Xu, B. London, and J. M. Nerbonne. 1999. "Molecular basis of transient outward K^+ current diversity in mouse ventricular myocytes." *J Physiol* 521:587–99.

Guo, W., H. Li, B. London, and J. M. Nerbonne. 2000. "Functional consequences of elimination of $I_{to,f}$ and $I_{to,s}$: Early afterdepolarizations, atrioventricular block and ventricular arrhythmias in mice lacking Kv1.4 and expressing a dominant-negative Kv4 alpha subunit." *Circ Res* 87:73–9.

Guo, W., H. Li, F. Aimond, et al. 2002. "Role of heteromultimers in the generation of myocardial transient outward K^+ currents." *Circ Res* 90:586–93.

Guo, W., W. E. D. Jung, C. Marionneau, et al. 2005. "Targeted deletion of Kv4.2 eliminates $I_{to,f}$ and results in electrical and molecular remodeling with no evidence of ventricular hypertrophy or myocardial dysfunction." *Circ Res* 97:1342–50.

Hamill, O. P., A. Marty, E. Neher, B. Sakmann, and F. J. Sigworth. 1981. "Improved patch-clamp techniques for high-resolution current recording from cells and cell-free membrane patches." *Pflug Arch Eur J of Physiol* 391:85–100.

Horn, R., and A. Marty. 1988. "Muscarinic activation of ionic currents measured by a new whole-cell recording method." *J Gen Physiol* 92:145–59.

Horváth, B., D. Kiss, C. Dienes, et al. 2021. "Ion current profiles in canine ventricular myocytes obtained by the 'onion peeling' technique." *J Mol Cell Cardiol* 158:153–62.

Johnson, E. K., S. J. Springer, W. Wang, et al. 2018. "Differential expression and remodeling of transient outward potassium currents in human left ventricles." *Circ Arrhythm Electrophysiol* 11:e005914.

Kuo, H. C., C. F. Cheng, R. B. Clark, et al. 2001. "A defect in the Kv channel-interacting protein 2 (KChIP2) gene leads to a complete loss of I_{to} and confers susceptibility to ventricular tachycardia." *Cell* 107:801–13.

Liu, J., K. H. Kim, B. London, et al. 2011. "Dissection of the voltage-activated potassium outward currents in adult mouse ventricular myocytes: $I_{to,f}$, $I_{to,s}$, $I_{K,slow1}$, $I_{K,slow2}$, and I_{ss}." *Basic Res Cardiol* 106:189–204.

London, B., A. Jeron, J. Zhou, et al. 1998. "Long QT and ventricular arrhythmias in transgenic mice expressing the N terminus and first transmembrane segment of a voltage-gated potassium channel." *Proc Natl Acad Sci U.S.A.* 95:2926–31.

London, B., W. Guo, X.-H. Pan, et al. 2001. "Targeted replacement of KV1.5 in the mouse leads to loss of the 4-aminopyridine-sensitive component of $I_{K,slow}$ and resistance to drug-induced QT prolongation." *Circ Res* 88:940–6.

Niwa, N., W. Wang, Q. Sha, and J. M. Nerbonne. 2008. "Kv4.3 is not required for the generation of functional $I_{to,f}$ channels in adult mouse ventricles." *J Mol Cell Cardiol* 44:95–104.

Sakmann, B., and E. Neher. 1984. "Patch clamp techniques for studying ionic channels in excitable membranes." *Annu Rev Physiol* 46:455–72.

Sun, and H. S. Wang. 2005. "Role of the transient outward current (Ito) in shaping canine ventricular action potential--a dynamic clamp study." *J Physiol* 564:411–9.

Xu, H., W. Guo, and J. M. Nerbonne. 1999a. "Four kinetically distinct depolarization-activated K^+ currents in adult mouse ventricular myocytes." *J Gen Physiol* 113:661–78.

Xu, H., D. M. Barry, H. Li, S. Brunet, W. Guo, and J. M. Nerbonne. 1999b. "Attenuation of the slow component of delayed rectification, action potential prolongation, and triggered activity in mice expressing a dominant-negative Kv2 alpha subunit." *Circ Res* 85:623–33.

10

Models of Ion Channel Gating

Frank T. Horrigan and Toshinori Hoshi

CONTENTS

10.1 Introduction

Models of ion channel gating are conceptual and mathematical constructs designed to *reproduce* and *predict* kinetic and steady-state relationships between the opening/closing processes of a channel ("gating," e.g., its open probability [P_o]) and relevant stimuli (e.g., membrane potential [V_m], ligand concentration, mechanical deformation, temperature). Successful models *reproduce* the known input-output (I-O) relationships or signal transduction properties of a channel, and also *predict* or simulate the properties of a channel not yet measured and sometimes practically impossible to do so.

Gating models can take many forms, from data descriptor functions providing a purely phenomenological description of I-O relationships to all-atom molecular dynamics (MD) simulations based on explicit atomic representations of channels and their surrounding environment. We may eventually see electron wave function-based quantum mechanics (QM) simulations in the future. However, currently more common are gating schemes or semi-mechanistic models based on simplified representations of molecular structure and

DOI: 10.1201/9781003096214-12

function such as kinetic states or functional domains like "sensors" and "gates." The logistics for developing models and level of detail incorporated depend on the goal and availability of functional and structural information. Much of the "art" of model building lies in deciding what level of detail or simplification is required to address a goal and obtain a meaningful answer.

Every model is a simplification, owing to incomplete information as well as philosophical tenants such as Occam's razor or the idea of parsimony; hypotheses that can explain an effect by making the fewest assumptions are preferred. A simpler model is preferred over a complex one, provided they cannot be distinguished based on the experimental observations. One caveat is that "simplest" does not have a unique or universal definition. Models with the fewest free parameters do not necessarily make the fewest mechanistic assumptions (Horrigan and Aldrich 2002).

This chapter serves as a starting point for those who are curious about approaches to develop ion channel gating models. The chapter utilizes voltage-gated K^+ (Kv) channels and Ca^{2+}- and voltage-gated Slo1 K^+ (BK) channels as exemplar systems to illustrate important concepts, principles and rationales in model building. We regret that space constraints prevent us from presenting many excellent models of other channels. Readers interested in specific channels are referred to other chapters in this volume.

10.2 The Goals and Benefits of Modeling

Channel gating models provide quantitative predictions to understand the role of channels in cell signaling, to test hypotheses about physicochemical mechanisms of gating or

FIGURE 10.1 (CONTINUED)

The goals of modeling. (A) A model of electrical excitability describing feedback between ion channels and membrane voltage (V_m). Models of V_m-dependent K^+, Na^+ and Ca^{2+} channels and nicotinic acetylcholine receptors (AChRs), must reproduce time-dependent relationships between stimuli (V_m, ACh) and open probability (P_o). The contribution of each channel to overall conductance for different ions (G_x) is defined by their P_o, number, single-channel conductance and ion selectivity. A circuit model incorporating ionic conductances and passive electrical properties of the cell (e.g., membrane capacitance and leak conductance) is used to predict V_m. (B) Steady-state log(P_o)–V_m relations for mouse Slo1 BK (mSlo1) channels measured at different [Ca^{2+}] are fit by the semi-mechanistic Horrigan–Aldrich (HA) model (depicted by curves), which describes gating in terms of voltage sensor (VS) activation, Ca^{2+}-binding, opening of the activation gate, and allosteric interactions among these processes (Figure 10.3G) (Horrigan and Aldrich 2002). The model provides a basis for interpolating sparsely sampled data, spanning a wide range of V_m and $[Ca^{2+}]_i$. (C) The mean P_o–V_m relation for mSlo1 in 0 Ca^{2+} (C_1) is fit (curves) by two alternative models of V_m-dependent gating (C_2) (Horrigan, Cui, and Aldrich 1999), which are indistinguishable based on macroscopic data (●), but diverge at low P_o. Data derived from single-channel recordings (▲) rule out scheme A, a sequential model that assumes channels cannot open until all four voltage sensors are activated, but are well fit by scheme B, an allosteric model that is equivalent to the HA model in 0 Ca^{2+}. (D) Sensitivity analysis: the HA model in 0 Ca^{2+} is defined by five parameters including the equilibrium constants for VS (J_0) and gate (L_0) at V = 0, the gating charge associated with these transitions (z_J, z_L), and the allosteric factor D. To determine which parameters influence V_m dependence, the maximal logarithmic slope of P_o–V_m relations predicted by the model was plotted as a function of the parameters J_0, L_0, D, or total gating charge $z_T = 4z_J + z_L$ (right panel), where both axes were normalized to the values in panel C, showing that maximal slope is sensitive to allosteric coupling (D) as well as charge (Ma, Lou, and Horrigan 2006). (E) Normalized P_o–V_m relations for wild-type (WT) and R207Q mutant mSlo1 channels were fit simultaneously by the HA model with only the VS equilibrium constant J_0 altered (Horrigan, Cui, and Aldrich 1999), suggesting that R207 in the S4 segment does not contribute to gating charge.

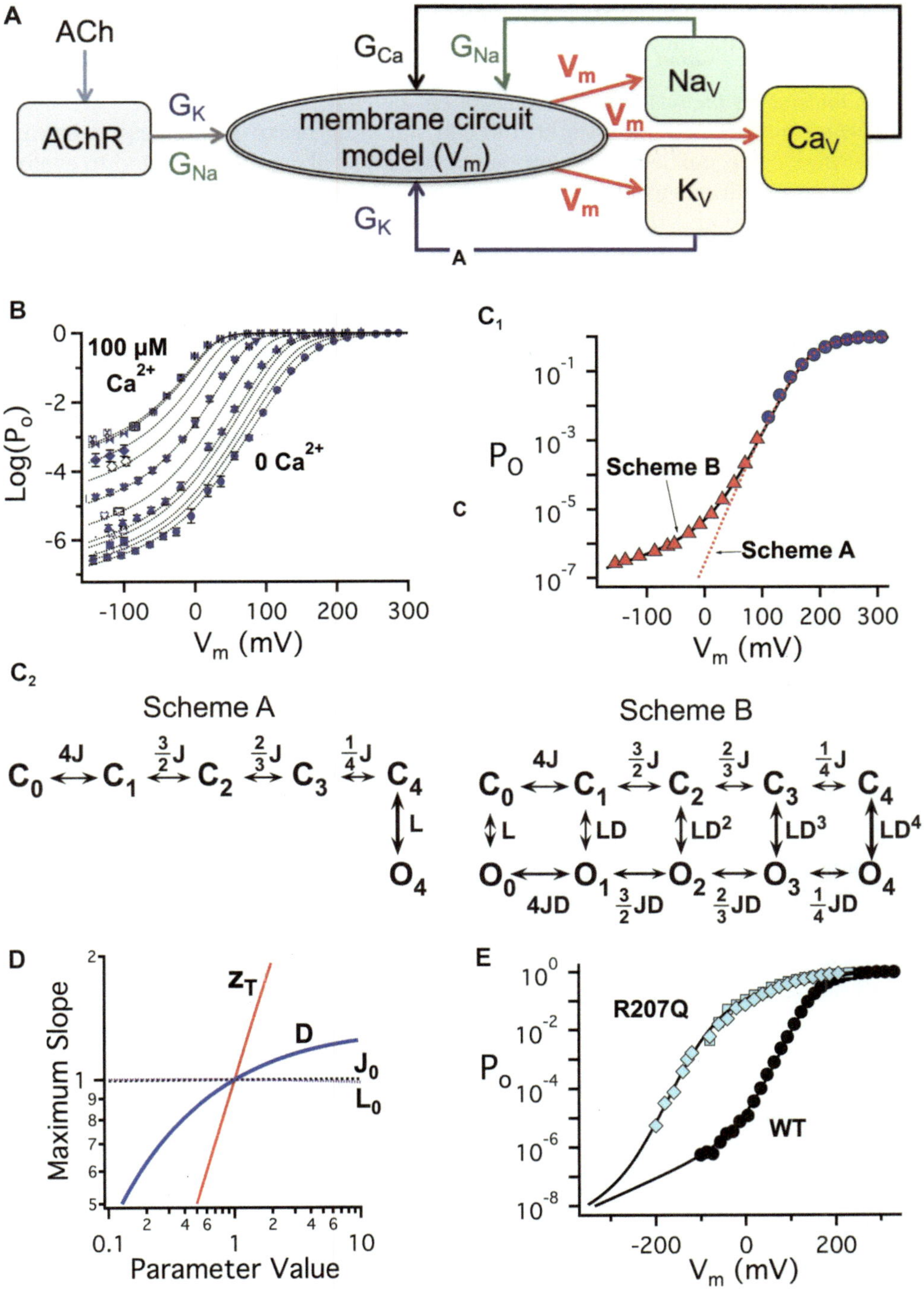
A
ACh
AChR
GK
GNa
GCa
GNa
GK
membrane circuit model (Vm)
Vm
Vm
Vm
NaV
CaV
KV
A
B
0
-2
-4
-6
Log(Po)
100 μM Ca2+
0 Ca2+
-100 0 100 200 300
Vm (mV)
C1
PO
Scheme B
Scheme A
C
-100 0 100 200 300
Vm (mV)
C2
Scheme A
Scheme B
D
Maximum Slope
ZT
D
J0
L0
0.1 1 10
Parameter Value
E
R207Q
WT
Po
-200 0 200
Vm (mV)

to infer information about elementary gating events based on function. These goals and the requirements that they place on models are discussed next.

1. *To understand the role of ion channels in cell signaling, including electrical excitability*: Gating models provide insight into cell signaling because changes in the number and/or properties of different channels in a cell can be simulated to evaluate their contributions to signaling or to mimic effects of channel regulation in physiology, disease and therapy. Electrical excitability involves feedback between ion channels and their effects on V_m (Figure 10.1A) that can, in some cases, be modeled using phenomenological descriptions of I-O relationships without a detailed understanding of channel gating but more frequently require gating schemes or semi-mechanistic models (see later) to predict how channels respond and contribute to signaling, especially when multiple variables such as V_m and Ca^{2+} are involved.
2. *To predict channel behavior under conditions that are difficult to measure experimentally*: Measuring the response of a channel to all relevant stimuli is often impractical. A gating model helps to predict a channel's response to combinations of conditions poorly sampled by the measured results (Figure 10.1B) or conditions that are difficult to achieve experimentally (e.g., high temperature). While most models can interpolate a channel's response to conditions spanned by the original data set, extrapolating predictions beyond the experimental conditions typically requires semi-mechanistic models based on an understanding of the physical mechanisms underlying V_m or ligand sensitivity. Likewise, predicting how multimodal channels integrate combinations of stimuli is facilitated by understanding how gating mechanisms are influenced by each stimulus and how those mechanisms interact (Figure 10.1B).
3. *To translate a mechanistic hypothesis of ion channel function into a testable prediction*: Discrimination of gating mechanisms specified by models provides insight into many processes including stimulus integration and drug action. Models also facilitate the design of experiments to distinguish between competing hypotheses by revealing experimental conditions where the hypotheses' predictions diverge (Figure 10.1C).
4. *To define the relationship between elementary processes and channel response*: *Sensitivity analysis* can determine the effect on channel function of systematic changes in parameters, thereby facilitating identification of elementary processes (e.g., rate-limiting steps, charge movement, or ligand-binding events) (Figure 10.1D) or defining experimental conditions where perturbations in such processes can best be detected. Similarly, models are used to derive estimates of important elementary parameters (e.g., gating charge, ligand K_D) by fitting the kinetic or steady-state properties of gating.
5. *To understand relationships between channel structure and function*: Modeling allows the role of individual structural components to be defined in terms of the gating mechanisms, including elementary processes or elementary parameters, that are perturbed by mutation (e.g., Figure 10.1E), thus providing insights into physicochemical mechanisms.

10.3 Model Generation

10.3.1 Should I Build a (New) Gating Model?

The benefits of developing a successful gating model are numerous. But in most cases, the effort required probably does not justify *de novo* development. This may come as a surprise, but development of a new model requires a rigorous effort to constrain parameter values (see later) and rule out alternative models. Using an ill-defined model offers little insight and may in fact be detrimental. If the channel of interest has a better-studied ortholog for which a good gating model already exists, adapting the existing model may require less effort and provide valuable comparative information.

10.3.2 What Kind of a Model?

The selection depends on the goal or the purpose of the model, and some of the possible model types are (a) phenomenological, (b) statistically parsimonious and (c) semi-mechanistic (Figure 10.2).

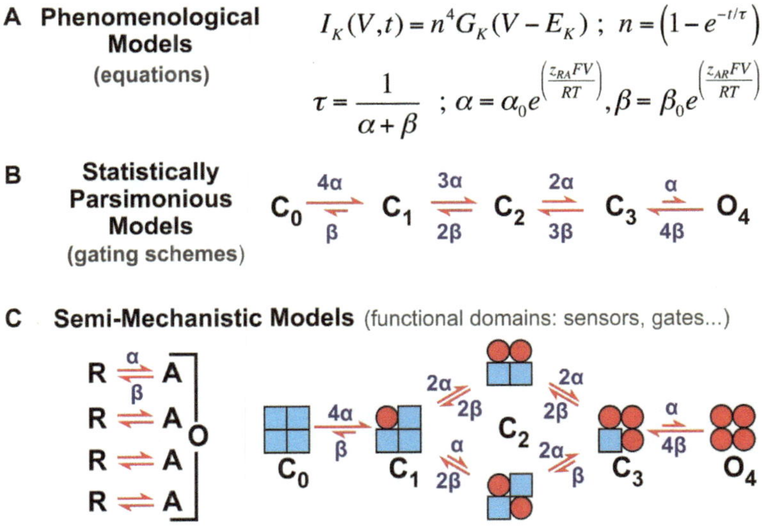

FIGURE 10.2
Three gating model types are used to represent the Hodgkin and Huxley (HH) model of K_V channels. (A) A phenomenological model describes the relationship between the K^+ current (I_K), voltage (V) and time (t) by a negative exponential function to the fourth power, where the time constant τ is defined by two V-dependent rate constants (α, β). (B) A statistically parsimonious HH scheme with four closed and one open state. (C) Two semi-mechanistic representations of the HH model assume the channel opens when all four independent and identical V-sensors transition from resting (R, ■) to activated (A, ●). The forward rate constant for the C_0–C_1 transition (4α) is four times that for voltage sensor (VS) activation because the C_0 state contains four VSs, any one of which may activate to generate C_1. Rate constants for other forward and backward transitions are similarly dependent on the number of VSs available to activate or deactivate (i.e., statistical factors). Potential degeneracy in the C_2 state is illustrated by two conformations that are not equivalent; but if assumed to be functionally indistinguishable, can be lumped into a single C_2 state, yielding a scheme equivalent to panel B.

10.3.2.1 Phenomenological Models

A phenomenological model is simply a set of equations that describes channel gating, which may or may not have a realistic physical meaning. An extreme example is to use polynomial equations to describe ionic current kinetics. A polynomial fails to provide any physical or mechanistic interpretation. Other functions are more intimately associated with "physical realities." For example, a *negative exponential function* of the form $I(t) = A(1-e^{(-t/\tau)})$ or $I(t) = Ae^{(-t/\tau)}$, where $I(t)$ is the current size as a function of time t, A is a scaling factor and τ is the time constant, results from any system where the rate of change depends only on the present condition. A negative exponential function may be used to represent activation kinetics of an ionic current without delay on presentation of a stimulus. A *product* of multiple negative exponential functions, for example, $I(t)=A(1-e^{(-t/\tau)})^n$, is often used to describe the "sigmoidal" time course or activation with a delay, as in the Hodgkin and Huxley model (HH model) of voltage-gated Na^+ and K^+ currents (Figure 10.2A) (Hodgkin and Huxley 1952). Multiplication of the equations implies that the underlying systems behave probabilistically in an independent manner. Another function used frequently in phenomenological models is a *Boltzmann function* of the form $P_o(V_m) = \frac{1}{1+e^{\left(\frac{V_{0.5}-V_m}{SF}\right)}}$, where $P_o(V_m)$ is P_o at the membrane potential V_m; $V_{0.5}$ is the half-activation V_m, where $P_o = 0.5$; and SF is a factor that determines the steepness of the curve. This equation would be considered *mechanistic only for the simple two-state closed–open channel.* For more complex gating with more than two states, the formulation is largely phenomenological; alterations in multiple aspects of channel gating can lead to changes in $V_{0.5}$ and SF. The preceding equation remains frequently utilized for comparative reasons, but the parameters are *phenomenological data descriptors* for most channel types.

10.3.2.2 Statistically Parsimonious Gating Schemes

To develop a statistically parsimonious gating scheme, one relies on statistical criteria (e.g., residual sum of squares, maximum likelihood ratios) to determine how many components, such as kinetic states, to include and how they are arranged. The resulting scheme is statistically parsimonious, balancing the overall goodness-of-fit/badness-of-fit and the cost of additional free parameters. However, the model is parsimonious only for the data set analyzed, ignoring infrequent transitions, and may not be so for other conditions (see Section 10.3.2.3); thus, their general applicability or extensibility may be limited.

The simplest gating scheme for a channel is obviously the two-state model with one closed state and one open state. In the statistically parsimonious gating scheme approach, additional states and transitions are added to describe the available observations only if a certain criterion is met. This approach is thus data-driven and requires high-quality experimental results. Excellent experimental tips are found in Sakmann and Neher (1995).

The statistically parsimonious gating scheme approach (and the semi-mechanistic approach) almost always makes the assumption that ion channel function is a time-homogeneous Markov process with a finite number of states; (a) an ion channel has a relatively small number of *kinetically distinguishable states* and (b) the transitions among these discrete states can be described by time-invariant rate constants. An ion channel protein undoubtedly has numerous conformational states or free-energy minima, as suggested in part by different classes of single-particle cryo-EM structures and all-atom MD simulations; however, these free-energy minima are grouped into a smaller number of *functionally distinguishable states*. Electrophysiological methods typically detect events lasting from ~20 μs to hundreds of seconds, and up to a dozen or so *kinetically distinguishable states* have been reported (see later). The assumption that rate constants are time-invariant relates to the idea that the rate of leaving a state *under a given condition* is "memoryless" and does not depend on how long the channel has stayed in that state. Consider, for example, the two-state scheme with one closed state (C) and one open state (O). The rate constant k_{CO} describes the mean number of transitions from C to O per unit time and k_{OC} is that for the reverse transition.

$$C \underset{k_{OC}}{\overset{k_{CO}}{\rightleftharpoons}} O \qquad \text{Scheme 10.1}$$

According to the time-homogeneous Markov postulate, provided that the *experimental condition remains the same*, the values of k_{OC} and k_{CO} remain invariant whether the channel has been open (or closed) for 1 ms or 1 s. Consequently, the dwell times within each state are distributed according to single negative exponential functions with time constants for closed and open of $1/k_{CO}$ and $1/k_{OC}$, respectively. Consistent with the Markov postulate, experimentally measured single-channel dwell times are generally well described by a negative exponential function or a sum of negative exponential functions (Sakmann and Neher 1995). Some caveats are that dwell times shorter than ~100 μs may be poorly resolved and require correction for missed events, and that it is uncertain theoretically whether rate constants remain "memoryless" on time scales approaching the molecular vibrational range (<1 ps).

How many states? At least two states of an ion channel are relatively easily recognized based on their conductance: closed and open. However, there may be additional kinetically distinct states that share very similar single-channel current amplitudes, such as the four closed states predicted by the HH model (Figure 10.2B).

In general, the following guidelines may be used to estimate the *minimum* number of states that a gating model must contain to be consistent with kinetic data. First, if a channel visits N states, whether closed or open, during gating, there should be in theory N–1 negative exponential components in ionic current kinetics. Similarly, if a channel traverses *m* open states and *n* closed states, its open durations and closed durations are described, in theory, by sums of *m* and *n* negative exponential components, respectively. In practice, it may be difficult to detect all of the aforementioned components because their amplitudes may be very small and/or their time constants too similar to distinguish (Magleby and Weiss 1990; Colquhoun and Hawkes 1995b; Colquhoun and Hawkes 1995a); but performing the preceding analysis under different stimulus conditions may be helpful. It is important to note that time constants do not correspond directly to the mean dwell time in any state, except for the simplest models (closed–open, open–closed–open, and closed–open–closed). Detailed treatments of single-channel dwell-time analysis are found in Horn and Lange (1983), Magleby and Weiss (1990), Colquhoun and Sigworth (1995), and Colquhoun and Hawkes (1995b).

Additional kinetic analysis may be incorporated to infer the number of states and how they are arranged. In a typical analysis of single-channel dwell times, information regarding the sequential order in which individual events of different durations occur is discarded (see Chapter 8). *Conditional probability analyses* often attempt to utilize aspects of this "sequence" information and may take on many different forms. For example, adjacent open and closed durations may be analyzed (Colquhoun and Hawkes 1987; Magleby and Song 1992). A more general approach, pioneered by Horn and colleagues, calculates the likelihood that an observed sequence of idealized single-channel closed and open events is generated by different models, and then selects the most parsimonious model based on statistical criteria (Horn and Lange 1983; Horn 1987). This approach has been adapted and improved, incorporating the hidden Markov concept so that noisy and/or multichannel data can be analyzed (Chung et al. 1990; Klein, Timmer, and Honerkamp 1997). Macroscopic ionic currents generated by a large number of homogeneous channels can be analyzed a similar manner (Milescu, Akk, and Sachs 2005). Free software is available to implement various gating-scheme building processes and simulations on multiple operating system platforms (e.g., https://qub.mandelics.com, http://www.ucl.ac.uk/Pharmacology/dc-bits/dcwinprogs.html).

Additional measurement types provide information complementary to that of ionic currents that can further constrain statistically parsimonious gating schemes and provide insight regarding the underlying molecular transitions. For example, gating currents, associated with the movement of charged structural elements in voltage-dependent channels, are valuable for probing transitions among states with the same conductance and may eliminate certain models from consideration (Zagotta, Hoshi, and Aldrich 1994; Schoppa and Sigworth 1998) (see also Chapters 2 and 7). Optical methods (e.g., patch fluorometry) can potentially be applied to any channel type to monitor conformational changes associated with gating (see Chapter 11). But, while the statistical approach holds great appeal as an "objective" model building strategy and can in principle be applied to any single data type, one unresolved question is *how to weigh qualitatively different data.* Each data type has its own unique error and noise characteristics making them difficult to compare statistically. In addition, different data types can exhibit different sensitivity to changes in the same model parameters, and therefore differ in their ability to constrain certain transitions or distinguish between models that contain those transitions. If gating current analysis favors one model and single-channel analysis favors another, what criterion does one use to choose one over the other? For such conditions, the increasingly popular global fitting process is also of very limited usefulness. No satisfactory solution to this important and nagging issue has been put forward to date.

Stimulus dependence of rate constants? Stimulus-induced transitions among different states of an ion channel, beyond those by the prevailing thermal energy, are driven by changes in rate constants. For example, voltage-gated ion channels change their P_o because the values of some rate constants are influenced by changes in membrane potential (ΔV_m). To develop a successful gating scheme, the dependence of the rate constants on the relevant stimuli must be described. The stimulus dependence could be described by any function, but some are more readily understood using well-known physical principles. Many of the formulations used in the gating scheme approach have ready physical interpretations and are also applicable in the semi-mechanistic approach.

Rate constants in gating schemes are often given molecular and physicochemical connotations using the Arrhenius activation energy/transition state formulation $k = f' e^{(-\Delta G/RT)}$, where k is the rate constant (1/s); ΔG is the Gibbs free energy difference between the state in which the channel resides and the transition state or the energy barrier (herein referred to as the activation energy); R and T have their usual meanings; and f' is known as the frequency factor, vibrational/collisional factor or pre-exponential factor. The value of f' is often quoted to be roughly 10^9 to 10^{12} /s, but *it is poorly defined and it has no direct experimental validation.* However, ΔG can often be defined in terms of physical processes underlying stimulus-dependent rate constants. The aforementioned Arrhenius formulation can be extended to make a rate constant dependent on V_m, ligand concentration and/or temperature. More details and practical examples are discussed in the previous version of this chapter (Horrigan and Hoshi 2015).

10.3.2.3 Semi-Mechanistic Models

Truly mechanistic models of ion channel gating are built on descriptions of the elementary processes that give rise to functionally and kinetically distinct states. Mechanistic models ideally involve explicit simulation of protein conformational changes but more commonly are based on the common view that ion channels are modular proteins whose function can be described in terms of the collective activity of distinct functional domains such as sensors and gates. Mechanistic models thus typically describe the movement or "activation" of functional domains and their interaction with each other as well as combinatorial factors relating to the number of functional domains that can be activated within a channel. Such models are best described as "semi-mechanistic" because they incorporate simplifying or parsimonious assumptions to describe the activation of individual functional domains. Nonetheless, semi-mechanistic models offer several important advantages over phenomenological models and statistically parsimonious models as summarized next.

a) *Incorporation of diverse data types, a priori concepts, and nonquantifiable pieces of information.* Semi-mechanistic models incorporate a variety of information that cannot, or cannot easily, be integrated into phenomenological models or statistically parsimonious models. While some theoretical concepts, such as those defining stimulus dependence of rate constants, can be implemented in statistically parsimonious gating schemes, others describing how functional domains interact (e.g., allosteric factors; see Section 10.4.3) or the relationship between rate constants defined by independent functional domains (i.e., statistical factors; see Figure 10.2B, C) cannot be imposed on gating schemes without a mechanistic framework. Likewise, nonquantifiable structural information including the symmetry of homotetrameric channels and the presence or absence of direct physical contact between functional domains can place important constraints on semi-mechanistic models that cannot be implemented in a purely statistically parsimonious gating scheme approach. Semi-mechanistic approaches capable of incorporating difficult-to-quantify structural information are therefore considered by many as conceptually elegant. Because the semi-mechanistic models incorporate various prior information and concepts, such models may be considered to be Bayesian heuristic in nature.
b) *Predictive power.* Semi-mechanistic models are more likely than phenomenological models or statistically parsimonious models to be extensible or applicable beyond conditions where the model was originally tested. One reason is that semi-mechanistic models describe the effects of stimuli in terms of well-understood biophysical processes that are valid over a wide range of stimulus conditions. Not only do they include *stimulus-dependent rate constants,* as described earlier for statistically parsimonious models, but also a number and arrangement of states and rate constants consistent with, for example, the number of sensor domains. Statistically parsimonious gating schemes tend to exclude states that are rarely occupied under the testing condition (e.g., with all sensors activated) and may therefore fail to provide accurate predictions when extended to conditions where those states are more frequently occupied.
c) *A framework for structure/function analysis.* Interpretations of the effect of mutagenesis in terms of their impact on functional domains and elementary processes typically require a semi-mechanistic model. This point cannot be overemphasized.

Statistically parsimonious gating schemes often fail to provide useful comparative information when perturbations are large enough that the "best" statistically parsimonious gating scheme for mutant and wild-type channels are different (e.g., due to differences in occupancy).

d) *Application to related channels.* Homologous and orthologous channels are expected to contain the same number and type of functional domains. Therefore, a semi-mechanistic model for one channel should provide a reasonable framework for related channels. By contrast, statistically parsimonious models may not be applicable to homologous/orthologous channels if, as in the case of mutation noted earlier, the distribution of occupied states is different.

One disadvantage of semi-mechanistic models is that they tend to sacrifice some ability to fit functional data compared to statistically parsimonious models with an equivalent number of free parameters. One reason is that a semi-mechanistic model is constrained to "fit" observations that are not necessarily addressed by the statistically parsimonious model. Thus, a statistically parsimonious model may be preferred if the goal is to faithfully reproduce a particular functional response. But semi-mechanistic models may be preferred for a variety of other reasons as outlined.

Semi-mechanistic models and statistically parsimonious gating schemes are not necessarily mutually exclusive. In principle, statistically parsimonious gating scheme approaches should/could converge on gating schemes that reflect underlying biophysical mechanisms and may be used to infer or test mechanistic assumptions. Likewise, constraints imposed on gating scheme architecture by semi-mechanistic models, such as the connectivity of states or relationships between rate constants, may be maintained in building and ranking gating schemes. However, in practice, low occupancy of some states and/or errors in the estimation of rate constants limit the utility of such approaches. Exceptions are cases where the gating mechanism is simple enough and experimental conditions appropriate to adequately sample all *accessible* states and transitions. Typically this involves the use of stimulus conditions that limit the number of accessible states, such as voltages that fully activate voltage sensors (e.g., to study inactivation), saturating ligand concentrations, or even the use of channels that have been mutated to reduce the number of functional sensors and simplify possible gating mechanisms (Shelley et al. 2010).

10.4 Examples of Semi-Mechanistic Models and Assumptions

An advantage of phenomenological models or statistically parsimonious models is that systematic and prescribed approaches exist to construct them (see Horrigan and Hoshi 2015). Semi-mechanistic models, by contrast, can be more difficult to refine because they are built upon mechanistic assumptions that may be difficult to validate directly or quantitatively. Furthermore, if a semi-mechanistic model fails to reproduce some features of the data, it cannot be "fixed" simply by extending the original hypothesis. Rather, alternative hypotheses must be formulated, which can be a complex process involving the synthesis of diverse information, identification of mechanisms that are consistent with the data, and translation of biophysical principles into simplified models. *Therefore, it is useful to review some examples of semi-mechanistic models, both to illustrate the types of mechanistic assumptions that have been successful and the types of data and observations that are taken into account in the process of formulating new hypotheses.*

10.4.1 The Hodgkin and Huxley (HH) Model

The *HH model* was developed originally as a phenomenological model (Figure 10.2A) but also given a mechanistic interpretation (Hodgkin and Huxley 1952), and has served as a foundation for many subsequent semi-mechanistic models. Regarding their K^+ current formulation, one accurate assumption/implication was (in modern terms) that voltage-dependent K^+ channels contain four identical voltage sensor (VS) domains that activate largely independently (see Volume II, Chapter 4). This idea was based on the observation that the delayed activation of K^+ conductance in squid axon is adequately described by fourth-order kinetics (but see later), whereas deactivation exhibits a single exponential time course. A simple explanation is that channels are open only when all four identical subunits, each with a VS, are activated. Activation of each VS is in turn described by a two-state voltage-dependent transition between resting (R) and activated (A) states (Figure 10.2C). The HH model predicts a sequential K^+ channel gating scheme (Figure 10.2B) with four closed states (C_0–C_3) and a single open state (O_4), defined by the number of activated VSs (0–4), and also predicts a characteristic relationship between rate constants and equilibrium constants for different transitions determined by the number of VSs available to activate or deactivate (i.e., statistical factors 4, 3, 2 and 1 in Figures 10.3B and 10.4A).

10.4.2 Modern Models of K_V Channel Gating

Although Hodgkin and Huxley correctly predicted the number of VSs in K^+ channels, this was in fact an estimate based on parsimony (including computational limitations of the time) rather than direct evidence, like most assumptions in their model. In subsequent decades, more complicated models were proposed to describe ionic and gating currents in squid axon. However, it was not until molecular biology and the patch-clamp technique were applied to K^+ channels that many mechanistic assumptions could be properly tested. In the 1990s, models of the first cloned Kv channel, Shaker, were developed (Zagotta, Hoshi, and Aldrich 1994; Schoppa and Sigworth 1998) based on high-resolution patch-clamp recordings of single-channel and macroscopic K^+ currents as well as gating currents and structural information. A critical piece of information was that Kv channels (but not Nav or Cav channels) are indeed composed of four identical subunits, each with a single VS domain (VSD). This placed absolute constraints on the architecture and symmetry of any proposed model that together with more detailed functional measurements allowed other assumptions of the HH model to be tested and modified. The Zagotta–Hoshi–Aldrich (ZHA) model of Shaker as modified by Ledwell and Aldrich (1999) (Figure 10.3B) has been used to describe a variety of Kv channels and their perturbations by mutagenesis. The key mechanistic assumptions are, briefly:

(a) *The channel contains four independent and identical VSs that activate in two steps (R1 → R2→ A).* That VSs activate in at least two steps, with some hints of cooperativity among them, is required to describe multiexponential gating current kinetics (Zagotta, Hoshi, and Aldrich 1994; Schoppa and Sigworth 1998).

(b) *Channel opening (C-O) represents a concerted conformational change that follows but is distinct from VS activation.* That subunits can undergo both independent transitions

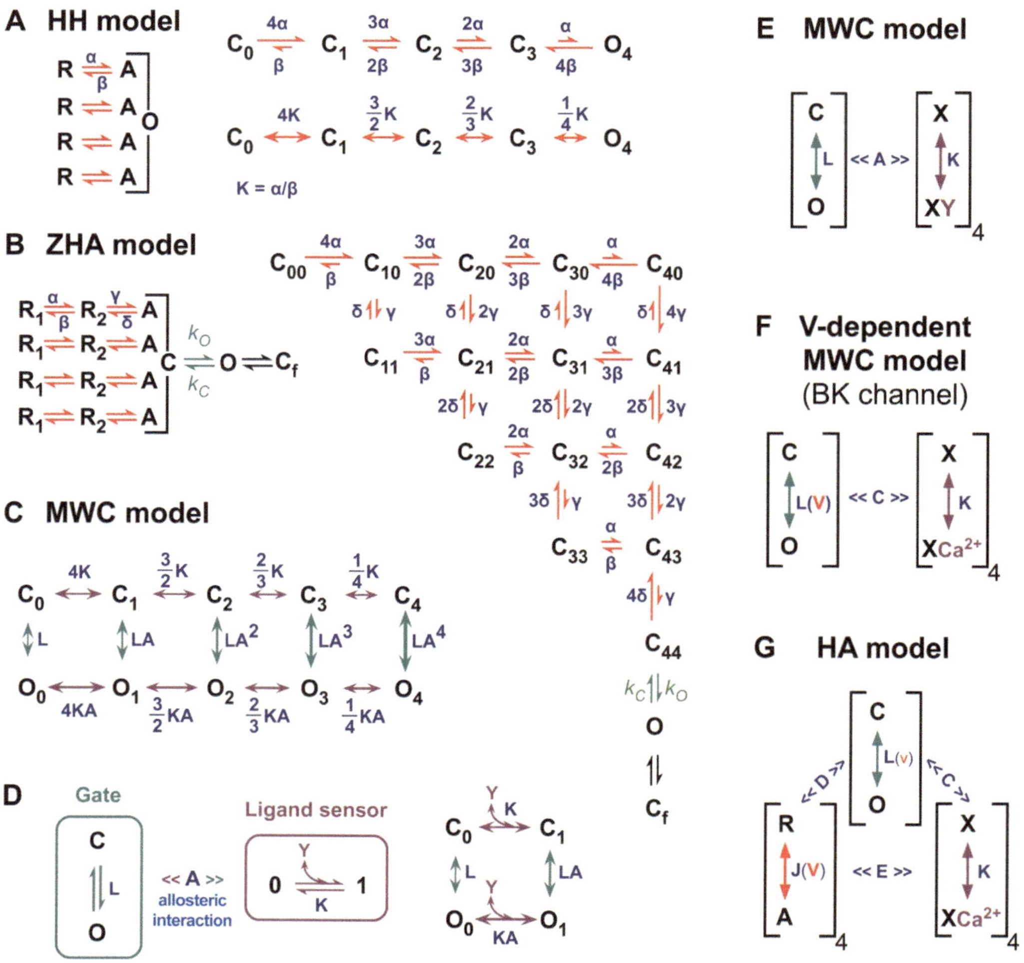

FIGURE 10.3
Semi-mechanistic model examples. (A) The HH model (left panel), as described in Figure 10.2, generates a sequential scheme (right panel) with characteristic rate constants (upper scheme) or equilibrium constants (lower scheme) representing the ratio of forward to backward rate constants. (B) A modified ZHA model of Kv channel activation (left panel) assumes that VSs activate in two steps (R1 to R2 to A) and opening occurs after all VSs are activated. The C_f state is needed to account for brief "flicker" closed events observed when single channels are maximally activated by voltage. The equivalent gating scheme (right panel) contains 15 closed states ($C_{m,n}$) in addition to C_f, where m, n represent the number of VSs in R_2 and A conformations, respectively. (C) MWC allosteric model of ligand-dependent activation for a tetrameric channel can be represented by a gating scheme with five closed and five open states, with subscripts indicating the number of activated ligand sensors. The unliganded channel can open with equilibrium constant L. The equilibrium constant for activating a single ligand sensor is $K = [Y]/K_D$ when channels are closed and KA when open, where [Y] is the ligand concentration and A is an allosteric factor. (D) A simple MWC-type mechanism with one ligand sensor can be described (left panel) by a gate transition [C-O], sensor activation [0-1], and an allosteric interaction (A) or (right panel) by an equivalent gating scheme. (E) The MWC model can be represented by allosteric interaction of one gate with four ligand sensors. (F) A V-dependent MWC-type model of BK channel activation assumes an allosteric interaction between one gate and four Ca^{2+} sensors, with the equilibrium constant for the C-O transition L(V) being V-dependent (Cox, Cui, and Aldrich 1997). (G) The HA model of BK channel activation describes gating in terms of allosteric interactions between one gate, four Ca^{2+} sensors and four VSs (Horrigan and Aldrich 2002).

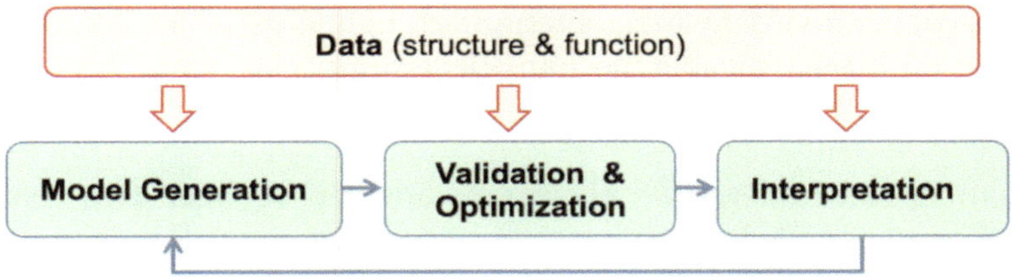

FIGURE 10.4
A typical model development workflow, including different types of data that may be taken into consideration during the iterative process of generating, testing and modifying a model.

associated with voltage sensing and concerted transitions associated with opening was supported by a variety of evidence including the relationship between ionic and gating currents in WT and mutant channels (Zagotta, Hoshi, and Aldrich 1994; Ledwell and Aldrich 1999). While the exact number of steps involved in opening may be subject to question (Schoppa and Sigworth 1998), the fundamental concept that the VSs and ion conduction gate represent structurally and functionally distinct domains whose activation in Kv channels is tightly coupled is well supported by the atomic structures of Kv channels (Long et al. 2007) and by experiments designed to probe conformational changes in these different domains (Broomand and Elinder 2008; Mannuzzu, Moronne, and Isacoff 1996).

10.4.3 Allosteric Mechanisms and Models

Allostery, often defined broadly as "action at a distance," should play a role in the function and regulation of most ion channels because channel proteins are modular with structurally distinct functional domains (Colquhoun and Lape 2012). The parts of the channel that detect stimuli (sensor domains) or those that interact with regulatory proteins or drugs are distal from the pore–gate domain that regulates opening and closing. Allosteric mechanisms have been incorporated in many semi-mechanistic models predominantly to describe ligand-gated channels or drug action but also for channels that respond to other stimuli such as V_m as discussed later.

The term "allosteric" was originally coined to describe the idea that two different ligands can interact with a protein and influence each other's apparent binding affinity indirectly through a reversible protein conformational change, or "allosteric transition," as opposed to directly competing via steric hindrance for a common binding site. This concept was extended to describe mechanisms of cooperative binding in oligomeric proteins. To account for the cooperative binding of oxygen to hemoglobin, Monod, Wyman and Changeux (1965) devised the MWC allosteric model, which has had a significant impact on the understanding of ligand action comparable to that of the HH model on membrane excitability, and has influenced or formed the basis for many ion channel models.

10.4.3.1 The MWC Model

Like the HH model, the MWC model is based on extensive experimental evidence including measurements of oxygen binding as well as structural and biochemical evidence that hemoglobin is a tetramer whose tertiary and quaternary conformation and subunit–subunit interactions are altered by ligand binding. The MWC approach can be applied directly to describe the cooperative activation of tetrameric ligand-gated ion channels (see Chapter 3) and is based on three plausible mechanistic assumptions, which stated for tetrameric channels are:

1. The channel can undergo an allosteric transition between two conformations, closed (C) and open (O), involving a concerted change in all four subunits.
2. The affinity of ligand for its receptor in all four subunits is the same but is greater in the open than the closed conformation (i.e., $K_{DC} > K_{DO}$).
3. Opening can occur whether or not a ligand is bound.

These assumptions define a two-tiered gating scheme (Figure 10.3C) with five closed and five open states, representing the two allosteric conformations (C, O) with different numbers of ligands bound (0–4). These assumptions also define the relationship between equilibrium constants indicated in the model and explain why ligand binding induces channel opening and why ligand binding is cooperative. To understand these principles, consider a simpler MWC-type model with one ligand-binding site, which generates a four-state gating scheme with two closed and two open states (Figure 10.3D, right panel). The equilibrium constant between the two closed states (C_0–C_1) is defined in terms of ligand concentration and dissociation constant as $K = [Y]/K_{DC}$. Similarly, the equilibrium constant between the two open states (O_0–O_1) is defined as KA, where A is an allosteric factor representing the relative affinity of the ligand for the open and closed conformations ($A = K_{DC}/K_{DO}$). Consequently, if the transition between closed and open in the absence of ligand (C_0–O_0) is assigned an equilibrium constant L, then the equilibrium constant for opening when ligand is bound (C_1–O_1) must be LA. If opening favors ligand binding ($A > 1$), then ligand binding must favor opening ($LA > L$). This is required simply by thermodynamic principles that demand that the equilibrium constant between two states, which is the product of all the equilibrium constants for the intervening transitions, be independent of path (e.g., $C_0 \Leftrightarrow C_1 \Leftrightarrow O_1$ vs. $C_0 \Leftrightarrow O_0 \Leftrightarrow O_1$).

Several additional features of Figure 10.3D are worth noting, both in terms of the MWC model and gating schemes in general. First, the aforementioned mechanistic assumptions only constrain the equilibrium constants but not the rate constants in the model because ligand-binding affinity or K_D is an equilibrium concept that relates the ratios of the forward and backward rates constants for a ligand-binding step. Additional assumptions are necessary to include rate constants in the model (Horrigan and Hoshi 2015). Equilibrium models often serve a critical role in the development of semi-mechanistic and kinetic models because fewer mechanistic assumptions and free parameters are required to constrain the model and because equilibrium constants are more readily linked to free-energy differences than rate constants. That said, kinetic models are constrained by the same thermodynamic principles noted earlier. In particular the principle of microscopic reversibility or detailed balance requires in a cyclical scheme like Figure 10.3D that the product of all the rate constants around a cycle in one direction (e.g., $C_0 \rightarrow C_1 \rightarrow O_1 \rightarrow O_0 \rightarrow C_0$) must equal the product of rate constants in the opposite direction ($C_0 \rightarrow O_0 \rightarrow O_1 \rightarrow C_1 \rightarrow C_0$). This constraint applies to any cyclical gating scheme: allosteric models, statistically parsimonious models and semi-mechanistic models. Finally, it is important to note that not all cyclic models are allosteric. For example, if the affinity of a ligand for closed and open conformation were identical, its interaction with a channel would be described by a ten-state model like MWC but with $A = 1$: an independent action model.

The scheme in Figure 10.3D can be extended to four ligand-binding sites (Figure 10.3C) to illustrate why the MWC model predicts cooperative binding. The ten-state MWC scheme exhibits a characteristic relationship between horizontal equilibrium constants, which is identical to that in the HH model (Figure 10.3A) reflecting that sensors, in this case ligand-binding sites, in different subunits are independent and identical. However, the equilibrium constants for transitions among open and closed conformations differ by the allosteric factor A. As a consequence, to satisfy detailed balance, the equilibrium constant for channel opening must increase by a factor A for each ligand bound. In other words, the effects of ligand binding on opening are energetically additive. This illustrates why binding of ligands to all four sites can produce a large (A^4-fold) increase in the equilibrium constant for opening, despite a much smaller A-fold allosteric affinity change for each site. Cooperative binding arises because binding promotes opening, and open channels in turn can bind additional ligand molecules with higher affinity than closed channels. The equilibrium properties of the MWC model can be described by a semi-mechanistic representation (Figure 10.3E), which includes a concerted conformational change to open the channel [C-O], ligand-binding in each of four identical subunits $[X\text{-}XY]_4$, where Y is a ligand, and the interaction between these processes is described by an allosteric factor A.

Alternative models of allostery and cooperative binding in hemoglobin and other proteins have been proposed such as the sequential KNF model (Koshland, Nemethy, and Filmer 1966), which assumes individual subunits change conformation when ligand binds. However, MWC-type models have been widely applied in ion channel models because the key allosteric transitions in channels (e.g., ion conduction gate opening) are thought to be concerted or nearly concerted. Thus, this model provides an elegant and parsimonious mechanism to account for the cooperative response of many ion channels to ligands and other stimuli.

Allosteric mechanisms in voltage-dependent gating relate to the distinction between obligatory and nonobligatory coupling between VSs and the ion conduction gate. In ligand-gated channels described by MWC-type models, the ligand sensors and gate are coupled in a nonobligatory manner in that opening does not require ligand binding; ligand binding and opening are distinct processes but influence each other (Figure 10.3E). In comparison, for the voltage-dependent Kv channel models in Figure 10.4A and B, the coupling processes are obligatory because channels cannot open unless all VSs are activated. One possible explanation for such a mechanism is that the resting VS prevents gate opening by steric hindrance, analogous to direct (nonallosteric) competition between two ligands. By contrast, an allosteric mechanism would allow channels to open whether or not VSs are activated and could involve indirect interaction between the VS and gate.

10.4.3.2 The HA Model of BK Channels

BK potassium channels (aka Slo1, Maxi K, $K_{Ca}1.1$, KCNMA1; see Volume II, Chapter 7) are activated by both depolarization and intracellular Ca^{2+}. The response to Ca^{2+} was first studied in detail at the single-channel level. McManus and Magleby (1991) showed through analysis of native channels in different $[Ca^{2+}]$ and constant V_m that BK channels can occupy at least three open and five closed states arranged in two (closed, open) tiers, as expected for a channel that can open and close with multiple Ca^{2+} ions bound and can bind Ca^{2+} whether open or closed. The authors noted that the arrangement of states in their statistically parsimonious model resembled a subset of those predicted by the MWC-type model. Cox, Cui, and Aldrich (1997) subsequently proposed a voltage-dependent MWC-type model to describe the kinetic and steady-state properties of macroscopic K^+ current from heterologously expressed mouse Slo1 (mSlo1) channels. To account for the effects of V_m they assumed that the C-O transition in the MWC model was voltage-dependent (Figure 10.3F). Consistent with the assumptions that voltage- and Ca^{2+}-dependent transitions were distinct, they observed that BK channels could be fully activated by V_m depolarization in the absence of Ca^{2+} binding (Cox, Cui, and Aldrich 1997). Nonetheless, a two-state model of voltage-dependent gating was difficult to reconcile with the tetrameric nature of BK channels and inconsistent with evidence that gating currents in BK channels activate much more rapidly than ionic currents. Indeed a detailed analysis of ionic and gating currents over a wide range of conditions indicated that voltage-dependent gating in BK channels, like Ca^{2+}-dependent gating, could best be described in terms of an allosteric mechanism as formulated in the Horrigan–Aldrich (HA) model (Horrigan and Aldrich 2002) (Figure 10.3G). The HA model has been used to interpret the effects of many modulators and mutations on BK channels, has been adapted to the proton-sensitive BK channel homolog Slo3 (Zhang, Zeng, and Lingle 2006), and has served as a guide for developing semi-mechanistic models of other multimodal voltage- and ligand-gated channels (e.g., HCN, TRPV1). The key mechanistic assumptions of the HA model in terms of voltage-dependent activation are:

(a) *Channels contain four independent and identical VSs whose activation can be described by a two-state process* ($[R\text{-}A]_4$ *in the model).* Two-state activation is consistent with the kinetics and steady-state activation of gating currents (unlike K_V channels). The assumption that VSs can activate independently is consistent with the relationship between the voltage dependence of ionic and gating current but may be considered a simplifying assumption because there is also some evidence for non-independent activation of VSs in the presence of Ca^{2+} (Horrigan and Aldrich 2002; Shelley et al. 2010).

(b) *VS activation and channel opening are distinct processes.* This is obvious in BK channels because the kinetics of gating currents and ionic currents differ by almost two orders of magnitude; the majority of gating charge moves before channels open.

(c) *VSs promote channel opening through an allosteric interaction* (D *in the model).* This is one of the crucial differences between models of BK and Kv channel gating but is firmly supported by several lines of evidence. First, BK channels can activate in a nearly voltage-*in*dependent manner at negative V_m where VSs are in a resting state, but P_o is greatly increased when VSs are activated. Second, VSs can activate whether channels are open or closed according to a variety of evidence, including gating current, "limiting-slope" analysis, delays in activation and deactivation of ionic current (Horrigan and Aldrich 2002), and single-channel analysis indicating that multiple open and closed states can be occupied in 0 Ca^{2+} or saturating Ca^{2+} (Talukder and Aldrich 2000; Rothberg and Magleby 2000; Shelley and Magleby 2008). These observations together support the notion that both voltage- and Ca^{2+}-sensor activation promote but are not required for channel opening.

(d) *Voltage- and Ca^{2+}-sensors interact* (E *in the model).* An allosteric interaction between sensors is evidenced by the ability of Ca^{2+} to shift the voltage dependence of gating charge movement while channels are closed (Horrigan and Aldrich 2002).

10.4.4 Semi-Mechanistic Models Represent a Balanced Approach

The HH model, which started out as a phenomenological model in the mid-20th century, is still in use today. Despite some known shortcomings, the HH model has stood the test of time for multiple reasons. Implementation of the HH model is relatively straightforward, but perhaps more significantly, it has proven largely consistent with the modular structural nature of voltage-gated ion channels, transforming itself from a phenomenological model to a semi-mechanistic model. The Kv channel models of Shaker of Zagotta et al. (ZHA model) and Schoppa and Sigworth were developed from the start to balance statistical parsimony and structural information. A similar developmental approach was employed for the HA model of BK channel gating and the resulting model has been applied to other channel types. The longevity and wide applicability of select models are due in part to their semi-mechanistic approach where the parameters were well constrained by the experimental observations and yet difficult-to-quantify structural information is incorporated.

10.5 Simulations

A gating scheme or a semi-mechanistic model is used to simulate time-dependent state occupancy probabilities, ionic currents, or other functional properties during the model building process (Figure 10.4 and also in later "application" phases [Figure 10.1]) (see also Chapter 8). The most general approach to calculate state occupancy probabilities for any Markov model is the Q matrix method (Colquhoun and Hawkes 1995a), which can also be used to determine open and closed dwell-time distributions. State occupancy probabilities can also be calculated by numerical integration, which is readily implemented with commercially available data analysis/mathematical packages. Macroscopic ionic currents can be simulated based on the occupancy and conductance of different states as well as the number of channels; and gating currents can be simulated in a similar manner based on changes in occupancy and charge movement associated with different state transitions. Single-channel ionic currents can be stochastically simulated using a pseudorandom number generator and taking into account the conductance of the currently occupied state and rate constants for transitions leaving it. The resulting idealized current trace may be processed through an analog or a digital filter and combined with "noise" matching the instrumentation/background noise of the recording system. Such simulated records can then be analyzed in the same way as experimental records as a way to optimize the model parameter values (Figure 10.4). Assuming that the channels work independently of each other, multichannel results, including macroscopic currents, can be simulated by simply summing the results of multiple one-channel trials. For a more detailed description of these procedures, interested readers are referred to Horrigan and Hoshi (2015).

10.6 Parameter Optimization, Validation and Interpretation

As outlined in Figure 10.4, parameter optimization represents an important step. This is evidenced by the observation that those gating models where the parameters are well constrained by direct experimental observations (i.e., models with high degrees of *parameter identifiability*) tend to find wide use. The importance of parameter optimization and constraint cannot be overemphasized. During model development, validation and interpretation is not possible unless parameters can be constrained. "Validation" involves determining whether a model can fit the data, how well it fits and possibly testing predictions that would help distinguish it from alternative models. Discrepancies between the model prediction and data must then be "interpreted" to construct improved models. But if parameters are poorly constrained, then a model is also poorly defined and difficult to test or improve.

10.7 Summaries, Conclusions and Future Developments

We presented three classes of gating models – phenomenological models, statistically parsimonious gating schemes and semi-mechanistic models. Each has its unique strengths and weaknesses, related to the goal of model building. Phenomenological models are relatively easy to develop but typically offer little mechanistic insight. Statistically parsimonious gating schemes provide detailed information about the channels in the data sets analyzed and some schemes may be given limited mechanistic and/or structural connotations. However, protein structural information is difficult to incorporate into statistically parsimonious gating schemes and they frequently suffer from the lack of extensibility or applicability to results other than the original data sets such as new stimulus conditions, mutants, and channel homologs. Semi-mechanistic models incorporate difficult-to-quantify information including structural organizations of the multimeric channel complexes with multiple functional sensors and gates, which may interact in an allosteric manner. Semi-mechanistic models probably offer the best predictive power and extensibility, acting as elegant conceptual frameworks under a variety of conditions.

Even semi-mechanistic models of channel gating contain simplifying assumptions including the Markov postulate that ion channels contain discrete kinetically distinguishable states connected by memoryless rate constants. This premise may appear in contrast with the numerous and short-lived protein conformations observed in MD simulations. However, many of the conformations in the MD simulations are probably functionally indistinguishable. Undoubtedly, the fields of gating model development and structural simulations will converge eventually. In theory, structural simulations based on *physiologically relevant atomic structures* have great promise to provide insight into the conformational changes that underlie gating and the mechanisms governing ion permeability. However, computational power has only recently advanced to the stage that such structural models can start to simulate single-channel behavior near time scales relevant to channel gating (i.e., microseconds to milliseconds) (Jensen et al. 2012). Further, the force-field equations used in MD simulations are continuing to evolve, for example, by incorporating polarizability of atoms. The relative merits of various semi-mechanistic models and MD simulations are currently difficult to compare.

Gating models, especially semi-mechanistic models, will continue to be utilized not only to study ion channel structure–function issues but also to assess functional contributions of different channels to cell signaling. For example, the dynamic clamp method (Prinz, Abbott, and Marder 2004) to investigate functional roles of various ionic currents relies on a real-time interface between computer and cell such that ionic currents are simulated in response to the cell's V_m and injected back into the cell to determine their effect on V_m. Simulations of cell function required for systems physiology and neuroscience will be greatly aided by further development of ion channel gating models.

Acknowledgments

We regret that space constrain forces us to limit our discussion to only a few select gating models. Our apologies extend to numerous authors whose gating models were not discussed.

Suggested Readings

Broomand, A., and F. Elinder. 2008. "Large-scale movement within the voltage-sensor paddle of a potassium channel-support for a helical-screw motion." *Neuron* 59 (5):770–7.

Chung, S. H., J. B. Moore, L. G. Xia, L. S. Premkumar, and P. W. Gage. 1990. "Characterization of single channel currents using digital signal processing techniques based on Hidden Markov Models." *Philos Trans R Soc B: Biol Sci* 329 (1254):265–85.

Colquhoun, D., and A. G. Hawkes. 1987. "A note on correlations in single ion channel records." *Proc R Soc B: Biol Sci* 230 (1258):15–52.

Colquhoun, D, and A. G Hawkes. 1995a. "A Q-matrix cookbook: how to write only one program to calculate the single-channel and macroscopic predictions for any kinetic mechanism." In *Single-channel recording*, edited by B. Sakmann, and E. Neher, 589–633. New York: Plenum.

Colquhoun, D., and A. G. Hawkes. 1995b. "The principles of the stochastic interpretation of ion-channel mechanisms." In *Single-channel recording*, edited by B. Sakmann and E. Neher, 397–482. New York: Plenum.

Colquhoun, D., and R. Lape. 2012. "Perspectives on: conformational coupling in ion channels: allosteric coupling in ligand-gated ion channels." *J Gen Physiol* 140 (6):599–612. doi:10.1085/jgp.201210844.

Colquhoun, D., and F. J. Sigworth. 1995. "Fitting and statistical analysis of single-channel records." In *Single-channel recording*, edited by B. Sakmann and E. Neher, 483–587. New York: Plenum.

Cox, D. H., J. Cui, and R. W. Aldrich. 1997. "Allosteric gating of a large conductance Ca-activated K^+ channel." *J Gen Physiol* 110 (3):257–81.

Hodgkin, A. L., and A. F. Huxley. 1952. "A quantitative description of membrane current and its application to conduction and excitation in nerve." *J Physiol* 117:500–44.

Horn, R. 1987. "Statistical methods for model discrimination: applications to gating kinetics and permeation of the acetylcholine receptor channel." *Biophys J* 51 (2):255–63. doi:10.1016/S0006-3495(87)83331-3.

Horn, R., and K. Lange. 1983. "Estimating kinetic constants from single channel data." *Biophys J* 43 (2):207–23. doi:10.1016/S0006-3495(83)84341-0.

Horrigan, F. T., and R. W. Aldrich. 2002. "Coupling between voltage sensor activation, Ca^{2+} binding and channel opening in large conductance (BK) potassium channels." *J Gen Physiol* 120 (3):267–305.

Horrigan, F. T., J. Cui, and R. W. Aldrich. 1999. "Allosteric voltage gating of potassium channels I. mSlo ionic currents in the absence of Ca^{2+}." *J Gen Physiol* 114 (2):277–304.

Horrigan, F. T., and T. Hoshi. 2015. "Models of ion channel gating." In *Handbook of ion channels*, edited by J. Zheng, and M. C. Trudeau, 83–101. Boca Raton, FL: CRC Press, Taylor & Francis Group, LLC.

Jensen, M. Ø., V. Jogini, D. W. Borhani, A. E. Leffler, R. O. Dror, and D. E. Shaw. 2012. "Mechanism of voltage gating in potassium channels." *Science* 336 (6078):229–33. doi:10.1126/science.1216533.

Klein, S., J. Timmer, and J. Honerkamp. 1997. "Analysis of multichannel patch clamp recordings by hidden Markov models." *Biometrics* 53 (3):870–84.

Koshland, D. E., Jr., G. Nemethy, and D. Filmer. 1966. "Comparison of experimental binding data and theoretical models in proteins containing subunits." *Biochemistry* 5 (1):365–85.

Ledwell, J. L., and R. W. Aldrich. 1999. "Mutations in the S4 region isolate the final voltage-dependent cooperative step in potassium channel activation." *J Gen Physiol* 113 (3):389–414.

Long, S. B., X. Tao, E. B. Campbell, and R. MacKinnon. 2007. "Atomic structure of a voltage-dependent K^+ channel in a lipid membrane-like environment." *Nature* 450 (7168):376–82.

Ma, Z., X. J. Lou, and F. T. Horrigan. 2006. "Role of charged residues in the S1-S4 voltage sensor of BK channels." *J Gen Physiol* 127 (3):309–28.

Magleby, K. L., and L. Song. 1992. "Dependency plots suggest the kinetic structure of ion channels." *Proc Biol Sci* 249 (1325):133–42. doi:10.1098/rspb.1992.0095.

Magleby, K. L., and D. S. Weiss. 1990. "Estimating kinetic parameters for single channels with simulation: a general method that resolves the missed event problem and accounts for noise." *Biophys J* 58 (6):1411–26.

Mannuzzu, L. M., M. M. Moronne, and E. Y. Isacoff. 1996. "Direct physical measure of conformational rearrangement underlying potassium channel gating." *Science* 271 (5246):213–16.

McManus, O. B., and K. L. Magleby. 1991. "Accounting for the Ca^{2+}-dependent kinetics of single large-conductance Ca^{2+}-activated K^+ channels in rat skeletal muscle." *J Physiol* 443:739–77.

Milescu, L. S., G. Akk, and F. Sachs. 2005. "Maximum likelihood estimation of ion channel kinetics from macroscopic currents." *Biophys J* 88 (4):2494–515. doi:10.1529/biophysj.104.053256.

Monod, J., J. Wyman, and J.-P. Changeux. 1965. "On the nature of allosteric transitions: a plausible model." *J Mol Biol* 12:88–118.

Prinz, A. A., L. F. Abbott, and E. Marder. 2004. "The dynamic clamp comes of age." *Trends Neurosci* 27 (4):218–24. doi:10.1016/j.tins.2004.02.004.

Rothberg, B. S., and K. L. Magleby. 2000. "Voltage and Ca^{2+} activation of single large-conductance Ca^{2+}-activated K^+ channels described by a two-tiered allosteric gating mechanism." *J Gen Physiol* 116 (1):75–99.

Sakmann, B., and E. Neher. 1995. *Single-channel recording*. 2nd ed. New York: Plenum.

Schoppa, N. E., and F. J. Sigworth. 1998. "Activation of Shaker potassium channels III - An activation gating model for wild-type and V2 mutant channels." *J Gen Physiol* 111 (2):313–42.

Shelley, C., and K. L. Magleby. 2008. "Linking exponential components to kinetic states in Markov models for single-channel gating." *J Gen Physiol* 132 (2):295–312. doi:10.1085/jgp.200810008.

Shelley, C., X. Niu, Y. Geng, and K. L. Magleby. 2010. "Coupling and cooperativity in voltage activation of a limited-state BK channel gating in saturating Ca^{2+}." *J Gen Physiol* 135 (5):461–80. doi:10.1085/jgp.200910331.

Talukder, G., and R. W. Aldrich. 2000. "Complex voltage-dependent behavior of single unliganded calcium-sensitive potassium channels." *Biophys J* 78 (2):761–72.

Zagotta, W. N., T. Hoshi, and R. W. Aldrich. 1994. "Shaker potassium channel gating. III: evaluation of kinetic models for activation." *J Gen Physiol* 103 (2):321–62.

Zhang, X., X. Zeng, and C. J. Lingle. 2006. "Slo3 K^+ channels: voltage and pH dependence of macroscopic currents." *J Gen Physiol* 128 (3):317–36.

11

Investigating Ion Channel Structure and Dynamics Using Fluorescence Spectroscopy

Rikard Blunck

CONTENTS

11.1 Introduction

The function of ion channels, as electrogenic transport proteins, can be directly measured with electrophysiological techniques (Chapters 7–9). In this chapter, we will discuss how fluorescence techniques link the dynamics of the functional electrophysiological data to high-resolution structures (Chapters 12–13) and how to obtain structural information in a native environment to compare it directly to high-resolution structures or use it as restraints for structural (Chapters 14–15) and kinetic (Chapter 10) modeling.

Historically, the first use of fluorescence to study ion channel structure was to monitor the intrinsic fluorescence of tryptophan and tyrosine. By exploiting intrinsic fluorescence, no additional labeling technique was required, making these experiments possible before the cloning of ion channels. Using synthetic bungarotoxin and tetrodotoxin derivatives, it became possible to move from intrinsic fluorescence to fluorescently labeling specific

DOI: 10.1201/9781003096214-13

positions in the channels. Fluorescence recordings of channel activity picked up significantly with the development of fluorescent calcium indicators in the early 1980s. But it was not until the cloning and recombinant expression of ion channels in the late 1980s to early 1990s that the molecular biology tools as well as the structural information became available to specifically target important regions in the channels in order to follow their conformational changes. Accordingly, the first voltage-clamp fluorometry measurements, simultaneous measurements of site-directed fluorescence and electrophysiology, followed after just a few years and were later extended to a wide variety of channels and other transport proteins. Single-channel tracking in mammalian cells became possible by fusion with green fluorescent protein or one of its derivatives, and investigation of ion channels by fluorescence spectroscopy was further improved by higher sensitivity down to the single-molecule level and different labeling techniques.

11.2 Modulation of Fluorescence

How information on conformational changes of proteins is detected by the fluorophores can be derived from the general concept of fluorescence. Each fluorophore has characteristic excitation and emission spectra, which give the relative probability of absorbing and emitting a photon of a specific wavelength, respectively. When a fluorophore is excited, it absorbs a photon with the correct energy to lift an electron to a higher electronic state (typically from S0 to S1; Figure 11.1a). The excitation occurs into a higher vibrational level; the vibrational energy is rapidly dissipated such that the electron will assume the S1 ground state. From the S1 ground state, the electron may return to the S0 state by emitting a photon corresponding to the energy difference (fluorescence, f). This transition again occurs to the higher vibrational states of S0 (Figure 11.1b) such that, during excitation, the vibrational energy is added to the energy difference between S0 and S1 ground states, whereas it is subtracted from it during emission. Consequently, emission is shifted to wavelengths longer than excitation (Stokes shift). The continuous excitation and emission spectra result from superposing the transitions to each vibrational state (Figure 11.1b).

Not every photon is absorbed by the fluorophore, and not every excited state results in emission of a photon. The absolute probability to absorb a photon from the ground state and the probability to emit a photon from the excited state are given by the *extinction coefficient* (ε)* and the *quantum yield* (QY), respectively. When no photon is emitted during return of the electron to its ground state, the energy may be dissipated into the molecule or surrounding (nonradiative decay, k_{NR}). If one considers the excitation/emission cycle in a kinetic model (Figure 11.1d), then the number of photons emitted per excitation cycle, the quantum yield (QY), is given by the rate constant for the emission of a photon (f) normalized to the sum of all pathways to the ground state:

$$QY = \frac{f}{f + k_{NR}} \tag{11.1}$$

* More precisely, the extinction coefficient is the probability normalized to the concentration and pathway. The actual probability is given by the absorption cross section σ, which is related to the extinction coefficient by $\sigma = 1000 \cdot \ln(10) \cdot \varepsilon / N_A$.

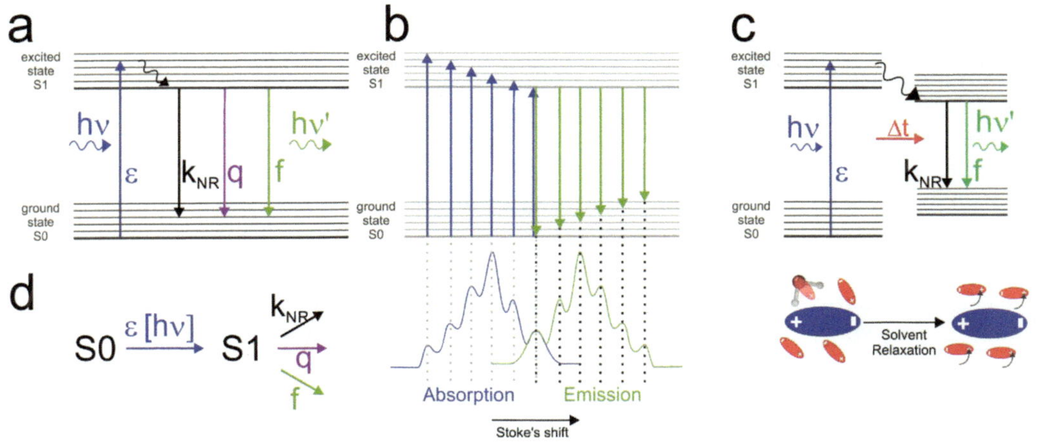

FIGURE 11.1
(a) Jablonski diagram of typical fluorescence process. After excitation of a photon, the excited electron can return to the ground state either by emitting a photon (fluorescence, f), by nonradiative decay (k_{NR}) or by quenching (q) if a quencher is present. (ε, extinction coefficient). (b) Emission and excitation spectra result from excitations and emission from the ground vibrational to higher vibrational states. (c) At the moment of excitation, the dipoles of the solvent are aligned to the S0 dipole moment. This leads to a higher energy of the S1 state, whose dipole moment is generally different. After the dipoles of the solvent realigned to the S1 state (solvent relaxation), the energy of S1 is reduced, whereas the energy of S0 is increased, as the dipoles are now misaligned to S0. As a consequence, the emission spectrum is shifted to longer wavelengths. (d) Kinetic model of the fluorescence process.

where k_{NR} is the rate constant for nonradiative decay. The *lifetime* of the fluorophore denotes the time constant τ of the exponential decay related to leaving the excited state and equals the inverse of the sum of the rate constants leaving the excited state:

$$\tau = \frac{1}{f + k_{NR}}. \tag{11.2}$$

The fluorophore interacts with and emits electromagnetic radiation due to its dipole moment. The delay of the alignment of the solvent's dipoles (e.g., H_2O) to the fluorophore dipole moment changing during excitation (solvent relaxation) lowers the energy of the excited state, but at the same time increases the energy of the S0 ground state since the dipoles are now aligned to S1 (Figure 11.1c). The lower energy gap between S1 and S0 leads to a red-shift of the emission spectrum. Hence, if a fluorophore is moved from a hydrophilic to a hydrophobic environment, e.g., during protein movement, emission will be shifted to the blue.

More commonly, fluorescence is modulated by *quenching*. During quenching, the energy of the excited state is dissipated through interaction with a different molecule, similar to nonradiative decay mentioned earlier. Presence of a quencher offers an additional, very fast pathway back to the ground state (Figure 11.1a). If we assume that the rate constant for quenching is given by q (Figure 11.1d), the QY changes to

$$QY = \frac{f}{f + k_{NR} + q} \tag{11.3}$$

and the lifetime τ changes proportionally to

$$\tau = \frac{1}{f + k_{NR} + q}. \tag{11.4}$$

In proteins, one has to consider two different types of quenchers: those that are present in the surrounding solvent and those that are attached to the surrounding protein. The first group includes oxygen in the aqueous solution, or any quenchers intentionally added to the solution. The second group of quenchers contains residues of the surrounding protein. In particular the aromatic residues are well suited to quench fluorescence upon contact; of all natural amino acids, tryptophan is the strongest quencher followed by tyrosine. Phenylalanine, despite featuring an aromatic ring, is much less effective as a quencher. For instance, tetramethylthodamine (TMR) is quenched by 57%, 34% and 11% in the presence of 30 mM tryptophan, tyrosine and phenylalanine, respectively. Other amino acids such as methionine, histidine, glutamate or aspartate also quench fluorescence, albeit at much higher concentrations. The quenching efficiency depends also on the nature of the fluorophore. Concentrations of course are difficult to define when both fluorophore and quencher are attached to a protein and thereby restrained in their movement. For quenching to occur, the van der Waals radii of fluorophore and amino acid have to overlap, be it by collision (concentration) or by constriction to a small volume in a protein. While the fluorescence of the dye Cy5 is reduced by only 3% and 9% in the presence of 6 and 30 mM tryptophan, respectively, the quenching increases to 41% when Cy5 and tryptophane are attached to a small peptide, suggesting that the "local" or "effective" concentration is higher than 30 mM (Marme et al. 2003).

11.2.1 Förster Resonance Energy Transfer

Förster resonance energy transfer (FRET) is a dipole–dipole interaction that allows the transfer of an excited state from a donor fluorophore to the acceptor one. One may picture the donor as an emission antenna and the acceptor as a receiver antenna. If the dipole field of the donor reaches the acceptor, the donor induces the acceptor to oscillate and the acceptor may be excited, as if the oscillation was induced by a photon. The excitation of the acceptor leads in turn to a de-excitation of the donor, effectively transferring the excited state from donor to acceptor (Figure 11.2a). For this to occur, the emission spectrum of the donor fluorophore must overlap with the excitation spectrum of the acceptor fluorophore (Figure 11.2b).

The energy transfer efficiency (*ET*) is the fraction by which the quantum yield of the donor is decreased by the presence of the acceptor. As in the case of quenching discussed earlier, presence of an acceptor also leads to a new pathway to leave the excited state with

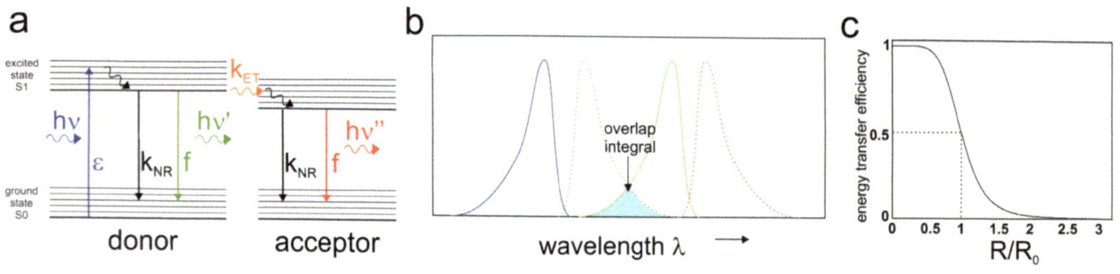

FIGURE 11.2
(a) Jablonski diagram for Förster resonance energy transfer (FRET). (b) In order for FRET to occur, transition energy differences have to coincide, i.e. the emission spectrum of the donor has to overlap with the excitation spectrum of the acceptor (donor, black; acceptor, gray; excitation spectrum, solid line; emission spectrum, dashed line). (c) Distance dependence of the energy transfer efficiency (*ET*) as a function of distance between the donor and acceptor normalized to the R_0.

the rate constant k_{ET}, which may be determined from the properties of donor and acceptor pair and their environment:

$$k_{ET} = \frac{161.9 \cdot f \cdot \kappa^2}{\pi^5 N_A n^4} \cdot \frac{1}{R^6} \int_0^\infty Em_D(\lambda) \cdot \varepsilon_A(\lambda) \cdot \lambda^4 d\lambda, \tag{11.5}$$

where R is the distance between donor and acceptor, Em_D is the normalized emission spectrum of the donor, ε_A the absorption spectrum of the acceptor, λ the wavelength, N_A the Avogadro constant, n the refractive index of the surrounding, and κ an orientation factor between donor and acceptor dipoles. The orientation factor κ is required as energy transfer is dependent on the arrangement of the two dipoles relative to one another. Full energy transfer efficiency is only possible if donor and acceptor are aligned in parallel, whereas a perpendicular orientation prevents energy transfer.

With the rate constant k_{ET}, the energy transfer efficiency becomes

$$ET = \frac{k_{ET}}{f + k_{NR} + k_{ET}} = 1 - \frac{f + k_{NR}}{f + k_{NR} + k_{ET}} = 1 - \frac{QY_{DA}}{QY_{DO}} \tag{11.6}$$

where QY_{DA} and QY_{DO} (Equation 11.1) are the quantum yields of the donor in the presence and absence of the acceptor, respectively:

$$QY_{DA} = \frac{f}{f + k_{NR} + k_{ET}}. \tag{11.7}$$

Accordingly, also the fluorescence lifetime of the donor decreased:

$$\tau_{DA} = \frac{1}{f + k_{NR} + k_{ET}}. \tag{11.8}$$

which simply reflects that the additional pathway to leave the excited state lowers the dwell time therein.

If k_{ET} equals the sum of the other rate constants ($f + k_{NR}$), then 50% of the energy is transferred to the acceptor ($k_{ET,0}$). With $k_{ET} \sim R^{-6}$, Equation 11.6 can be written as

$$ET = \frac{1}{1 + \left(\frac{R}{R_0}\right)^6} \tag{11.9}$$

with R_0 the distance between the donor and acceptor, where 50% of the energy is transferred to the acceptor. It can be calculated from Equation 11.5:

$$R_0 = \left[\frac{161.9 \cdot QY_D \cdot \kappa^2}{\pi^5 N_A n^4} \int_0^\infty Em_D(\lambda) \cdot \varepsilon_A(\lambda) \cdot \lambda^4 d\lambda \right]^{1/6} \tag{11.10}$$

R_0 typically ranges between 25 and 60 Å. Normalized to R_0, the distance dependence of ET is shown in Figure 11.2c.

11.3 Labeling Techniques

An important issue for the study of ion channels using fluorescence is the appropriate labeling technique. One has to consider specificity, color (spectra) and brightness as well as orientation of the dipole moment (FRET) and orthogonality with other labeling whenever two compounds (e.g., donor and acceptor) are required.

11.3.1 Thiol-Reactive Chemistry

A commonly used technique is the reaction of the fluorophores with the sulfhydryl groups of cysteines. The fluorophores (or other compounds) are covalently bound to a thiol-reactive linker. These include maleimide, methanethiosulfonate and iodoacetamide linkers (Figure 11.3a). Most fluorophores are commercially available with at least one of these linkers.

Iodoacetamide and maleimide covalently bind to the thiol group of a cysteine, whereas the disulfide bridge formed by the MTS linker can be separated by a reducing agent, e.g., dithiothreitol (DTT). This is frequently used when modifying with chemical compound to vary properties like size or charge of a residue. To improve labeling efficiency, cysteines should be reduced prior to labeling using dithiothreitol or tris(2-carboxyethyl)phosphine (TCEP).

11.3.2 Fluorescently Labeled Ligands or Toxins

Toxins and ligands have the immense advantage that they are highly specific for the ion channel in question with typical affinity constants in the range of 1 to 300 nM for toxins and in the micromolar range for ligands. Therefore, very low concentrations of toxin or ligand allow in situ labeling. The labeled ligands or toxins can act as the donor or acceptor for energy transfer measurements or act directly as the optical readout of ligand binding.

11.3.3 Genetically Encoded Fluorescent Labels

11.3.3.1 Fluorescent Proteins

The most commonly used genetically encoded fluorophores are the fluorescent proteins (FPs). Starting from the green fluorescent protein (GFP), a large number of derivatives are known today with a variety of spectral and chemical properties. FPs have been developed in all spectral colors and are utilized as chemical sensors, e.g., for pH, or – in the form of photoswitchable FPs – for superresolution imaging. Splitting GFPs into two components or in FRET pairs allows them to be triggered by certain stimuli, e.g., as membrane potential sensors. For the investigation of structural aspects of ion channels, FPs have been used to track conformational changes, but for this scenario their use is limited to areas of the protein where there is no steric hindrance by the surrounding protein due to their large size.

11.3.3.2 Ligand-Binding Domains

For some variations of the FRET systems, transition metals (Ni^{2+}, Cu^{2+} and Co^{2+}) or luminescent lanthanides (Tb^{3+}, Eu^{3+}) can act as acceptors and donors, respectively. The transition metals coordinate between two or more cysteines or histidines. By introducing double *cys* or *his*, for instance at positions n and n+4 in an α-helix, the transition-metal ions will

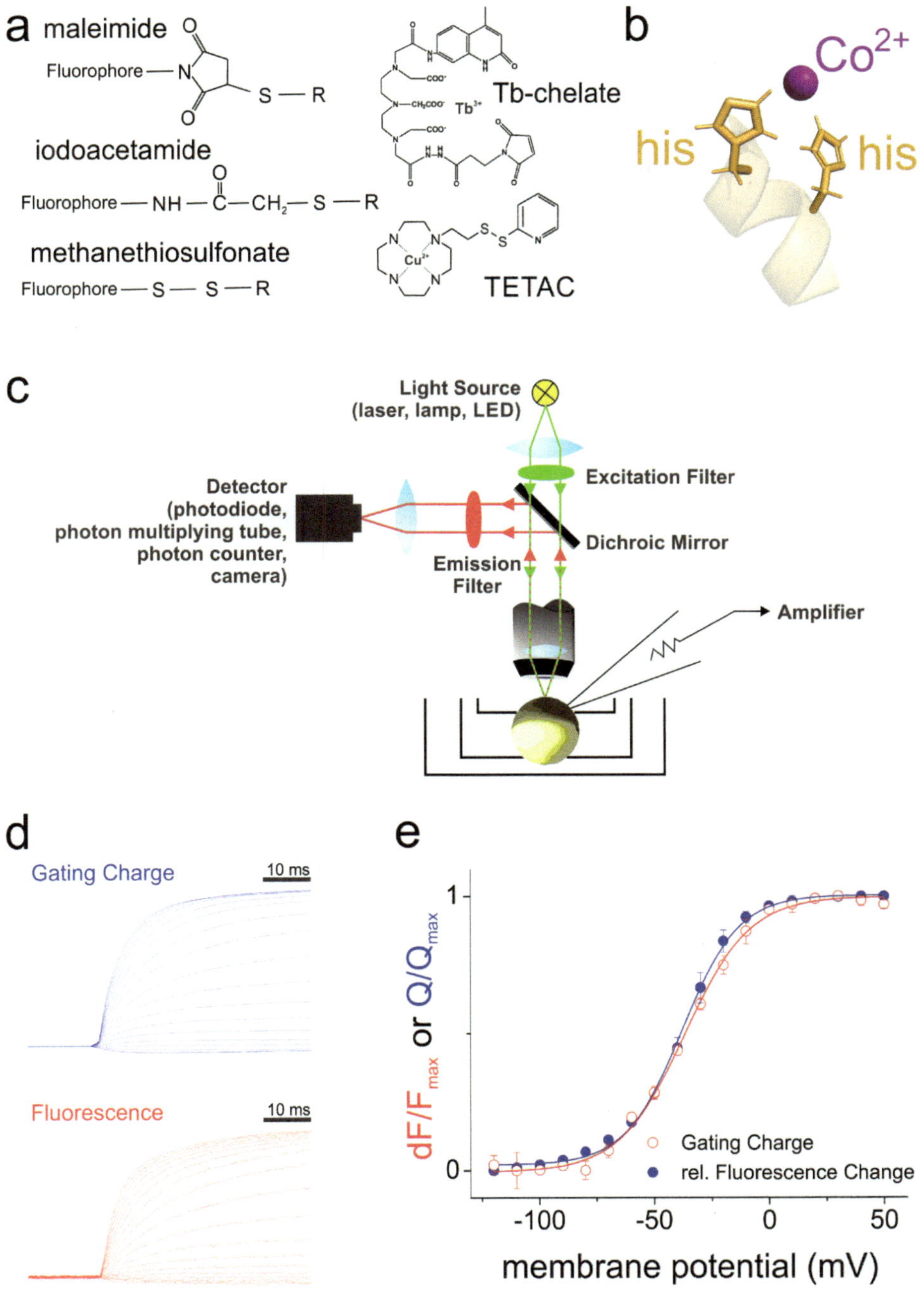

FIGURE 11.3
(a) Chemical structures of a fluorophore bound to the thiol group of a cysteine (R–S) via maleimide (top), iodoacetamide (center) and methanethiosulfonate (bottom). (Right) Tb-chelate and TETAC for chelation of Tb^{3+} and Cu^{2+}. (b) Coordination of Co^{2+} by a di-histidine in an α-helix. (c) Principal voltage-clamp fluorometry setup in cut-open oocyte voltage-clamp configuration. As light source, a low-noise lamp, laser or LED can be utilized. The emission is detected by a photodiode, a photomultiplying tube or a camera dependent on the configuration. For clarity, not all electrodes are shown. (d–e) Time courses (d) and voltage relations of gating charge and fluorescence response (e) of Shaker H4IR-W434F-A359C to depolarizing pulses from a holding potential of –90 mV. The channel was expressed in *Xenopus* oocytes and recordings performed in the cut-open oocyte voltage-clamp configuration. The fluorophore is linked to position A359C at the N-terminal end of the S4 and reports movements of the S4 during gating.

bind reversibly with high affinity to this site, requiring a sufficiently high concentration during the experiment (Figure 11.3b). Alternatively, Cu^{2+} can be chelated by thiol-reactive 1-(2-pyridin-2- yldisulfanyl)ethyl)-1,4,7,10-tetraazacyclododecane (TETAC).

A similar principle is used in the lanthanide-binding tags. Lanthanides can be chemically bound to the protein by a chelating compound via a thiol-reactive linker (Figure 11.3a). Genetically encoded, a high-affinity lanthanide-binding tag similar to an EF-hand motif can be formed by a short 17aa motif introduced into the protein. These ligand-binding tags are, however, large in comparison to transition-metal bridges or cysteine replacements so that restrictions apply as to where in the protein they can be introduced.

Both methods, the transition metals and ligand-binding tags, have the advantage that the acceptor and donor, respectively, are linked to the protein with a different chemistry than the FRET partner. Thus, the donor and acceptor can be more specifically directed to sites even within the same subunit without the need of stoichiometric labeling.

11.3.3.3 Fluorescent Unnatural Amino Acids

The most versatile genetically encoded fluorophores are unnatural or noncanonical amino acids (fUAAs), small genetically encoded fluorescent tags substituting the endogenous amino acid. As genetically encoded fluorescent tags, they have the immense advantage that they can be directed to any position in the protein and do not produce unspecific labeling. Due to the fUAAs' small size – just slightly larger than a tryptophan – they can be inserted almost anywhere in the protein. In addition, the site does not have to be accessible because they are incorporated during synthesis and therefore do not require posttranslational modification. Despite their small size, some fUAAs still have photochemical properties similar to organic fluorophores from which they are derived. To incorporate unnatural amino acids (UAAs) into the protein of interest, a nonsense codon is introduced at the position of interest and the corresponding tRNA–UAA complex provided to be inserted into the position in question. There are two principal manners of generating the aminoacylated tRNAs, pioneered by Peter G. Schultz (Scripps Research Institute, San Diego, California). The first way is to synthesize directly the aminoacylated tRNA and coexpress it together with the protein RNA. While expression levels achieved are typically lower with this method, it does not restrict the size of the UAA, allowing typical organic dyes, such as bodipy, Cy5 and coumarin derivates, to be incorporated. In the second approach, the tRNA-UAA complex is assembled in the expression system by coexpression with an orthogonal tRNA/tRNA–synthetase pair. The experimenter has to only provide the synthetic UAA to the cells. While the use of aminoacylated tRNA allows more flexibility in the chemical structure of the UAA, using an orthogonal pair of tRNA/tRNA–synthetase typically yields higher expression levels. With this method, Anap has been introduced and has found widespread application also in FRET systems. Anap is about the size of a tryptophane but does not possess the same brightness as organic dyes.

11.4 Obtaining Structural and Dynamic Information from Fluorescence Measurements

In this section, we will discuss how modulations in the fluorescence properties of the specifically attached fluorophores or fluorophore pairs are applied to obtain structural and dynamic information on membrane proteins.

11.4.1 Kinetics of Local Structural Rearrangements

One of the most powerful uses of fluorescence is the detection of the kinetics of local structural rearrangements. This is particularly useful when the function of the channel is simultaneously monitored using electrophysiology, a technique called voltage-clamp fluorometry (VCF). In VCF, the fluorescence changes are measured simultaneously with the electrophysiological recordings so that structural and functional data can be directly correlated to one another. This is achieved by mounting the electrophysiological setup on a fluorescence microscope (Figure 11.3c). VCF has been used in combination with any electrophysiological method. In order to monitor local conformational rearrangements, the channel is site-directedly labeled with a small fluorophore such that the properties of the fluorophore are determined by its immediate environment. When during stimulation of the channel, the tagged part of the channel undergoes a conformational change, the fluorescence will likely be modulated as explained earlier. The fluorophore either becomes accessible to the surrounding solution, which would lead to a red-shift of the emission spectrum, or it becomes quenched by residues of the surrounding protein. The first effect can be used to probe for water accessibility of regions of the channel proteins; a fluorophore is attached to various positions and the center wavelength of the emission spectrum determined. Water-accessible residues will show red-shifted emission spectra due to solvent relaxation. In case of (un-) quenching, the fluorescence intensity is altered.

The fluorescence signal resulting from movement of the protein – spectral shift and quenching – will be an exponential or multi-exponential intensity change, whose kinetics and amplitude will depend on the strength of the stimulus (e.g., voltage or ligand concentration; Figure 11.3b). It is important to realize that the time course of the fluorescence change reflects the kinetics of the transition that is monitored by the fluorescence and does not directly follow the conformational change of the protein itself. In other words, you know when the protein moves, not how fast.

Let us assume that the observed transition can be described as a two-state process A ↔ B. The fluorophore will be in a different environment in each of the two positions and accordingly fluoresce with different intensities F_A and F_B. The total fluorescence intensity observed F_{total} will be

$$F_{total} = Prob(A)F_A + Prob(B)F_B = F_A + Prob(B)\Delta\ F \qquad (11.11)$$

where A and B are the occupancies of both states with $Prob(A) + Prob(B) = 1$ and $\Delta\ F = F_B - F_A$.

The fluorescence signal reports local rearrangements. These local rearrangements can be mapped onto the channel by attaching the label at different positions throughout the protein. Their kinetics can also be correlated with the functional data such as ionic current or gating currents (Figure 11.3d). For example, if the probe is attached to the top of the S4 in voltage-gated ion channels, the fluorescence signal will follow the gating charge equilibrium (Figure 11.3e). Best practice is to always compare kinetics within the same mutant, for instance, fluorescence change versus current development. It is also possible to attach two fluorophores with different labeling techniques and obtain two conformational changes simultaneously.

Indirect estimations on the movement of the protein may be made based on the fact that quenching requires fluorophore and quencher to collide. If a quenching signal is present after labeling, a tryptophane – as the strongest quencher – in its proximity in the 3D structure can be removed to test whether the signal disappears. If no signal is present, a tryptophane can be introduced to provoke quenching. To determine exact distance changes, the fluorophore can be linked via C-linkers of various length, providing a lower limit for the distance.

11.4.2 Intra- and Intermolecular Distance Measurements

The best way to determine distances is FRET due to its strong distance dependence. The distances within a protein complex are typical in the range up to 70 Å and intramolecular movements in the range up to 15 Å. For these distances, FRET has proven very efficient. FRET is most efficient in the donor–acceptor distance range between $0.5R_0$ and $1.5R_0$. Beyond that interval, the distance dependence flattens (Figure 11.2c). Since the distance dependence of the FRET efficiency is normalized to R_0, the smallest measurable distance changes are proportional to R_0 – the higher R_0, the lower the resolution. In other words, a small distance change cannot be detected with a long R_0. Typical donor–acceptor pairs are listed in Table 11.1.

Technically the easiest way to extract a distance from FRET measurements are the so-called three-cube method and the spectral FRET method. In the three-cube method, emission in the donor range upon donor excitation and emission in the acceptor range upon both donor and acceptor excitation are determined. In the spectral FRET method, the emission spectrum is recorded both upon donor and acceptor excitation. From these measurements, the apparent FRET efficiency can be determined from the ratio *FR* of sensitized emission (number of acceptors excited by FRET) to the direct acceptor excitation (number of acceptors present).* Before that the sensitized emission needs to be corrected for donor fluorescence ("leak") detected in the sensitized emission and acceptor fluorescence directly excited by the light source. The required correction factors can be predetermined for the setup and filters used. A detailed derivation can be found in Erickson et al. (2001).

The limitation of the intensity FRET is that only an apparent energy transfer can be determined – with the exception of single-molecule FRET (Figure 11.5) – as seldom all donors are paired exactly with one acceptor. In the three-cube method and the acceptor photobleaching method, the number of acceptors and donors, respectively, may be overestimated since both a lower FRET efficiency (longer distance) and more unpaired

TABLE 11.1

Specific Donor–Acceptor Pairs Most Suited for Specific Case Scenarios

Scenario	Name	Donor	Acceptor	R_0 (Å)	Type
Distance measurement	LRET	Tb-chelate	TMR	60	Lifetime
			Alexa-488	43	
			Lucifer-Yellow	23	
			ABD	15	
Small distances	tmFRET	Anap	Co^{2+}	12	Intensity
Protein assembly	BRET	RLUC[a]	GFP	75	Intensity
			eYFP	45	
Depth in membrane	FRET	TMR	DPA[b]	35	Intensity
		Fluorescein		45	

[a] RLUC, *Renilla* luciferase. As RLUC is bioluminescent, i.e., emits photons in response to a chemical reaction in the absence of excitation light, less background is produced.[b] DPA, dipicrylamine. This amphiphatic negatively charged absorber acts as acceptor, although it is not fluorescent itself. It distributes between inner and outer membrane leaflet as a function of membrane potential.

* To be exact, the ratio would have to be corrected by the ratio of the extinction coefficients of the donor and acceptor, which determines the number of excited states in sensitized emission and direct excitation assuming equal excitation intensities.

donors and acceptors would lead to a lower apparent FRET efficiency. If, for instance, any unpaired donors were present, Equation 11.6 would transform to

$$ET = \frac{1}{1-a}\left(1 - \frac{QY_{DA}}{QY_{DO}}\right) \tag{11.12}$$

where a is the fraction of unpaired donors. If the distribution among the populations is unknown or even the subject of investigation, distances can no longer be derived from analyzing the fluorescence intensities. However, the populations or conformations may be separated by determining the *fluorescence lifetimes* of the donor and acceptor. The lifetime of a fluorophore is the time constant τ of the fluorescence decay given by Equation 11.2. It can experimentally be determined by exciting the fluorophore with a light pulse and observe the subsequent exponential decay. A mixture of two lifetimes shows as two distinct time constants in the fluorescence decay, providing the two correct energy transfer efficiencies (Figure 11.4a).

Assuming several populations of donors i – each with a different distance to the acceptor – coexist, they will have different transfer rates k_{ETi}. According to Equation 11.8, each will also have a different time constant in the presence of the acceptor τ_{DAi} of

$$\tau_{DAi} = \frac{1}{f + k_{NR} + k_{ETi}}. \tag{11.13}$$

Upon an excitation pulse, the fluorescence decay will be a superposition of all intensities

$$I_{DA}(t) = \sum_i I_{DAi}(t) = \sum_i I_{DAi}(0) \cdot \exp(-t / \tau_{DAi}). \tag{11.14}$$

Combining Equations 11.1 and 11.2 gives $QY = f\tau$ so that the energy transfer efficiency becomes

$$ET = 1 - \frac{\tau_{DA,i}}{\tau_{DO}}. \tag{11.15}$$

and

$$R_i = R_0 \cdot \left(\frac{\tau_{DO}}{\tau_{DA,i}} - 1\right)^{-1/6}. \tag{11.16}$$

Thus, we can calculate the FRET efficiency and the distances directly from the experimentally obtained lifetimes. This has the additional advantage that lifetimes are independent of concentration and can be compared between different preparations. Please note that a donor with two acceptors in different distances will still decay with a single but distinct time constant.

11.4.2.1 Linking Distances to Structures and Models

When following movements in a protein, one has to consider their geometry since only the distance between two points can be determined by FRET. Homomers are typically arranged symmetrically, and because of this symmetry some movements are not detected by FRET. For homotetrameric ion channels, for instance, identical positions are generally

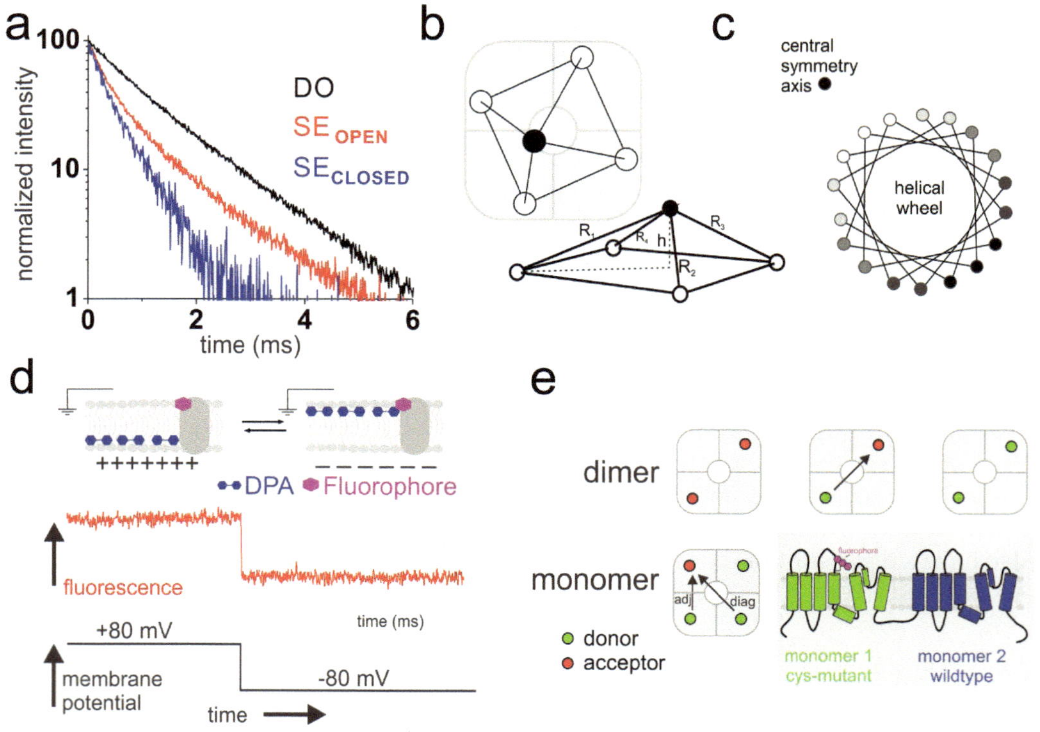

FIGURE 11.4
(a) Example of LRET lifetime measurements using terbium chelate. Shown are the luminescent decays after excitation with a pulse at 337 nm for donor only (DO) and sensitized emission in the closed and open state. Shorter distance leads to higher LRET efficiency and faster decay times. (Reproduced from Faure et al. 2012.) (b) For trilateration, donors are positioned at symmetric positions in the protein with a donor positioned off the central symmetry axis. From the four distances, the position in space of the donors can be reconstructed. (c) The position of a helix in space can be reconstructed from the distance difference of the residues that are located closer (lighter) or farther (darker) relative to the central symmetry axis. (d) The FRET acceptor is added to the membrane while the donor is linked to the protein of interest. At positive membrane potential, the negatively charged acceptor (dipicrylamine, DPA) is located at the inner leaflet, leading to high fluorescence. Upon reversal of the potential, the acceptor translocates to the outer leaflet, resulting in lower fluorescence due to energy transfer. (e) Tetramers formed as dimers of dimers arrange with the identical subunits across the pore so that only dimers with each one acceptor and one donor will result in FRET signals.

arranged in a square (Figure 11.4b). Therefore, any movement vertical to the membrane surface or circular around the central pore causes no distance changes when identical positions are labeled. To make movements in these directions visible, one must choose a different fixpoint in the channel, for instance, a labeled toxin or the membrane (Figure 11.4d). In general, one chooses the distance vector (i.e., the line between donor and acceptor) as parallel as possible to the expected movement and avoid a situation where both donor and acceptor move synchronous parallel to this direction since this would cancel out the signal.

In homooligomers, one also has to consider the labeling stoichiometry. Donors transferring to more than one acceptor will decay with a different time constant. Donors can also be placed adjacent or diagonal to its acceptor (Figure 11.4e). To prevent complex analysis, concatemers can be constructed that result in a single distance (Figure 11.4e).

In an era where research increasingly relies on atomistic models of protein structures, the question arises how the FRET distances may be linked to structures or models thereof. If a structure is available, the obtained distances can directly be compared with the structure to verify its validity in a more native environment. The power of the fluorescence measurements, however, lies in the possibility to obtain dynamic data. For the distances, this means that changes in distances in different states of the channel can be obtained. These distances can then be used as harmonic constraints in molecular dynamics simulations to "pull" a known structure into a new state (see Chapter 15). Similarly, homology models based on a known structure of a related channel can be validated and refined using FRET distances.

But the distances can also directly provide structural information in the absence of a structure by triangulation using the symmetry of homomers. For triangulation, the symmetric positions in a homomer are labeled with the donor, and a single acceptor is positioned off-center at a known position in the protein (Figure 11.4b).

Even if only distances relative to the central axis are determined, they can be used to recreate the movement of a known secondary structure. The position of any secondary structure (e.g., an α-helix) in space can be reconstructed from a minimum of three distances (more is better). From this, movements of rigid motifs during transition from one state to another can be predicted (Figure 11.4c).

11.4.2.2 Lanthanide-Based RET and Transition-Metal FRET

In the calculation of FRET distances, the orientation between donor and acceptor molecules enters in Equation 11.10 as the orientation factor κ. However, fluorophore movement might be impaired when fluorophores are attached to a protein. Any restriction can be measured in the fluorescence anisotropy. In the case of LRET (lanthanide-based RET) and tmFRET (transition-metal FRET), an ion with appropriate photophysical properties replaces the donor and acceptor, respectively. Since ions do not show a constant dipole moment, both orientation factor $\kappa (= 2/3)$ and the distances can be calculated exactly.

In LRET, the lanthanides terbium or europium act as donor. They absorb light in the near-ultraviolet range and emit luminescence in the visible wavelength range. As single atoms, the emission occurs in distinct lines (Figure 11.5c). The terbium ion is bound to the protein via a chelate with a reactive linker (polyaminocarboxylate-carbostyril-maleimide; Figure 11.5a) or a genetically encoded chelator. Terbium luminesces with a lifetime in the range of 1–2 ms (Figure 11.5b); this long lifetime allows using standard recording equipment, and, more importantly, it allows easily separating donor and acceptor lifetimes.

tmFRET also uses an ion but as acceptor. This has the advantage that the acceptor itself does not need to be fluorescent but merely to absorb in the range of donor emission. Typical transition metals used for this method are cobalt, copper or nickel. The great advantage of tmFRET is that the acceptor ions bind with high affinity to a genetically introduced di-histidine motif (see earlier), allowing for the donor-only measurement to be performed first, and then adding the acceptor to the solution to obtain the donor in the presence of acceptor signal. In combination with a genetically encoded donor such as the fluorescent unnatural amino acid Anap (R_0 Anap-Co = 12 Å), a completely genetically encoded system with independent donor and acceptor labeling can be accomplished. The unspecific absorption or energy transfer to transition-metal ions in bulk solution has to be corrected by determining the effect in the absence of an acceptor-binding motif. The distance is finally calculated via the normalized fluorescence changes:

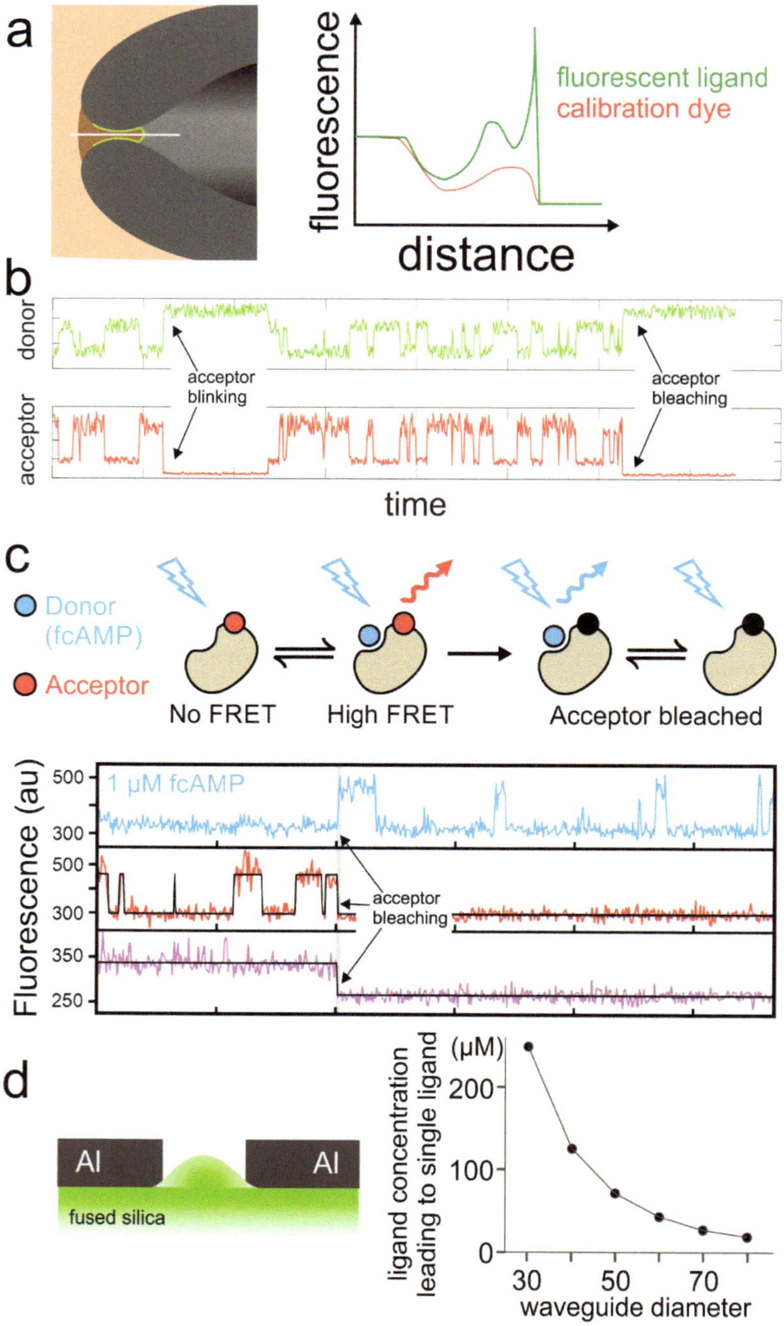

FIGURE 11.5
(a) Excised patches expressing ligand-gated channels are imaged. The fluorescent ligand leads to intensity in the bulk solution and the membrane (green), whereas a red calibration dye in the solution is recorded using different filters (red). The difference is the membrane-bound ligand concentration (Reproduced from Biskup et al. 2007). (b) Donor and acceptor fluorescence intensity in single-pair FRET. The donor adopts its donor-only intensity if the acceptor enters blinking or is photobleached. (c) spFRET signal of fluorescently labeled cGMP to isolated cyclic-nucleotide binding domains of HCN channels isolated by zero-mode waveguides (ZMW). (Original figure from Goldschen-Ohm et al. 2016.) (d) Principle of ZMW and relation between size of the nanopores and detection volume (Levene et al. 2003).

$$ET = \frac{F_{noHH} - F_{HH}}{F_{noHH} \cdot F_{HH} + F_{noHH} - F_{HH}}$$

where the F_i represent the ratio by which fluorescence decreased by addition of the acceptor in the presence (*HH*) or absence (*noHH*) of a di-histidine motif.

11.4.3 Ligand Binding

One aspect of ligand-gated ion channels is the binding of the ligands to their respective binding sites and how this process is coupled to channel activation. Fluorescence techniques have proven extremely helpful to readout the binding step to the ligand-binding site. To address the problem, two principal approaches are feasible: first, to monitor the conformational change of the binding site related to the ligand binding as discussed earlier.

In the second approach the ligands themselves are fluorescently labeled. By simultaneously measuring current and fluorescence in excised patches (*patch-clamp fluorometry*), binding of the fluorescently labeled ligand to the channels in the patch leads to increased fluorescence intensity at the membrane. In principle, the "local concentration" of the ligand at the membrane is increased (Figure 11.5a). This increase is not observed in the absence of channels, indicating that no unspecific labeling of the membrane occurred. As ligands remain in the solution, background fluorescence is removed by normalization to a second dye in the bulk solution.

An improved method to exclude the background from the labeled ligands in the solution is to reduce the detection volume such that only a single molecule is present. This has been achieved using zero-mode waveguides (see later for more details). While this method increases the information that is gained on the labeling step, the information on the current is lost.

11.4.4 Single-Channel Fluorescence

Certain aspects of structural changes are hidden in the ensemble fluorescence measurements such as the movement of single subunits relative to one another or processes not occurring in a synchronized fashion, e.g., in transporter proteins. Similarly, only the rate-limiting step in a sequence of events can be resolved. For this reason, single-channel fluorescence recordings entered the focus of interest. When recording single-molecule fluorescence, a high signal-to-noise ratio is required, as the absolute fluorescence changes will naturally be very small. This is achieved with the advancements in detection optics and high-sensitivity cameras typical for single-molecule fluorescence. The signals themselves are generated by quenching or (single pair, sp) FRET, as described earlier. As single proteins alternate between different conformations, single-channel fluorescence data will follow the stochastic fluctuations resembling single-channel electrophysiology. Accordingly, similar techniques need to be used to analyze the data, including amplitude and dwell-time histograms as well as direct fits to Markov models. However, single-molecule fluorescence has additional properties that need to be taken into account: photobleaching and photoblinking (Figure 11.5b).

Photobleaching is the chemical destruction of a fluorophore upon prolonged or repeated excitation. Thus, photostable dyes should be selected. Blinking is the more important problem; during photoblinking, the fluorophores enter temporarily a triplet state and do not fluoresce during this time. To reduce photobleaching and triplet state, triplet state quenchers and oxygen scavengers can be added to the solutions. Often blinking occurs in a similar time scale as the actual gating events. In this case, interpretation is easier using spFRET because donor and acceptor fluorescence change antisymmetrically.

The data also has to be correlated with known characteristics of the channel. While macroscopic properties may be called upon, it would be best to directly correlate single-channel fluorescence and current. Currently, most single-channel fluorescence results were obtained from purified proteins reconstituted in supported bilayers so that no electrophysiological data was available. But single-channel fluorescence was also simultaneously recorded with macroscopic currents in *Xenopus* oocytes. spFRET and single-channel recording was achieved simultaneously from gramicidin toxin channels.

Let us return to the problem of detecting ligand binding, where two effects counter each other: detection of a single molecule while maintaining a sufficiently high concentration for binding. To achieve that, a single ligand resides – in average – in the detection fluorescence volume despite a micromolar concentration that is required for efficient binding. The detection volume has to be reduced to 10^{-19} L, which corresponds to a cube of 65 nm side length, about a tenth of a wavelength of visible light. Such a reduction is achieved in zero-mode waveguides (ZMWs; Figure 11.5d). Here, a glass substrate is covered with a metal coating containing holes with diameters in the range of 30–80 nm. This basically creates a Faraday cage with very small mesh size. The boundary conditions on the metal surface prevent electromagnetic waves with wavelengths larger than the diameter of the holes to exist and they are reflected. As in total internal reflection fluorescence (TIRF) microscopy, this reflection generates an exponentially attenuated electromagnetic field in the entry to the hole. A hole size of 50 nm would lead to approximately 1 molecule in the detection volume at 50 μM concentration. Using fluorescently labeled cAMP, as described earlier, spFRET measurements follow binding and conformational changes on the single-molecule level (Figure 11.5c).

11.5 Conclusions

With ever-improving imaging hardware and a plethora of fluorophores with various properties, fluorescence became an indispensable tool for ion channel research, be it in structural biology, molecular neuroscience or cardiac research. Fluorescence recordings are mainly limited by the ability to direct the label to a specific position. This aspect is steadily improved in particular with the development of novel detection assays and increased availability of genetically encoded tags. The advantages of an optical readout include, first, that fluorophores are sensitive to their environment and are thus modulated by a variety of factors; and, second, that optical recordings are spatially resolved. Furthermore, the optical recordings do not interfere with electrophysiology so that both can be recorded simultaneously. The spatial resolution permits parallelization of the systems, which plays a crucial role in the development of high-throughput assays or in recording from complex systems such as neuronal tissue. In this chapter, we concentrated on how to exploit fluorescence spectroscopy in order to obtain structural information. However, calcium, pH or voltage sensors are also common uses for optical readouts that can be used in high-throughput systems. For instance, the temperature sensors in TRPV1 channels have been found by random mutagenesis and subsequent screening using a high-throughput calcium-influx assay on transiently-transfected HEK293 cells.

The spatial resolution may also be exploited to optically excite ion channels in a specified region and thereby induce neuronal activity. By linking either a ligand or a blocker via a linker, whose length is photoswitchable (azobenzene), to an ion channel, the channel can be turned on and off using light of different wavelengths.

Suggested Readings

Biskup, C., J. Kusch, E. Schulz, V. Nache, F. Schwede, F. Lehmann, V. Hagen, and K. Benndorf. 2007. "Relating ligand binding to activation gating in CNGA2 channels." *Nature* 446 (7134):440–3. doi:10.1038/nature05596.

Blunck, R. 2015. "Investigation of ion channel structure using fluorescence spectroscopy." In *Handbook of ion channels*, edited by J. Zheng, and M. C. Trudeau, 113–33. Boca Raton, FL: CRC Press.

Blunck, R. 2021. "Determining stoichiometry of ion channel complexes using single subunit counting." *Methods Enzymol* 653:377–404. doi:10.1016/bs.mie.2021.02.017.

Borisenko, V., T. Lougheed, J. Hesse, E. Fureder-Kitzmuller, N. Fertig, J. C. Behrends, G. A. Woolley, and G. J. Schutz. 2003. "Simultaneous optical and electrical recording of single gramicidin channels." *Biophys J* 84 (1):612–22. doi:10.1016/S0006-3495(03)74881-4.

Cha, A., G. E. Snyder, P. R. Selvin, and F. Bezanilla. 1999. "Atomic scale movement of the voltage-sensing region in a potassium channel measured via spectroscopy." *Nature* 402 (6763):809–13

Dai, G., and W. N. Zagotta. 2017. "Molecular mechanism of voltage-dependent potentiation of KCNH potassium channels." *Elife* 6. doi:10.7554/eLife.26355.

Erickson, M. G., B. A. Alseikhan, B. Z. Peterson, and D. T. Yue. 2001. "Preassociation of calmodulin with voltage-gated Ca(2+) channels revealed by FRET in single living cells." *Neuron* 31 (6):973–85.

Faure, E., G. Starek, H. McGuire, S. Berneche, and R. Blunck. 2012. "A limited 4 A radial displacement of the S4-S5 linker is sufficient for internal gate closing in Kv channels." *JBC* 287:40091–8. doi:10.1074/jbc.M112.415497.

Goldschen-Ohm, M. P., V. A. Klenchin, D. S. White, J. B. Cowgill, Q. Cui, R. H. Goldsmith, and B. Chanda. 2016. "Structure and dynamics underlying elementary ligand binding events in human pacemaking channels." *Elife* 5. doi:10.7554/eLife.20797.

Gordon, S. E., M. Munari, and W. N. Zagotta. 2018. "Visualizing conformational dynamics of proteins in solution and at the cell membrane." *Elife* 7. doi:10.7554/eLife.37248.

Heron, A. J., J. R. Thompson, B. Cronin, H. Bayley, and M. I. Wallace. 2009. "Simultaneous measurement of ionic current and fluorescence from single protein pores." *J Am Chem Soc* 131 (5):1652–3. doi:10.1021/ja808128s.

Kalstrup, T., and R. Blunck. 2017. "Voltage-clamp fluorometry in Xenopus oocytes using fluorescent unnatural amino acids." *J Vis Exp* 123. doi:10.3791/55598.

Leisle, L., F. Valiyaveetil, R. A. Mehl, and C. A. Ahern. 2015. "Incorporation of non-canonical amino acids." *Adv Exp Med Biol* 869:119–51. doi:10.1007/978-1-4939-2845-3_7.

Levene, M. J., J. Korlach, S. W. Turner, M. Foquet, H. G. Craighead, and W. W. Webb. 2003. "Zero-mode waveguides for single-molecule analysis at high concentrations." 299 (5607):682–6. doi:10.1126/science.1079700.

Luo, L., Y. Wang, B. Li, L. Xu, P. M. Kamau, J. Zheng, F. Yang, S. Yang, and R. Lai. 2019. "Molecular basis for heat desensitization of TRPV1 ion channels." *Nat Commun* 10 (1):2134. doi:10.1038/s41467-019-09965-6.

Mannuzzu, L. M., M. M. Moronne, and E. Y. Isacoff. 1996. "Direct physical measure of conformational rearrangement underlying potassium channel gating." *Science* 271 (5246):213–16.

Marme, N., J. P. Knemeyer, M. Sauer, and J. Wolfrum. 2003. "Inter- and intramolecular fluorescence quenching of organic dyes by tryptophan." *Bioconjug Chem* 14 (6):1133–9.

Pantazis, A., K. Westerberg, T. Althoff, J. Abramson, and R. Olcese. 2018. "Harnessing photoinduced electron transfer to optically determine protein sub-nanoscale atomic distances." *Nat Commun* 9 (1):4738. doi:10.1038/s41467-018-07218-6.

Taraska, J. W., and W. N. Zagotta. 2010. "Fluorescence applications in molecular neurobiology." *Neuron* 66 (2):170–89.

Turro, N. J., V. Ramamurthy, and J. C. Scaiano. 2010. *Modern molecular photochemistry of organic molecules*. Sausalito, CA: University Science Books.

Wang, S., J. B. Brettmann, and C. G. Nichols. 2018. "Studying structural dynamics of potassium channels by single-molecule FRET." *Methods Mol Biol* 1684:163–80. doi:10.1007/978-1-4939-7362-0_13.

12

Ion Channel Structural Biology in the Era of Single-Particle Cryo-EM

Jianhua Zhao and Yifan Cheng

CONTENTS

12.1 Introduction

Until a few years ago, most ion channel structures were determined by X-ray crystallography (see Chapter 13) and a few by electron crystallography. Examples of success include voltage-gated potassium channels (MacKinnon, 2003), voltage-gated sodium channels (Payandeh et al., 2011) and water channels (Fujiyoshi et al., 2002; Harries et al., 2004). As well-established and advanced structure-determination methods, both crystallographic methods enabled the determination of atomic structures of many ion channels and contributed significantly to our understanding of ion channels in general. However, obtaining well-ordered crystals of sufficient sizes of a target ion channel is a prerequisite and thus a bottleneck. For many ion channel families, particularly those with large soluble domains,

DOI: 10.1201/9781003096214-14

such as transient receptor potential (TRP) channels, ryanodine receptors (RyRs) and the IP3 receptor, crystallizing the whole channel has not been accomplished.

Single-particle cryo-electron microscopy (cryo-EM) bypasses the requirement of crystallization and relies on computationally aligning a large number of projection images of individual molecules captured by an electron microscope from molecules in near native form embedded in a thin layer of vitreous ice (Cheng et al., 2015). A three-dimensional (3D) reconstruction of a target protein can be determined by iteratively determining and refining the angular relationship between all images (i.e., refinement; see Section 12.5.6) and by sorting out a homogeneous subset of particle images (i.e., classification; see Section 12.5.7) so that the resolution can be progressively improved, and hopefully to a level that is sufficient for de novo atomic model building. Technological breakthroughs in the early 2010s enabled atomic resolution structure determination by this method. The atomic structure determination of the TRPV1 ion channel by single-particle cryo-EM (Cao et al., 2013; Liao et al., 2013), the first high-resolution ion channel structure determined by this method without the need of growing any crystals, marked the beginning of a new era for the structural biology of ion channels (Cheng, 2018).

12.2 Why Single-Particle Cryo-EM?

The main advantage of single-particle cryo-EM, by definition, is the ability to determine structures without the need of growing crystals of any kind. Recognizing this advantage, single-particle cryo-EM has long been used for structural studies of some ion channels, particularly those with very large soluble domains, including structural studies of RyRs (Serysheva et al., 2005), IP3 receptors (Ludtke et al., 2011) and TRPV1 (Moiseenkova-Bell et al., 2008). However, resolutions of these early attempts were limited to low resolution, bringing the nickname of "blobology" to the method. Embarrassingly, early structures of some ion channels determined by single-particle cryo-EM were proven to be completely wrong (Mio et al., 2007; Sato et al., 2004; Serysheva et al., 2003). Recognizing the methodological shortfalls motivated the field to develop rigorous criteria and procedures to prevent potential future mistakes (Ludtke et al., 2011). Among major technological hurdles that limit the resolution and throughput is the image recording medium, either photographic film or charge-coupled device (CCD) cameras. Photographic film is incompatible with automation, and CCD cameras do not detect electrons directly but uses a thin layer of scintillator to convert the electron signal to photons for camera detection. The real breakthrough came from the development of modern direct electron detection cameras, which became commercially available in early 2010s. Among early direct detection cameras, the K2 camera was the first one featuring single-electron counting capability (Li et al., 2013), which was revolutionary, and was quickly adapted by all other manufacturers.

The first breakthrough of applying single-particle cryo-EM to ion channels was the determination of the atomic structure of the TRPV1 ion channel (Liao et al., 2013), a target that evaded crystallization despite extensive efforts over a decade from many crystallography laboratories around the world. This initial success sparked tremendous attention on cryo-EM and marked the beginning of a new era in ion channel structural biology. Without the restriction of needing to form crystals and with the capability of achieving atomic resolution, single-particle cryo-EM was demonstrated to be an excellent tool for determining the atomic structure of ion channels, which has since revolutionized the ion channel structural biology field. Nowadays, thanks to the continuous methodology development, single-particle cryo-EM has become a dominant methodology in structure determination of ion channels. In the following, we will describe some basic experimental and computational procedures used in single-particle cryo-EM structure determination of ion channels, aiming to provide some conceptual understanding of how the method is applied to ion channel structural studies.

12.3 Sample Preparation of Ion Channels for Single-Particle Cryo-EM

Biochemical sample preparation for single-particle cryo-EM is quite similar to those used for crystallization. Ion channels, either from an endogenous source or from recombinant overexpression systems, were first solubilized in detergent followed by purification. For recombinant proteins, the purification is often performed with affinity purification followed by size exclusion chromatography. The goal of purification optimization is to produce highly purified stable protein with high yield. For recombinant proteins, various affinity tags are inserted in the expression vector to enable affinity purification. To optimize expression and the choice of detergent, a green florescent protein (GFP) is often inserted to the expression vector to enable florescent-based size exclusion chromatography (FSEC), which allows rapid screening of detergent and expression vector design (Hattori et al., 2013). The biochemical criteria of a successful purification for single-particle cryo-EM are quite similar as those used for X-ray crystallography, such as the quality of the gel-filtration peak, the purity of the sample and the thermal stability of the purified channels. Although it is always true that a better biochemically behaved sample produces better structures at higher resolution, the requirement of sample quality for single-particle cryo-EM is somewhat less stringent than for X-ray crystallography. The capability of computational classification makes it possible to derive high-resolution structures from a less homogeneous sample. For membrane proteins, particularly for ion channels, one advantage for single-particle cryo-EM is that one can reconstitute detergent-solubilized channels back into a lipid bilayer environment, using lipid nanodiscs (Gao et al., 2016) or saposin nanoparticles (Frauenfeld et al., 2016), and study structures of ion channels in a lipid bilayer environment. Single-particle cryo-EM structure determination is compatible with the use of these membrane mimetic systems. Purified ion channels, either in detergent or in lipid bilayer mimetic systems, are then used to prepare cryo-EM grids by following well-established plunge-freezing procedures. Optimization is also needed to produce cryo-EM grids with suitable ice thickness and optimal particle distributions.

12.4 Cryo-EM Experiment: Data Acquisition and Interpretation

Cryo-EM data acquisition is nowadays carried out by automated procedures from cryo-EM grids loaded into an electron microscope (Cheng et al., 2021). Nowadays, a typical state-of-the-art setup includes an electron microscope equipped with a field emission electron source that produces a highly coherent electron beam to image frozen hydrated biological samples, an automated sample loading device that allows rapid and efficient sample exchange, and a high-end direct electron detection camera. Cryo-EM data acquisitions are now carried out by using automation programs, such that once it is set up properly, it will collect cryo-EM images continuously for tens of hours without human intervention. Recorded images are pipelined to computer workstations for immediate data processing, often on the fly during the data acquisition.

12.5 Foundations of Microscopy and Data Processing

A major effort of single-particle cryo-EM structure determination is placed in image processing. There are several popular pipeline packages that make the image processing pipeline relatively easy to follow for a novice. However, conceptually, it helps tremendously to learn some basic concepts that are embedded in the process of image analysis. In the following, we briefly introduce some basic image analysis concepts, as well as parameters that are related to these concepts and often included in publications for evaluating the quality of the structures being published.

12.5.1 Fourier Theory

Image processing of cryo-EM data is based heavily on Fourier theory, which involves describing images and objects using sine and cosine waves. A signal $f(x)$ can be described by

$$f(x) = a1 \cdot \sin(b1 \cdot x + c1) + a2 \times \sin(b2 \cdot x + c2) + a3 \cdot \sin(b3 \cdot x + c3) + \ldots. \quad (12.1)$$

where $a1$, $a2$, $a3$, ... are the *amplitudes* of each wave; $b1$, $b2$, $b3$, ... are the *frequencies* of each wave; and $c1$, $c2$, $c3$, ... are the *phases* of each wave (Figure 12.1a).

In some cases, such as estimating resolution in cryo-EM (see later), it's much more convenient to talk about the signal in terms of the amplitude, frequency and phase. Let $f(x)$ represent the signal and $f'(b) = [a,c]$ represent the signal in terms of amplitude a, frequency b, and phase c (Figure 12.1b). Example #1 demonstrates how a sine wave can be described by a single point in this representation, while Example #2 demonstrates how even a complicated function in *real space* (i.e., in x) can be described by only two points in *frequency space* (i.e., in b).

Mathematically, switching between $f(x)$ and $f'(b)$ can be achieved by the *Fourier transform* (FT) and the *inverse Fourier transform* (FT^{-1}):

$$FT\left[f(x)\right] = F(b) = \int f(x) e^{-i \cdot x \cdot b} dx \quad (12.2)$$

$$FT^{-1}\left[F(b)\right] = f(x) = \int F(b) e^{i \cdot x \cdot b} db \quad (12.3)$$

$$FT^{-1}\left[F(b)\right] = FT^{-1}\left\{FT\left[f(x)\right]\right\} = f(x) \quad (12.4)$$

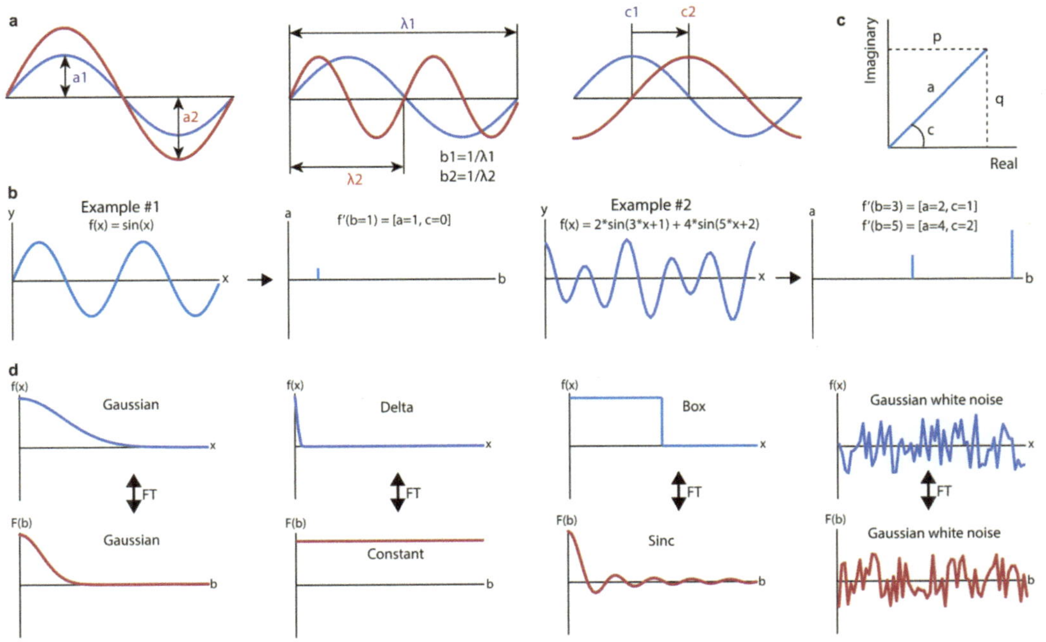

FIGURE 12.1
Fourier theory basics and examples. (a) Sine and cosine waves are described by the amplitude (*a*), frequency (*b*), and phase (*c*) of the wave. (b) Examples of real-space and corresponding frequency-space representations of a signal. (c) Graphical depiction of a Fourier component. (d) Useful examples of functions and their Fourier transforms. Note that the power spectrum is the square of the Fourier transform ($|F|^2$).

where $i=\sqrt{-1}$ and $e^{i\cdot b} = \cos(b) + i\cdot\sin(b)$). Typically, a function $F(b) = \mathrm{FT}[f(x)]$ is represented by its real (p) and imaginary (q) *Fourier components* (Figure 12.1c) where

$$F(b) = [p,q] = p + i\cdot q = a\cdot\{\cos(c) + i\cdot\sin(c)\} \tag{12.5}$$

It follows that the amplitude a and phase shift c can be calculated from the Fourier components $[p,q]$ by

$$a = \sqrt{|p + i\cdot q|} = \sqrt{(p + i\cdot q)(p - i\cdot q)} = \sqrt{p^2 + q^2} \tag{12.6}$$

$$c = cos^{-1}\left(\frac{p}{a}\right) = sin^{-1}\left(\frac{q}{a}\right) \tag{12.7}$$

A plot of a^2 versus b is the *power spectrum* and is often the plotted form of the Fourier transform of a function or image in *Fourier space* or *frequency space*. Some useful examples of functions and their Fourier transforms are shown in Figure 12.1d.

Because the Fourier transform of a box function is a sinc function, one has to be careful with masking in cryo-EM. A tight mask can cut into the EM density, resulting in a box function that creates ripples in Fourier space and artificially increases the reported resolution by Fourier shell correlation estimation (see Section 12.5.10).

Expanding to two or three dimensions, we have

$$FT\left[f(x,y,z)\right] = F(bx,by,bz) = \int f(x,y,z)\cdot e^{-i\cdot(x\cdot bx+y\cdot by+z\cdot bz)}dx\,dy\,dz \quad (12.8)$$

$$b = \sqrt{bx^2 + by^2 + bz^2} \quad (12.9)$$

12.5.2 Central Slice Theorem

Cryo-EM imaging results in 2D projections of 3D molecules, which are then combined into a 3D map using the central slice theorem (Figure 12.2a). The central slice theorem posits that a 2D projection image in real space corresponds to a plane through the center of the 3D Fourier transform of the molecule. By combining many images of a molecule in different orientations, we can construct the 3D Fourier transform of the molecule and then perform an inverse Fourier transform to calculate the real-space map of the molecule. This procedure is equivalent to doing back projection in real space.

Samples in cryo-EM typically adopt a preferred orientation on a grid, resulting in poorly sampled or missing regions of Fourier space. Consequently, the resulting real-space map of the molecule may contain artifacts. A plot of particle image orientation parameters can be informative for identifying preferred orientation (Figure 12.2b), but it is inadequate as a quantifiable metric because misaligned particle images can appear as a uniform distribution of views. Furthermore, most data sets show some form of orientation bias, but it does

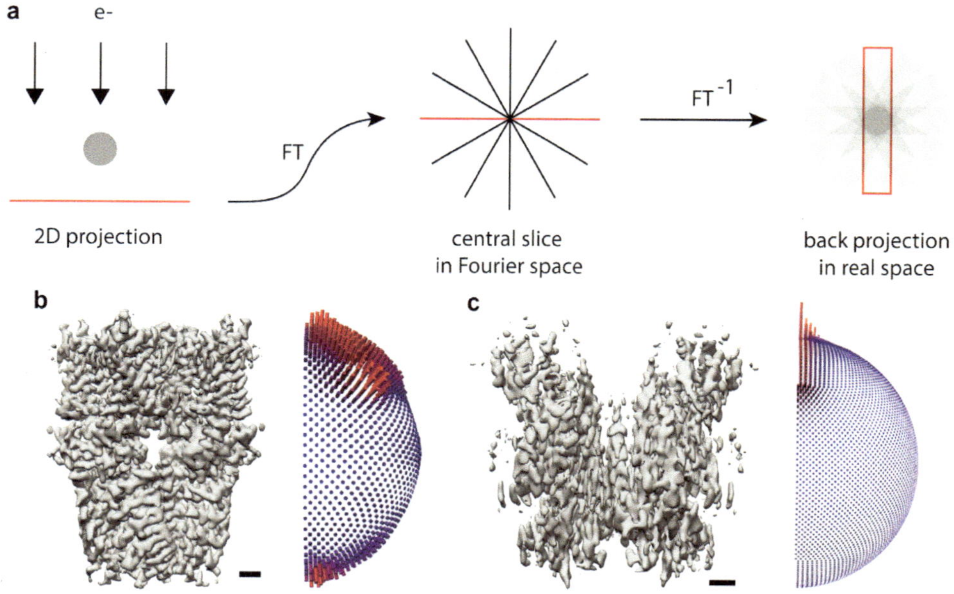

FIGURE 12.2
3D structures from 2D images. (a) Central slice theorem and back projection. (b) Cryo-EM map of TRPA1 (left) showing no obvious distortions in the map despite biased orientation distribution (right). Scale bar: 10 Å. (c) Cryo-EM map of TMEM16A (left) showing distortions in alpha-helical densities with biased orientation distribution (right). Scale bar: 10 Å.

not always cause distortions in the density map. To better quantify orientation bias, we can estimate the directional resolution of density maps (see Section 12.5.9.2).

12.5.3 Contrast Transfer Function

Biological samples imaged in cryo-EM are weak phase objects and most electrons go straight through the sample, resulting in low amplitude contrast. To better visualize samples, images are recorded out of focus to generate phase contrast. An image taken out of focus is the convolution of the defocused point spread function (PSF) with the object being imaged (Figure 12.3a):

$$Im_{defocused} = Im_{focused} * PSF \tag{12.10}$$

The images we get in cryo-EM are $g(x)$ and we want to remove the effects of the PSF to get back the original signal $f(x)$. To do so, we make use of the convolution theorem:

$$FT\left[f(x) * PSF\right] = FT\left[f(x)\right] \cdot FT\left[PSF\right] \tag{12.11}$$

Because it's easier to work with multiplication in Fourier space rather than convolution in real space, we typically think about correcting the images in Fourier space. The Fourier transform of the PSF is the contrast transfer function (CTF) and has the form

$$FT\left[PSF\right] = CTF = -\left(w_{phase}\right) \cdot \sin\left(\S\ \right) - \left(w_{amp}\right) \cdot \cos\left(\S\ \right) \tag{12.12}$$

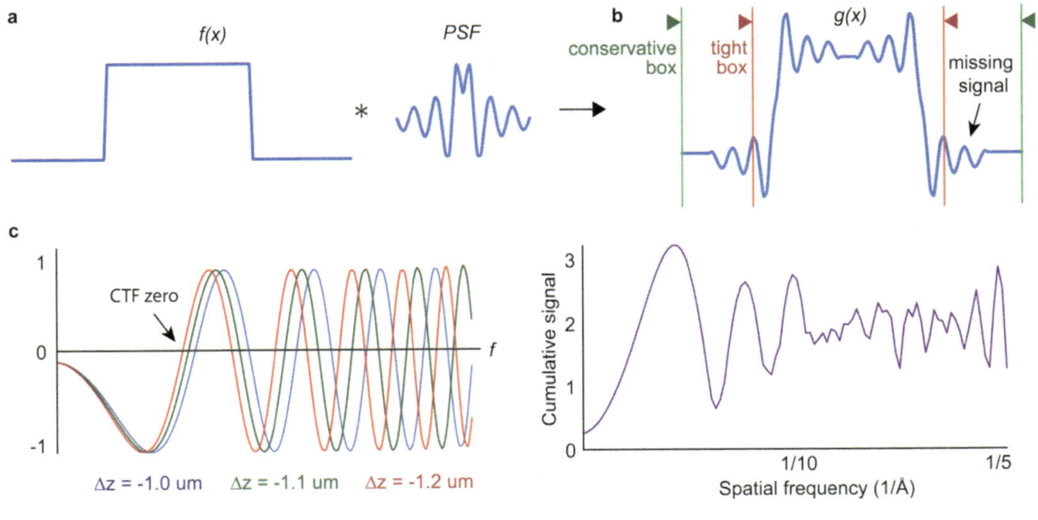

FIGURE 12.3
Contrast transfer function (CTF). (a) The defocused image $g(x)$ is the convolution of the original signal $f(x)$ with the point spread function (PSF). (b) Modulation by the CTF causes signal to "spread out," necessitating a large box when extracting particle images. (c) Zeros in the CTF necessitate images at different defocus to be combined to fill in the missing gaps in signal (left plot, simulated assuming 300 kV). Summing CTFs with defocus values ranging from –1.0 to –1.2 μm shows good coverage of the Fourier space signal (right plot), demonstrating that a defocus range of a few hundred nanometers is sufficient to ensure complete sampling of Fourier space.

$$w_{phase} = \sqrt{1-\left(w_{amp}\right)^2} \tag{12.13}$$

$$X = \left(\pi \cdot \lambda \cdot a^2\right) \cdot \Delta z - \left(\pi \cdot \lambda^3 \cdot a^4\right) \cdot C_s \tag{12.14}$$

where w_{amp} and w_{phase} are the fractional amounts of amplitude and phase contrast (typically, w_{amp} is set to ~0.07 in cryo-EM); λ is the wavelength of the electrons, ~0.02Å for 300 kV; a is the spatial frequency or resolution; Δz is the defocus of the image; and Cs is the spherical aberration, which is a fixed number for an electron microscope, typically ~2.5~2.7 mm. Because amplitude contrast is negative (from the cosine term), we typically take images underfocused rather than overfocused so that the phase contrast is also negative to maximize contrast.

CTF is one of the most important concepts in single-particle cryo-EM, or in general in electron microscopy. Because of the CTF, images recorded in an electron microscope are often not directly interpretable but require correction of the CTF prior to any further image analysis, such as image alignment or classification. CTF correction is a process of deconvolution, but most importantly, it is the correction of phase flipping, which happens whenever CTF goes through zero. Because of CTF zeros, a complete correction of CTF for a single image is not possible, but can be achieved by combining many images recorded with different defocuses:

$$FT\left(Im_{corrected}\right) = \frac{CTF_1 \cdot FT\left(Im_1\right) + CTF_2 \cdot FT\left(Im_2\right) + CTF_3 \cdot FT\left(Im_3\right)\ldots}{CTF_1^2 + CTF_2^2 + CTF_3^2 \ldots} \tag{12.15}$$

where Im_{sum} is the combined and corrected image and {CTF_1, CTF_2, CTF_3,...} are the CTFs of defocused images {Im_1, Im_2, Im_3, ...}.

12.5.4 Box Size

The signal in cryo-EM images taken under-focus is modulated by the CTF, causing the signal to "spread out" (Figure 12.3b). Consequently, when extracting particle images from micrographs, it is important to choose a generous box size that is much larger than the sample of interest to capture all the signal. It should be noted that some of the signal is likely below the level of the noise. As a rule of thumb, the box size should be at least twice the size of the protein diameter.

12.5.5 Defocus Range

Due to the form of the CTF, the signal in a defocused image drops to zero at certain spatial frequencies. To fill in these missing gaps, we collect images at different defocus values and combine them (Figure 12.3c). Typically, a defocus range of a few hundred nanometers is sufficient.

12.5.6 Refinement

Map refinement refers to the procedure of figuring out how to fit all the 2D particle images together to form a 3D map (Figure 12.2a). Refinement involves estimating angular and

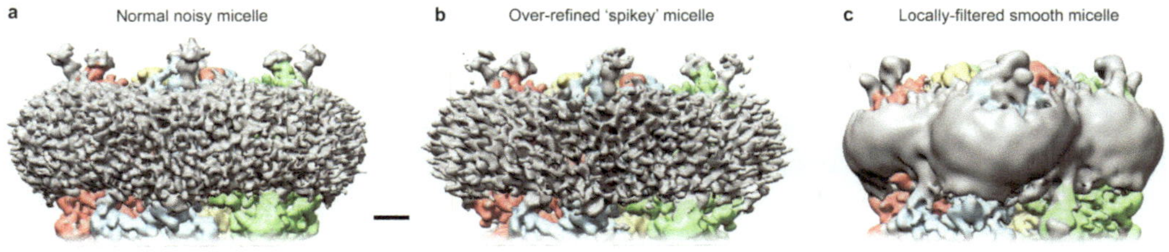

FIGURE 12.4
Map refinement. (a) A typical micelle in cryo-EM density maps looks noisy but relatively featureless. (b) A common sign of overrefinement in membrane proteins is the appearance of "spikes" in the micelle. Scale bar: 15 Å. (c) Masking and filtering select regions can improve refinement results by avoiding overrefinement of structured noise in the micelle and flexible regions of the protein. Map was obtained through nonuniform refinement in cryoSPARC.

translational shifts of an image relative to a reference map. Typically, it begins with a coarse global search over all angular and translational space at low spatial frequencies. As the resolution of the map improves over iterations, refinement switches to a fine local search over angular and translational shifts at higher spatial frequencies.

The signal-to-noise ratio is high at low spatial frequencies (i.e., the signal is much stronger than noise at low resolution). The inverse is true at high spatial frequencies. Therefore, it is crucial to be careful when using high spatial frequencies for map refinement. For membrane proteins, a typical sign of overrefinement is the appearance of structured noise features in the detergent micelle that look like "spikes" (Figure 12.4a). An overrefined map is not necessarily wrong, but one should be careful when interpreting the map. This is especially true for maps of proteins where some regions are more flexible or rigid compared to other regions.

12.5.7 Classification

Ion channels adopt different conformations (such as open or closed gate) and may interact with different proteins (i.e., with or without auxiliary subunits). Classification refers to sorting out a subset of particles within a data set that is sufficiently homogeneous in conformation and composition so that their angular and translational parameters can be refined to produce a 3D reconstruction at a higher resolution. Classification alone does not necessarily lead to a higher resolution structure. However, without sufficient homogeneity, the reconstruction calculated from a particle set cannot reach sufficient resolution. The more homogeneous the subset is, the fewer particles the subset contains that may limit the achievable resolution. In this regard, larger data sets are often necessary to sort out conformational heterogeneity, but demands more microscope time to collect the data and more computer resources to process the data. Classification is often performed in two ways: at the level of 2D images and at the level of 3D maps. 2D classification is often used to remove obvious junk images, such as images of ice. On the other hand, 3D classification can distinguish finer structural details. Classification and refinement are often iterated to achieve the final result.

12.5.8 Masking during Refinement

Specific regions in the protein can be targeted for refinement and classification by using a selective mask around the regions of interest. This can serve to avoid overrefinement of regions that may be flexible or not structured, such as the detergent micelle. In some programs (e.g., cisTEM; Grant et al., 2018), one can low-pass filter regions outside the mask instead of ignoring it completely, which can help image alignment. Alternatively, in other programs (e.g., cryoSPARC; Punjani et al., 2017), one can align images to a locally filtered map, which effectively filters flexible or nonstructured regions (Figure 12.4b). Both approaches serve to dampen high-resolution features in dynamic regions, which helps to avoid overrefinement in those regions.

12.5.9 Resolution Estimation

The resolution of a map is estimated by splitting a data set into two sets, constructing a map for each set and then calculating the Fourier shell correlation (FSC) between the two maps (Figure 12.5a). The FSC is calculated as

$$FSC(k) = \frac{\sum_i (H1_{ki} \cdot H2_{ki})}{\sqrt{\sum_i H1_{ki}^2} \cdot \sqrt{\sum_i H2_{ki}^2}} \tag{12.16}$$

where $H1$ and $H2$ are the Fourier transform of the two maps, k is the spatial frequency, and i represents the index of all the Fourier components at spatial frequency k. If the two maps were refined independently using two separate halves of the data set, then the FSC between these maps is referred to as the "gold standard FSC." The spatial frequency k at which the FSC falls below 0.143 is generally considered the resolution of the map (Figure 12.5a). This number corresponds in theory to a correlation of 0.5 between a perfect reference with the map from the combined dataset (Rosenthal and Henderson, 2003).

The FSC is a global resolution estimate because it calculates the correlation between two maps in spherical shells in Fourier space. Variations of the FSC calculation are available to measure different properties: the local FSC estimates the resolution in different regions of the map and the directional FSC estimates the resolution from different viewing directions of the map.

12.5.9.1 Local Resolution Estimation

The local resolution estimate can reveal regions of flexibility or disorder in a protein. While there are different ways to calculate local resolution, the local FSC can be calculated simply by splitting a map into different parts and then calculating the FSC for each section. The local resolution estimate of a map can inform which regions are better resolved and which regions might be more dynamic (Figure 12.5b). It is important during model building and interpretation to consider that some map regions are better defined than others.

12.5.9.2 Directional Resolution Estimation

The directional resolution estimate measures the completeness of Fourier space coverage and can reveal missing wedges in Fourier space due to sample orientation bias (Dang et al., 2017). The directional FSC (dFSC) can be calculated by evaluating the FSC in shells within conic sections rather than spherical sections (Figure 12.5c). A cryo-EM data set can exhibit biased orientation distributions (Figure 12.2b), but still have good coverage of Fourier space (Figure 12.5d). On the other hand, the dFSC can reveal missing wedges of data in Fourier space (Figure 12.5e) that results in distortion of the cryo-EM map (Figure 12.2c).

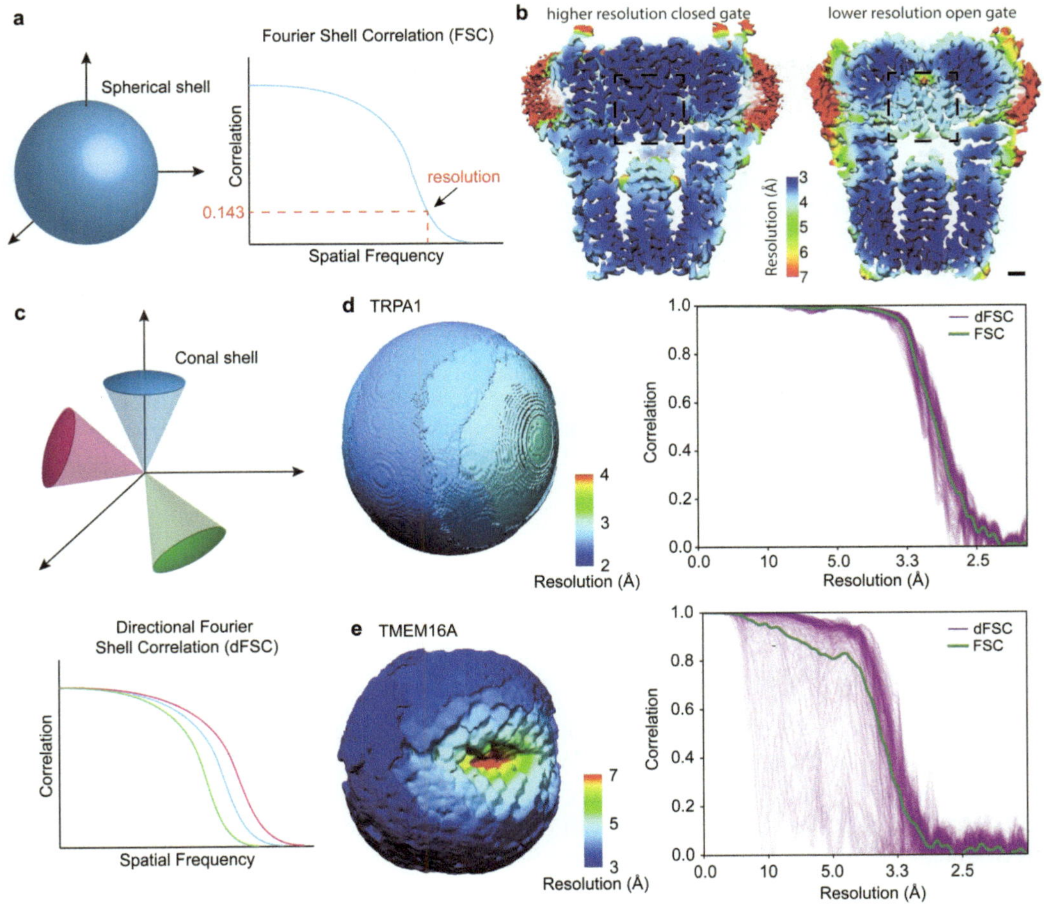

FIGURE 12.5
Resolution estimation in single-particle cryo-EM. (a) The Fourier shell correlation (FSC) calculates the correlation in spherical shells between two maps in Fourier space. The global resolution of a map is estimated at the point where the FSC between two half-maps drops below 0.143. (b) Local resolution estimation of two maps of TRPA1 in different conformations. The closed conformation of TRPA1 shows high resolution in the gate region (left), indicative of a more stable and rigid structure. The open conformation of TRPA1 shows lower resolution in the gate region (right), suggesting a more dynamic structure. Scale bar: 10 Å. (c) The directional Fourier shell correlation (dFSC) calculates the FSC in conal shells, providing resolution estimates in different directions. (d) dFSC of the TRPA1 example from Figure 12.2b shows similar resolution in different directions. (e) dFSC of the TMEM16A example from Figure 12.2c shows a large missing wedge in Fourier space.

12.5.10 Masking

The density of interest typically makes up only a small fraction of the cryo-EM map, with the rest of the map being featureless regions filled with noise. When estimating the resolution, we are only interested in the parts of the map that correspond to the biological sample. Therefore, regions outside the areas of interest are masked to flatten those values to zero. Done properly, this procedure provides a more accurate estimate of the resolution via FSC calculation of the region of interest. However, if a mask is chosen such that it cuts into the density where there is signal, this creates correlated sharp edges in both

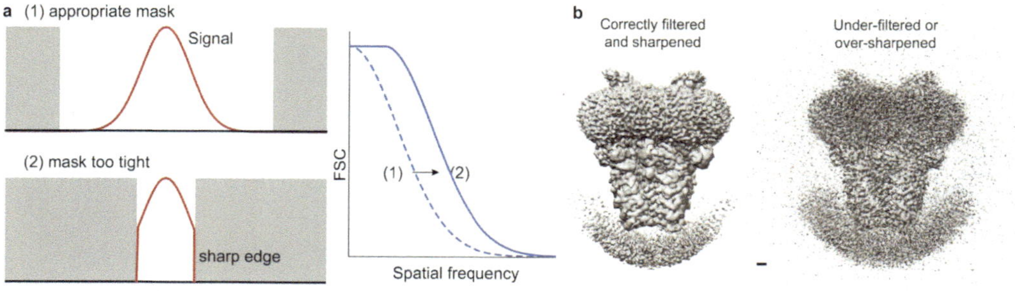

FIGURE 12.6
Postprocessing of cryo-EM maps. (a) Using a tight mask that cuts into density creates sharp edges that artificially boost the FSC between two half-maps. (b) Maps of TRPA1 sharpened with a low (left) and high (right) B-factor. Sharpening with a B-factor that is too high for the map results in noise "speckles" throughout the map (right). Filtering to a spatial resolution beyond the resolution of the map can also result in noise "speckles." Scale bar, 10 Å.

half-maps, which artificially boost the resolution estimate by FSC (Figure 12.6a). Therefore, a loose mask that includes the detergent micelle is recommended.

12.5.11 Filtering and Sharpening

Filtering a map involves dampening high spatial frequencies where the signal-to-noise ratio is low, as indicated by the FSC. When a map is filtered to the correct resolution, the map density should be smooth and continuous for well-defined regions (Figure 12.6b, left). On the other hand, various steps in imaging and processing can downweigh higher frequencies relative to lower frequencies. Sharpening attempts to correct the amplitude weighting and involves applying an inverse exponential function of the form e^{-ab}, where a is a spatial frequency dependent term and b is the B-factor. Sharpening with the appropriate B-factor should enhance high spatial frequency features while maintaining a smooth map with continuous density (Figure 12.6b, left). However, oversharpening can amplify noisy high spatial frequency features that result in "speckles" throughout the map (Figure 12.6b, right). Model refinement programs such as Phenix and Coot work better when maps are filtered and sharpened correctly, giving higher quality models with better statistics.

12.6 What Structural Information Can Cryo-EM Provide?

Since 2013, the number of ion channel structures determined by single-particle cryo-EM has been steadily increasing. By removing the obstacle of crystallization, the pace of structure determination has drastically increased. One result has been a dramatic increase in the number of high-resolution TRP channel structures from zero before 2013 to all subfamilies of this ion channel superfamily being determined in a matter of a few years. For some ion channels, such as NompC (Jin et al., 2017) and Piezo (Zhao et al., 2018), the shapes of these channels make one wonder if it would ever be feasible to crystallize them for X-ray crystallography.

Beyond determining the atomic structures of an ion channel, single-particle cryo-EM also facilitates rapid structural analysis of ion channels trapped in different functional states, most often by using known ligand, either agonists or antagonists, or in some cases by introducing specific mutations (Zhang et al., 2021; Zhao et al., 2020). Such structural analyses provide mechanistic insights into ion channel regulation in response to their physiological stimuli. In this regard, single-particle cryo-EM offers certain technological advantages, most often offered by the capability of computational classification allowing sorting out particles with heterogeneous conformations. Such capability is particularly suited to dissect the gating mechanism of ion channels.

However, it is worth noting that, in comparison with X-ray crystallography (see Chapter 13), there are also some technical disadvantages of single-particle cryo-EM. For example, in X-ray crystallography, the exact identity of ions can be determined by anomalous scattering. But there is no equivalent approach in single-particle cryo-EM. Identifying cations, such as Na^+ or Ca^{2+}, often relies on the geometry of cation coordination by surrounding residues (Autzen et al., 2018), which may not be as direct as the anomalous scattering method in X-ray crystallography.

12.7 Further Technological Advancement and Challenges ahead in Ion Channel Structural Biology

While the structural biology of ion channels has advanced significantly due to the technological breakthroughs in single-particle cryo-EM, there are still many challenges that will require further methodological advancements. For example, it is still difficult to determine the structure of voltage-gated ion channels under a defined voltage across the membrane. It is also difficult to study channels while keeping the environments in two sides of the membrane different. One example is to study ion channels under low pH extracellularly but neutral pH intracellularly. One plausible approach to address such challenges is to reconstitute the target ion channels into liposomes followed by structure determination directly from liposomes (Wang and Sigworth, 2009; Yao et al., 2020). In this way, it is possible to capture structures of ion channels while maintaining environmental differences, such as pH and ionic strength, across the membrane. In the foreseeable future, one can anticipate continuous technological advancements in single-particle cryo-EM that will enable answering many of the challenging questions in ion channel structural biology.

Suggested Readings

Autzen, H. E., et al. 2018. "Structure of the human TRPM4 ion channel in a lipid nanodisc." *Science* 359:228–32.

Cao, E., M. Liao, Y. Cheng, and D. Julius. 2013. "TRPV1 structures in distinct conformations reveal activation mechanisms." *Nature* 504:113–8.

Cheng, A., et al. 2021. "Leginon: new features and applications." *Protein Sci* 30:136–50.

Cheng, Y. 2018. "Membrane protein structural biology in the era of single particle cryo-EM." *Curr Opin Struct Biol* 52:58–63.

Cheng, Y., N. Grigorieff, P. A. Penczek, and T. Walz. 2015. "A primer to single-particle cryo-electron microscopy." *Cell* 161:438–49.

Dang, S., et al. 2017. "Cryo-EM structures of the TMEM16A calcium-activated chloride channel." *Nature* 552:426–9.

Frauenfeld, J., et al. 2016. "A saposin-lipoprotein nanoparticle system for membrane proteins." *Nat Methods* 13:345–51.

Fujiyoshi, Y., et al. 2002. "Structure and function of water channels." *Curr Opin Struct Biol* 12:509–15.

Gao, Y., E. Cao, D. Julius, and Y. Cheng. 2016. "TRPV1 structures in nanodiscs reveal mechanisms of ligand and lipid action." *Nature* 534:347–51.

Grant, T., A. Rohou, N. Grigorieff. 2018. "cisTEM, user-friendly software for single-particle image processing." *Elife* 7.

Harries, W. E., D. Akhavan, L. J. Miercke, S. Khademi, and R. M. Stroud. 2004. "The channel architecture of aquaporin 0 at a 2.2-A resolution." *Proc Natl Acad Sci U S A* 101:14045–50.

Hattori, M., R. E. Hibbs, and E. Gouaux. 2012. "A fluorescence-detection size-exclusion chromatography-based thermostability assay for membrane protein precrystallization screening." *Structure* 20:1293–9.

Jin, P., et al. 2017. "Electron cryo-microscopy structure of the mechanotransduction channel NOMPC." *Nature* 547:118–22.

Li, X., et al. 2013. "Electron counting and beam-induced motion correction enable near-atomic-resolution single-particle cryo-EM." *Nat Methods* 10:584–90.

Liao, M., E. Cao, D. Julius, and Y. Cheng. 2013. "Structure of the TRPV1 ion channel determined by electron cryo-microscopy." *Nature* 504:107–12.

Ludtke, S. J., et al. 2011. "Flexible architecture of IP3R1 by Cryo-EM." *Structure* 19:1192–9.

MacKinnon, R. 2003. "Potassium channels." *FEBS Lett* 555:62–5.

Mio, K., et al. 2007. "The TRPC3 channel has a large internal chamber surrounded by signal sensing antennas." *J Mol Biol* 367:373–83.

Moiseenkova-Bell, V. Y., L. A. Stanciu, I. I. Serysheva, B. J. Tobe, and T. G. Wensel. 2008. "Structure of TRPV1 channel revealed by electron cryomicroscopy." *Proc Natl Acad Sci U S A* 105:7451–5.

Payandeh, J., T. Scheuer, N. Zheng, and W. A. Catterall. 2011. "The crystal structure of a voltage-gated sodium channel." *Nature* 475:353–8.

Punjani, A., J. L. Rubinstein, D. J. Fleet, and M. A. Brubaker. 2017. "cryoSPARC: algorithms for rapid unsupervised cryo-EM structure determination." *Nat Methods* 14:290–6.

Rosenthal, P. B., and R. Henderson. 2003. "Optimal determination of particle orientation, absolute hand, and contrast loss in single-particle electron cryomicroscopy." *J Mol Biol* 333:721–45.

Sato, C., et al. 2004. "Inositol 1,4,5-trisphosphate receptor contains multiple cavities and L-shaped ligand-binding domains." *J Mol Biol* 336:155–64.

Serysheva, I. I., et al. 2003. "Structure of the type 1 inositol 1,4,5-trisphosphate receptor revealed by electron cryomicroscopy." *J Biol Chem* 278:21319–22.

Serysheva, I. I., S. L. Hamilton, W. Chiu, and S. J. Ludtke. 2005. "Structure of Ca2+ release channel at 14 A resolution." *J Mol Biol* 345:427–31.

Wang, L., and F. J. Sigworth. 2009. "Structure of the BK potassium channel in a lipid membrane from electron cryomicroscopy." *Nature* 461:292–5.

Yao, X., X. Fan, and N. Yan. 2020. "Cryo-EM analysis of a membrane protein embedded in the liposome." *Proc Natl Acad Sci U S A* 117:18497–503.

Zhang, K., D. Julius, and Y. Cheng. 2021. "Structural snapshots of TRPV1 reveal mechanism of polymodal functionality." *Cell* 184:5138–50 e5112.

Zhao, J., J. V. Lin King, C. E. Paulsen, Y. Cheng, and D. Julius. 2020. "Irritant-evoked activation and calcium modulation of the TRPA1 receptor." *Nature* 585:141–5.

Zhao, Q. et al. 2018. "Structure and mechanogating mechanism of the Piezo1 channel." *Nature* 554:487–92.

13

Protein Crystallography

Yoni Haitin and Moshe Giladi

CONTENTS

13.1 Ion Channels Physiology in the Era of Structural Biology

The idea of hydrophilic pores, allowing the translocation of, otherwise impermeable, charged particles across hydrophobic biological membranes, was introduced as early as the end of the 19th century (Bayliss 1915). During the 20th century, the concept of ion channels evolved through the groundbreaking development and use of various electrophysiological techniques, which provided the experimental framework for the visionary understanding of the inner workings of these specialized molecular machines (Hille 2001). Moreover, the development of molecular biology approaches, including cloning, mutagenesis and heterologous protein expression techniques, offered an unprecedented opportunity to deepen our knowledge of ion channel functional selectivity, gating mechanisms and overall architectures (Jan and Jan 1989). However, advancements in the field of membrane protein structural biology were required for the explicit description of key molecular mechanisms governing ion channel function. Indeed, by the end of the 20th century, these advancements culminated in the first structure of an intact potassium channel, resolved using X-ray crystallography (Doyle et al. 1998). Since then, the crystal structures of intact channels and fragments thereof have been determined at an ever-increasing rate (Li et al. 2021), providing snapshots of ion channel proteins in discrete states. With this high-resolution information, basic molecular processes governing the rapid and selective translocation of ions across cellular membranes are inferred. Here, we will provide a contemporary overview of the principles of protein crystallization and structure determination using X-ray crystallography (Figure 13.1).

DOI: 10.1201/9781003096214-15

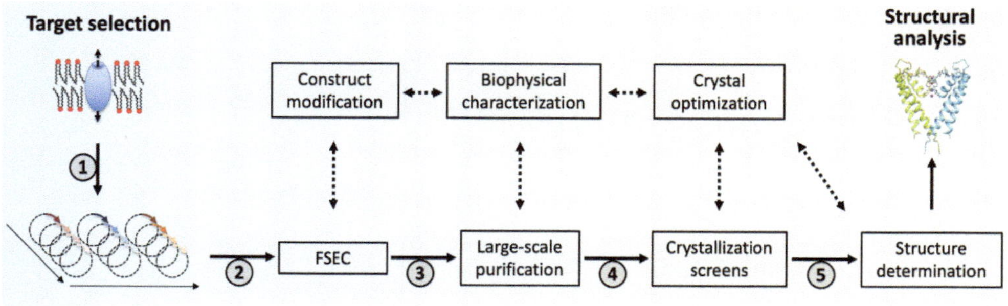

FIGURE 13.1
Protein structure determination workflow. (1) Following target selection, a library of constructs, covering different orthologs, isoforms or domains, are designed and subcloned enframe into a GFP-containing vector backbone. (2) These libraries are expressed in various systems and screened for monodispersed elution profiles by fluorescence-detection size-exclusion chromatography (FSEC) (see Box 13.1). If no suitable candidates are identified, the construct design should be revisited. (3) Positive hits are subjected to large-scale production. Samples are monitored for homogeneity and stability throughout the purification process using various biophysical characterization approaches. (4) Purified protein preparations (>95% purity at 200–250 μM) are subjected to initial crystallization trials. Initial crystal hits are optimized by varying the crystallization condition composition ("mother liquor") to yield reproducible single three-dimensional crystals. (5) Finally, optimized crystals are subjected to diffraction experiments, followed by structure determination.

13.2 Why Crystallization?

The field of protein X-ray crystallography has made unimaginable progress since the determination of myoglobin (Kendrew et al. 1958) and hemoglobin (Perutz et al. 1960) structures. However, the main bottleneck of structural studies using X-ray crystallography remains the process of obtaining well-diffracting crystals, built from a highly ordered repeated arrangement of the protein building blocks. Why are crystals needed to solve the structures of proteins (and other molecules) in the first place? In principle, the interaction of X-ray light with electrons results in the scattering of the incident photons. Since the wavelength of light in the X-ray band is comparable with atomic radii and covalent bond lengths, the high-resolution structure of the scatterer can be deduced from the emerging scattering pattern (see Box 13.2). However, as the scattering produced by a single molecule is far below the detection threshold, the periodic organization of the protein molecules in the crystal lattice provides signal amplification, which is proportional to the number of identical scattering molecules (Figure 13.2). With an increase in crystal size, decrease in water content and minimal conformational variation, the scattering power of a given crystal is enhanced, allowing one to obtain measurable data that contains the information needed to determine the structural details at higher resolution. In addition, several groundbreaking technological improvements enabled a significant decrease in the crystal sizes needed for obtaining a complete data set, from millimeter to micrometer on edge dimensions. Indeed, the field of protein structural biology using X-ray crystallography was revolutionized by (1) the worldwide availability of synchrotron radiation sources, which provide immensely bright X-ray radiation; (2) the use of cryogenic conditions, diminishing radiation damage during data collection; and (3) the introduction of highly sensitive detection technologies, increasing the data collection throughput (Kwan, Axford, and Moraes 2020). Yet, in order to overcome the bottleneck of obtaining well-diffracting single crystals, protein samples of high quality and sufficient quantities must be obtained first.

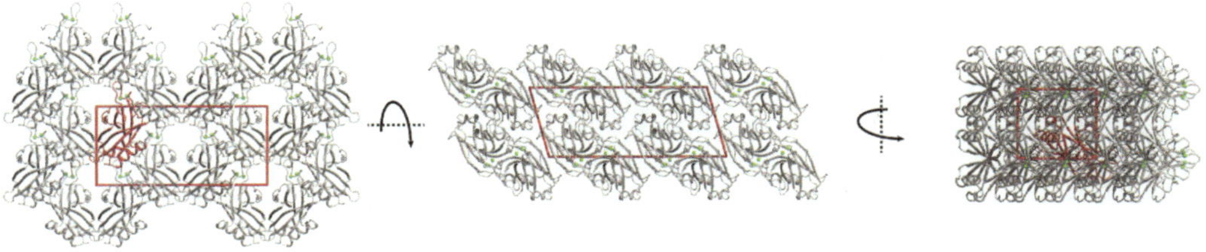

FIGURE 13.2
Crystal organization. The crystalline organization of the C2B domain of DOC2B (PDB 4LDC) is illustrated from three angles. The unit cell is demarcated by a red box, while a single protein molecule comprising the asymmetric unit (ASU) is colored red. In this case, the ASU corresponds to the functional, biological unit. Green spheres represent calcium ions.

13.3 Enhancing Crystallization Likelihood

Protein crystals are formed by regularly repeating molecular interactions in space (Figure 13.2). Importantly, the formation of such a specific spatial arrangement necessitates a highly homogeneous and monodisperse protein sample (see Box Figure 13.1). Thus, the crystallographer should also be a well-trained biochemist and molecular biologist in order to obtain highly pure, stable and homogeneous protein samples, in milligram quantities. Bioinformatic tools can be used during construct design to predict the likelihood of subsequent crystallization success. These predictions are based on the large number of available structures to date, and the experience gained by high-throughput structural genomics campaigns. For example, the *XtalPred* server (Slabinski et al. 2007) provides a comprehensive primary sequence analysis by combining the prediction of different protein properties from various web servers and integrating these characteristics into a single crystallization feasibility score.

As protein crystallization is a rare instance, in addition to meticulous construct design, it is advisable to perform pre-crystallization screens to characterize and optimize the stability and dispersity of the protein target, thereby enhancing the likelihood of success. Traditionally, these screens were performed using purified proteins, a lengthy and cumbersome procedure. In 2006, Kawate and Gouaux published their development of fluorescence-detection size-exclusion chromatography (FSEC), which is an analytical technique enabling rapid characterization of unpurified protein samples (Box 13.1). The most promising constructs are subjected to scaled-up expression and purification followed by biophysical analyses of their properties in solution (Box 13.2) (Niesen, Berglund, and Vedadi 2007; Borgstahl 2007). Together, thoughtful and careful construct design and optimization of sample preparation conditions are key for efficient and successful downstream structural investigations.

BOX 13.1 Fluorescence-Detection Size-Exclusion Chromatography (FSEC)

Size-exclusion chromatography (SEC) is commonly used as a purification procedure, where the proteins are separated according to their radius of hydration and are detected using UV absorbance. However, in unpurified samples, it is impossible to specifically identify the protein of interest due to the abundance of proteins sharing a similar size. Hence, FSEC was developed as a groundbreaking method allowing for fast, high-throughput screening of fluorescently labeled proteins to obviate the need for preliminary purification steps. FSEC is based on the genetic fusion of a green fluorescent protein (GFP) moiety to the protein of interest. Following expression of the fusion proteins, the cells are lysed and the cytosolic fraction (in the case of soluble proteins) or detergent-solubilized crude membrane fraction (in the case of integral membrane proteins) is subjected to SEC. Importantly, the protein of interest is specifically identified based on the fluorescence of the fused GFP, rather than its UV absorbance that is shared with all other proteins, using an inline fluorimeter. Samples are considered as well-behaving by exhibiting (1) an elution profile mainly consisting of a homogeneous population; (2) eluting at the expected column volume according to the estimated protein size; (3) showing minor populations of aggregates eluting at the column void volume (indicated by an arrow in Box Figure 13.1) and free GFP molecules (due to proteolysis if positioned at the C-terminus, or premature translation termination if fused at the N-terminus) (Box Figure 13.1, top). Conversely, an elution profile comprising of multiple peaks due to the coexistence of different oligomeric species, nonsymmetrical wide peaks representing nonhomogeneous dispersion, and large populations of aggregates or free GFP molecules are considered poorly behaving samples (Box Figure 13.1, bottom). Thus, FSEC provides crucial information on the expression level, integrity, stability and size dispersion of the target, but without the need for employing laborious purification steps. This allows the high-throughput screening of numerous constructs and buffer conditions (including detergents, if needed) amiable for downstream crystallization.

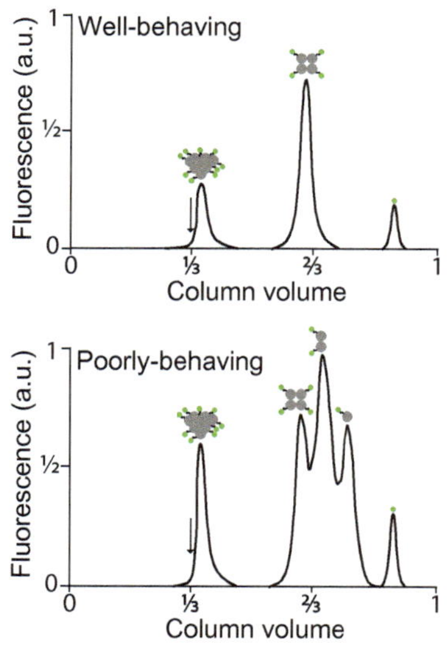

Box Figure 13.1

BOX 13.2 Biophysical Characterization of Sample Quality

Beyond meticulous construct design and protein purification, the crystallization tendency of a sample can be dramatically affected by its environment. For example, variables such as pH, small-molecule additives (e.g., glycerol, sucrose), ligands, and ionic composition or strength play a vital role in maintaining sample integrity and stability and should be systematically explored. Several approaches can be employed to characterize the purified protein sample under different buffer conditions. First, differential scanning fluorimetry (DSF) is a high-throughput screening method utilizing different fluorophores to assess sample stability in response to a gradual increase in temperature. Using commonly available real-time polymerase chain reaction (PCR) devices, DSF can report on the thermal denaturation melting temperature in the presence of different buffer conditions. Conditions resulting in an increased melting temperature are considered as stabilizing and are therefore favorable. In addition to the characterization of the sample stability using DSF, one should also scrutinize its size distribution state. While sample heterogeneity can be grossly assessed from the SEC profile, highly sensitive methods can be used to detect minor flaws that can be deleterious for crystallization. Dynamic light scattering (DLS) reports on the oligomeric species and aggregation state of the protein sample in solution with extremely high sensitivity (Box Figure 13.2A). Empirically, proteins that are monodisperse in undersaturated solutions readily crystallize (Box Figure 13.2B, C), while polydisperse proteins rarely do. Indeed, DLS provides a quantitative metric for sample polydispersity (Pd%), which is predictive of crystallization success. Together, as protein crystallization is driven by numerous variables, many of which are unknown, meticulous planning and a systematic approach are vital for the success in obtaining diffracting crystals (Box Figure 13.2D).

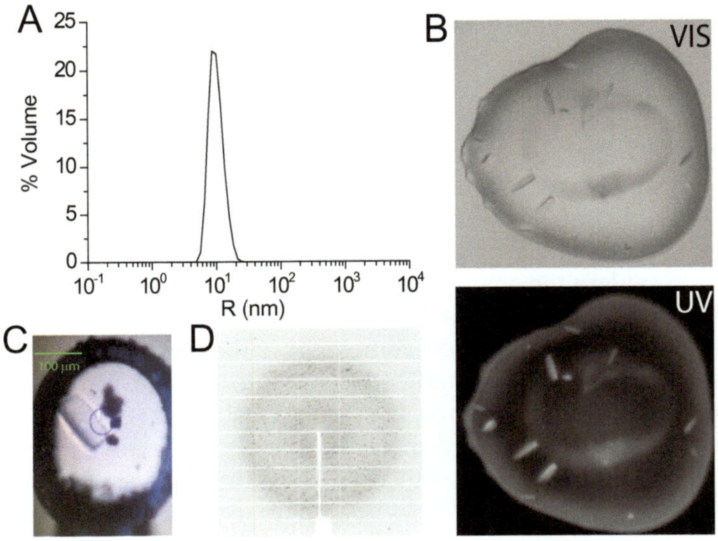

Box Figure 13.2

13.4 Protein Crystallization

The process of protein crystallization involves the phase transition of concentrated protein molecules, following the introduction of precipitating agents that drive the protein out of the solution. Briefly, in the crystallization experiment, the protein sample is brought to its solubility limit by mixing a sufficiently concentrated protein solution with precipitants that reduce protein solubility. The desired outcome is one where the solubility limit of the protein is surpassed and the supersaturated solution reaches a metastable state, where nucleation and the subsequent crystal growth can spontaneously occur. However, due to the unimaginable complexity of this system, the particular conditions that will yield protein crystals are impossible to predict. Therefore, a systematic exploration of protein and precipitant concentrations, precipitant types (e.g., salts, polyols, pH) and physical conditions (e.g., temperature and crystallization setup) is always required. To survey the infinite possible conditions, high-throughput approaches are routinely used.

The initial protein concentration at the beginning of the crystallization trials should be determined empirically. To this end, several protein concentrations are mixed with a solution of highly concentrated precipitants (i.e., 4M ammonium sulfate or 30% polyethylene glycol 5000) and precipitation is observed under a microscope. Concentrations in which the solution remains clear or contains heavy precipitate are deemed too low or high for crystallization, respectively. Next, commercial crystallization screening kits, based on data mining of solutions previously used successfully for protein crystallization, are usually used. The crystallization experiment can be further diversified at this point by employing different setups, such as sitting- or hanging-drop vapor diffusion platforms, protein-to-precipitant ratio, drop volume, and the incubation temperature used (Dessau and Modis 2010). The crystallization screens are periodically and routinely monitored using microscopy, and the state of each drop is scored and documented (i.e., clear, phase separation, amorphous precipitation, crystals and forms thereof). Importantly, components of the crystallization solution may form salt crystals. For example, mixing of calcium-containing protein solution with a phosphate-containing crystallization condition often results in the formation of $CaPO_4$ crystals. Hence, if a crystal hit is identified, its proteinaceous nature should be confirmed. This can be achieved using UV fluorescence (Box 13.2) or SDS-PAGE analysis of the crystals. The crystals should then be optimized by performing a systematic variation of the original condition and/or screening for crystallization-enhancing additives. The goal of this optimization step is to increase the likelihood of obtaining single crystals of appropriate size and to explore the emergence of different crystal forms. Ultimately, conditions yielding single crystals are identified and subjected to diffraction experiments.

As can be reasoned from the inherent complexity of the crystallization scheme described, it is not uncommon for initial crystallization trials to fail. Nevertheless, one should not be discouraged – perseverance is the key to success! Mitigating this difficulty can be approached by salvage of apparently unsuccessful crystallization trials or modification of the protein sample. Importantly, the outcome of the initial crystallization screens should be carefully examined. In many cases, while no clear crystal hits emerge at first glance, the screens contain conditions that give rise to crystalline material and can serve as initial seeds for further crystal growth. As crystal nucleation and growth may require different conditions, seeding enables bypassing the nucleation step by the mechanical separation and transfer of preformed nuclei to conditions that can support crystal growth, opening a new crystallization screening space to explore (Till et al. 2013). If unsuccessful, modification of the protein sample, at all production levels, may prove beneficial. Should the protein form a complex with known ligands (i.e., small molecules, peptides, DNA), their inclusion in the final crystallization mix may be favorable. Next, the protein sample may be rationally modified by demarcating the minimal domain core using limited proteolysis, followed by mass-spectrometric analysis and revision of construct design accordingly (Koth et al. 2003). In addition, as the purification tags and associated linkers are usually highly flexible and may interfere with ordered lattice formation, their removal during purification is advisable. Alternatively, if a soluble fusion tag is included in the construct, its presence may facilitate crystallization if rigidly fused to the protein of interest using a short linker (Koth et al. 2003). Similarly, the use of nanobodies, single-domain antibodies that can stabilize membrane proteins or protein complexes in specific conformations, may favor crystallization (Pardon et al. 2014). Finally, the use of mutagenesis to increase the overall protein stability and reduce surface entropy, followed by experimental validation using DSF, may be beneficial (Goldschmidt et al. 2007). Together, as protein crystallization is driven by numerous variables, many of which are unknown, meticulous planning and a systematic approach are vital for successfully obtaining diffracting crystals.

13.5 The Diffraction Experiment and Phasing Techniques

Protein crystals are formed by a periodic translational repetition of unit cells. Each unit cell, serving as the basic building block of the crystal, confines protein molecules, known as the asymmetric units (ASUs), which relate to one another by a set of symmetry operations (rotational and translational) (Figure 13.2). Thus, the unit cell is defined according to its symmetry and dimensions. Following irradiation by a coherent X-ray source, the photons are scattered by the electrons of the atoms forming the crystal, resulting in constructive and destructive wave interference events. These waves are described according to their wavelength, intensity and phase. While the wavelength of the photons remains unchanged, and the scattering intensity is directly recorded using 2D detectors, information regarding the phase is inherently lost during the experiment. Owing to the highly ordered crystalline organization, a discrete set of diffracted light spots, known as "reflections", is obtained (Box 13.2). By recording the diffraction pattern from many angles by crystal rotation, a 3D reconstruction of the scattering molecules within the unit cell can be reassembled, as discussed later. The resolution of a structure is determined by the maximal scattering angle achievable by the given crystal, detector and X-ray source. Indeed, the major advancements in X-ray detection technology, and the achievable photon flux in modern synchrotrons, allow us to obtain atomic resolution data sets from crystals previously deemed of insufficient quality (Karplus and Diederichs 2012). Moreover, the introduction of cryo-crystallography, where the entire experiment takes place at cryogenic temperatures, reduced the damage inflicted to the crystal by X-ray absorption of the high-energy photons. Collectively, these improvements enabled the collection of complete data sets from single crystals of ever-smaller dimensions.

In practice, the diffraction experiment is performed in two stages. First, the crystals are screened for diffraction quality, and their unit cell dimensions and symmetry are determined. This screening stage is performed by recording the diffraction pattern from a limited number of angles to minimize radiation damage and reduce the data processing load. Next, a data collection strategy is generated, usually using an automated pipeline, relying on the cell dimensions and symmetry parameters, as well as the maximal scattering angle achieved, which reflects the resolution limit of the given crystal. This strategy addresses important experimental parameters based on the initial characterization, including the expected overall radiation dose, the achievable data resolution limit, and the rotation angle (and the resulting number of frames) needed for obtaining maximal data completeness, which depends on the crystal symmetry. Finally, full data sets are obtained from the most promising crystals, according to the initial screening and strategy determination stages.

As the complete data set is recorded using a large number of frames, multiple and redundant observations of identical or symmetry-related reflections are made. Therefore, post-collection processing is required to apply various corrections (e.g., due to beam geometry and crystal orientation), integrate reflections dispersed over several frames, index each unique reflection throughout the data set, and finally, merge and scale identically indexed reflections. By performing such frame-by-frame processing, aberrations and technical anomalies can be identified, corrected, or excluded from the collected data sets early on. Moreover, the processing step provides several important quality metrics of the collected data sets, reported in the protein data bank (PDB) and published as part of the crystallographic table. These include the resolution range; the signal-to-noise ratio ($|I|/\sigma(I)$; the mean absolute reflection intensity divided by the standard deviation of the intensities); the number of unique reflections; data multiplicity (indicating the mean number of observations for each unique reflection); a merging R-value (representing the quality of the merging process); and data set completeness. These metrics are usually reported for the entire data set and at the high-resolution data shell.

A crucial remaining problem to address is the loss of the phase information during the diffraction experiment, as described earlier, known as the "phase problem." The missing information prevents the direct calculation of the electron density that scattered the photons. Over the years, several approaches for solving this problem were developed. The phases can be experimentally determined either by the binding of heavy atoms to the protein using crystal soaking or by incorporating them directly into the protein using selenomethionine residues in lieu of native methionines during protein expression (Koth et al. 2003). By exploiting the unique scattering properties of heavy atoms, their localization within the crystal can be determined, facilitating initial phase estimates for structure solution. More commonly, owing to the large number of available structures in the PDB (Berman et al. 2000), initial phases can be estimated using a method known as molecular replacement. Here, a structurally homologous protein, usually with ~30% or higher sequence identity, serves as a model that is systematically rotated and translated in the unit cell until its calculated diffraction best matches the observed data. This way the calculated phases from the known model can be used as an initial solution for the data set in question. In difficult cases, approaches harnessing the power of computational modeling and screening have been developed to extend the range of molecular replacement (Terwilliger et al. 2012; Jumper et al. 2021; Baek et al. 2021). These methods can improve the initial search probe, streamline the screen through numerous different homology models, or screen against a pool of conformers, thereby increasing search efficiency and likelihood for success.

13.6 Structure Determination and Quality Metrics

Following successful phasing, an initial electron density map is obtained. Depending on the phasing technique, an initial model may already be placed within the unit cell.

Alternatively, an initial model is built into the density, a process nowadays usually performed using automated tools (AutoBuild, ARP/wARP, Buccaneer) (Adams et al. 2010; Collaborative Computational Project 1994). At this stage, the initial model fits the electron density maps to a varying degree, depending on the quality of the data, the resolution range and the similarity between the model and crystallized protein. Thus, the structure and phases are gradually improved in an iterative refinement process (Figure 13.3). Early at this stage, only stable subregions within the protein, such as secondary structure elements and tightly bound ligands may be well resolved. However, peripheral and flexible regions should often be modified or rebuilt to fit the electron density. Thus, the crystallographer manually rebuilds the model to accommodate the electron density, followed by stereochemically restrained model improvement, and recalculation of the phases and electron density map. Concomitantly, ordered water and small molecules (e.g., from the crystallization solution or added ligands) are introduced during refinement. Next, the process repeats using the improved map to guide further model building, until the process reaches convergence, and no further improvements can be made. The improvement is quantified by the global linear residual (the *R*-value), which measures the discrepancy between the model and the data. Importantly, as refinement model building is prone to enhance model-induced biased interpretation of the data, due to the iterative nature of this process, an unbiased metric must be used to alleviate this concern. With the high number of observations made in the diffraction experiment, a small percentage (usually 5%–10%)

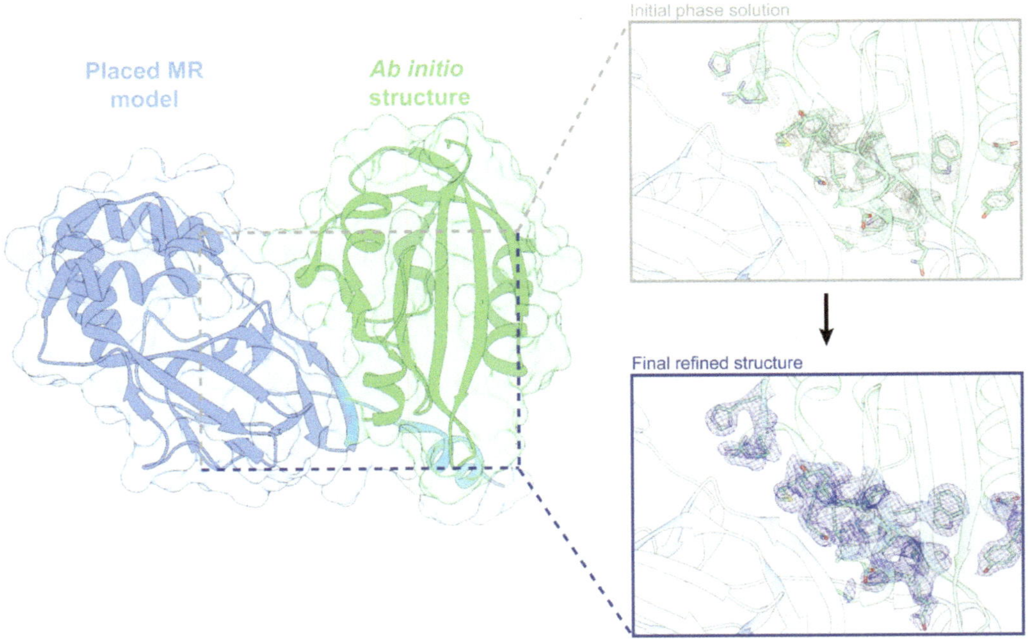

FIGURE 13.3
Comparison between initial and final electron density maps during structural solution using molecular replacement. The X-ray structure on the intracellular complex from the mouse EAG1 channel (mEAG1) (PDB accession 4LLO) was initially phased using the mEAG1 cyclic nucleotide binding-homology domain (blue; PDB accession 4F8A) as the search model. The resulting electron density map for the N-terminal PAS domain (green) emerged but was of low quality as reflected by multiple chain breaks and poorly resolved side chains (gray mesh). Following iterative model building and refinement, a highly detailed final map could be obtained (blue mesh). Maps are presented for interfacing residues of the PAS domain.

is set aside during the refinement process and serves for cross-validation. This results in two refinement metrics termed R_{work} (calculated from the reflections used for refinement) and R_{free} (calculated from the reflections used for cross-validation). Intuitively, R_{work} reaches a lower value compared to R_{free}. The two values should not diverge by more than ~5%. In addition to the agreement between the model and data, it should be remembered that every molecular atom arrangement must satisfy certain physicochemical rules. Thus, additional important quality metrics are the root mean square deviations (RMSD) of the bond lengths and angles. Moreover, the Ramachandran analysis, examining the φ and ψ torsion angles of every resolved residue in the protein backbone, should reveal that the vast majority of the residues are in favored or allowed regions of the Ramachandran plot. In most cases, no residues are present in disallowed regions of this plot, and if this is not the case a close examination of the structure is advised.

Importantly, the asymptotic nature of reaching a convergence between model and data implies that cessation of the refinement process is performed at the discretion of the crystallographer. Moreover, the expected values of the different quality metrics described earlier depend on data quality and resolution. Hence, a useful tool to determine whether the refinement process has run its course is to compare the quality metrics of the model to those of structures determined at a similar resolution (a feature implemented in crystallographic software suites such as PHENIX; Adams et al. 2010). Importantly, the quality metrics of each crystal structure deposited in the PDB are also compared to the entire bank and to structures of similar resolution, enabling a quick yet robust technical assessment of the structure by fellow scientists.

13.7 Structure Analysis and Visualization

The premise of structural biology is that a protein's structure defines its function. Therefore, the pinnacle of a crystallographic project is the analysis of the final structural model, which is aimed at gaining functional insights. As described in Section 13.6, the structure of the ASU is determined. However, the ASU does not necessarily coincide with the biological assembly, that is, the functional macromolecular assembly (Figure 13.2). For example, the ASU may consist of several copies of a biological assembly. This was the case in the structure of the complex between calmodulin and calmodulin-dependent protein kinase I (CaMKI) target sequence (PDB 1MXE) (Clapperton et al. 2002), where two biological units form the ASU. Conversely, a biological assembly may form between crystallographic symmetry-related adjacent ASUs. For example, in the structure of the NavAb voltage-gated sodium channel (Payandeh et al. 2011), the ASU consists of two polypeptide chains, while the biological assembly (the intact channel) is formed by four chains that are symmetry-related in the crystal (PDB 3RVY). To determine the composition and spatial organization of the biological assembly, it is useful to consult with the Proteins, Interfaces, Structures and Assemblies (PISA) server (Krissinel and Henrick 2007), which analyses all interfaces within the crystal and makes a prediction of the possible stable complexes. Moreover, PISA analysis provides valuable information regarding the mechanisms underlying subunit interactions and channel complex assembly mode. The structurally deduced oligomeric state and the role of specific interfaces should be corroborated with complementary biochemical and biophysical experimental approaches.

Once the composition of the biological unit has been determined, the structure can be annotated graphically to facilitate visual comprehension and to highlight key features. As protein structures are usually viewed as 2D images, a proper graphical representation is key for conveying the 3D information these structures entail. Over the years, several structural visualization software have been developed, with PyMol (http://www.pymol.org/) and UCSF Chimera (Pettersen et al. 2004) being most frequently used. The structure can be viewed using different representation modes, ranging from an all-atom ball-and-stick model, through cartoon representation highlighting secondary structure elements, to surface representation accentuating solvent-accessible regions (Figure 13.4). In addition, the structure can be colored according to various properties, such as coulombic potential (Konecny, Baker, and McCammon 2012), hydrophobicity (Kyte and Doolittle 1982) and evolutionary conservation scores (Ashkenazy et al. 2016) (Figure 13.4). Finally, different vantage points can be produced to enhance the 3D spatial grasp of the macromolecule.

A major driving force for determining the atomic resolution structures of ion channels is their ability to illuminate intricate nooks and crannies, underlying their gating, permeation and ion selectivity properties. Indeed, an important insight regarding the channel

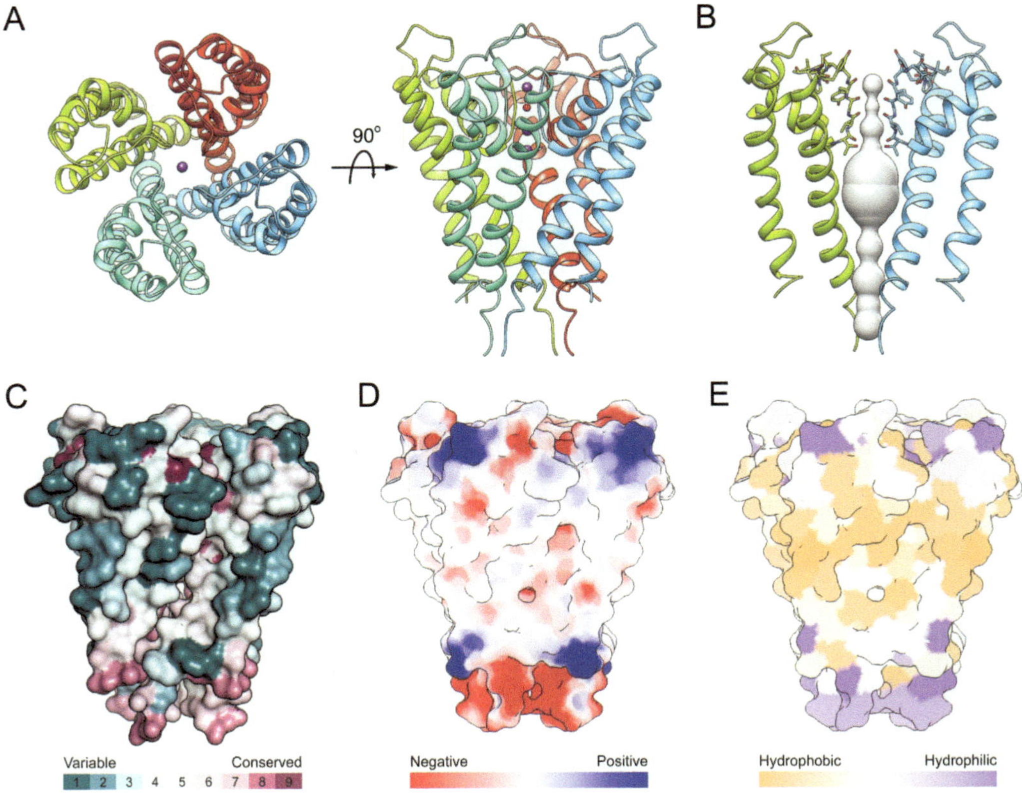

FIGURE 13.4
Common structural representations. (A) Illustration of the KcsA channel (PDB accession 1BL8), colored by chain, and viewed from the extracellular side (left panel) or the membrane plane (right panel). Potassium ions and a water molecule are represented as purple and red spheres, respectively. (B) Ion permeation path representation using the MOLE server (Pravda et al. 2018). For clarity, two diagonally positioned subunits were removed and the internal channel cavity is represented as a gray surface. (C–E) Surface representations of the channels colored according to evolutionary conservation score (C), as determined by the ConSurf server; electrostatic coulombic potential (D); and hydrophobicity (E).

state, as represented by the structure, can be inferred using methods examining their solvent-accessible surfaces and cavities. These tools provide various graphical representations, such as the pore diameter along the ion conduction path or a volumetric depiction overlayed onto the structural model under investigation (Smart et al. 1996) (Figure 13.4).

13.8 The Developing Role of X-Ray Crystallography in the Structural Biology Toolbox

Over the past decades, X-ray crystallography took center stage as the gold standard approach in structural biology. The recent "resolution revolution" of cryogenic electron microscopy (cryo-EM) (Kühlbrandt 2014), owing to the tremendous technological advancements achieved, is considered by some to obviate the need for protein crystallization en route for high-resolution structural investigations. Yet, the two methods can coexist and be used in a highly complementary fashion. For example, in cryo-EM, peripheral channel domains, which display spatial heterogeneity, usually offer diminished resolution compared with core structural elements. In such cases, the determination of a high-resolution crystal structure of the domain in question can be used to provide a holistic structure, which supersedes the molecular insights gained from either structural method.

While X-ray crystallography requires ample amounts of high-quality protein, its use as a central structural approach became routine in many labs worldwide, owing to the availability of synchrotron light sources and the highly streamlined data collection and analysis tools. Conversely, although cryo-EM requires much lower sample amounts and, importantly, bypasses the bottleneck of crystallization, it is currently suffering from high inequity in facility access, and the data collection and analysis procedures require considerable technical expertise. Finally, in contrast to X-ray crystallography, which can be applied to molecules of any size, the use of cryo-EM to solve the structures of small proteins is currently limited. Thus, for the foreseeable future, the use of cryo-EM will likely continue to grow, with X-ray crystallography continuing to serve alongside in providing a structural view of biology.

Suggested Readings

Adams, P. D., P. V. Afonine, G. Bunkóczi, V. B. Chen, I. W. Davis, N. Echols, J. J. Headd, et al. 2010. "PHENIX: a comprehensive python-based system for macromolecular structure solution." *Acta Crystallogr D Biol Crystallogr* 66 (2):213–21.

Ashkenazy, H., S. Abadi, E. Martz, O. Chay, I. Mayrose, T. Pupko, and N. Ben-Tal. 2016. "ConSurf 2016: an improved methodology to estimate and visualize evolutionary conservation in macromolecules." *Nucleic Acids Res* 44 (W1):W344–50.

Baek, M., F. DiMaio, I. Anishchenko, J. Dauparas, S. Ovchinnikov, G. R. Lee, J. Wang, et al. 2021. "Accurate prediction of protein structures and interactions using a three-track neural network." *Science* 373 (6557):871–6.

Bayliss, W. M. 1915. *Principles of general physiology*. New York: Longmans, Green, and Company.

Berman, H. M., J. Westbrook, Z. Feng, G. Gilliland, T. N. Bhat, H. Weissig, I. N. Shindyalov, and P. E. Bourne. 2000. "The protein data bank." *Nucleic Acids Res* 28 (1):235–42.
Borgstahl, G. E. O. 2007. "How to use dynamic light scattering to improve the likelihood of growing macromolecular crystals." *Methods Mol Biol* 363:109–29.
Clapperton, J. A., S. R. Martin, S. J. Smerdon, S. J. Gamblin, and P. M. Bayley. 2002. "Structure of the complex of calmodulin with the target sequence of calmodulin-dependent protein kinase I: studies of the kinase activation mechanism." *Biochemistry* 41 (50):14669–79.
Collaborative Computational Project, Number 4. 1994. "The CCP4 suite: programs for protein crystallography." *Acta Crystallogr D Biol Crystallogr* 50 (5):760–3.
Dessau, M. A., and Y. Modis. 2010. "Protein crystallization for x-ray crystallography." *J Vis Exp* 47 (January):e2285.
Doyle, D. A., J. Morais Cabral, R. A. Pfuetzner, A. Kuo, J. M. Gulbis, S. L. Cohen, B. T. Chait, and R. MacKinnon. 1998. "The structure of the potassium channel: molecular basis of K^+ conduction and selectivity." *Science* 280 (5360):69–77.
Goldschmidt, L., D. R. Cooper, Z. S. Derewenda, and D. Eisenberg. 2007. "Toward rational protein crystallization: a web server for the design of crystallizable protein variants." *Protein Sci* 16 (8):1569–76.
Hille, B. 2001. *Ion channels of excitable membranes*. Sunderland, Sinhauer, MA: Oxford University Press, Incorporated.
Jan, L. Y., and Y. Nung Jan. 1989. "Voltage-sensitive ion channels." *Cell* 56 (1):13–25.
Jumper, J., R. Evans, A. Pritzel, T. Green, M. Figurnov, O. Ronneberger, K. Tunyasuvunakool, et al. 2021. "Highly accurate protein structure prediction with AlphaFold." *Nature* 596 (7873):583–9.
Karplus, P. A., and K. Diederichs. 2012. "Linking crystallographic model and data quality." *Science* 336 (6084):1030–3.
Kawate, T., and E. Gouaux. 2006. "Fluorescence-detection size-exclusion chromatography for pre-crystallization screening of integral membrane proteins." *Structure* 14 (4):673–81.
Kendrew, J. C., G. Bodo, H. M. Dintzis, R. G. Parrish, H. Wyckoff, and D. C. Phillips. 1958. "A three-dimensional model of the myoglobin molecule obtained by x-ray analysis." *Nature* 181 (4610):662–6.
Konecny, R., N. A. Baker, and J. A. McCammon. 2012. "IAPBS: a programming interface to the adaptive poisson-boltzmann solver." *Comput Sci Discov* 5 (1):015005.
Koth, C. M., S. M. Orlicky, S. M. Larson, and A. M. Edwards. 2003. "Use of limited proteolysis to identify protein domains suitable for structural analysis." *Methods Enzymol* 368:77–84.
Krissinel, E., and K. Henrick. 2007. "Inference of macromolecular assemblies from crystalline state." *J Mol Biol* 372 (3):774–97.
Kühlbrandt, W. 2014. "The resolution revolution." *Science* 343 (6178):1443–4.
Kwan, T. O. C., D. Axford, and I. Moraes. 2020. "Membrane protein crystallography in the era of modern structural biology." *Biochem Soc Trans* 48 (6):2505–24.
Kyte, J., and R. F. Doolittle. 1982. "A simple method for displaying the hydropathic character of a protein." *J Mol Biol* 157 (1):105–32.
Li, F., P. F. Egea, A. J. Vecchio, I. Asial, M. Gupta, J. Paulino, R. Bajaj, et al. 2021. "Highlighting membrane protein structure and function: a celebration of the protein data bank." *J Biol Chem* 296:100557.
Niesen, F. H, H. Berglund, and M. Vedadi. 2007. "The use of differential scanning fluorimetry to detect ligand interactions that promote protein stability." *Nat Protoc* 2 (9):2212–21.
Pardon, E., T. Laeremans, S. Triest, S. G. F. Rasmussen, A. Wohlkönig, A. Ruf, S. Muyldermans, W. G. J. Hol, B. K. Kobilka, and J. Steyaert. 2014. "A general protocol for the generation of nanobodies for structural biology." *Nat Protoc* 9 (3):674–93.
Payandeh, J., T. Scheuer, N. Zheng, and W. A. Catterall. 2011. "The crystal structure of a voltage-gated sodium channel." *Nature* 475 (7356):353–9.
Perutz, M. F., M. G. Rossmann, A. F. Cullis, H. Muirhead, G. Will, and A. C. T. North. 1960. "Structure of hæmoglobin: a three-dimensional Fourier synthesis at 5.5-. resolution, obtained by x-ray analysis." *Nature* 185 (4711):416–22.

Pettersen, E. F., T. D. Goddard, C. C. Huang, G. S. Couch, D. M. Greenblatt, E. C. Meng, and T. E. Ferrin. 2004. "UCSF chimera – a visualization system for exploratory research and analysis." *J Comput Chem* 25 (13):1605–12.

Pravda, L., D. Sehnal, D. Toušek, V. Navrátilová, V. Bazgier, K. Berka, R. S. Vařeková, J. Koča, and M. Otyepka. 2018. "MOLEonline: a web-based tool for analyzing channels, tunnels and pores (2018 update)." *Nucleic Acids Res* 46 (W1):W368–73.

Slabinski, L., L. Jaroszewski, L. Rychlewski, I. A. Wilson, S. A. Lesley, and A. Godzik. 2007. "XtalPred: a web server for prediction of protein crystallizability." *Bioinformatics* 23 (24):3403–5.

Smart, O. S., J. G. Neduvelil, X. Wang, B. A. Wallace, and M. S. P. Sansom. 1996. "HOLE: a program for the analysis of the pore dimensions of ion channel structural models." *J Mol Graph* 14 (6):354–60.

Terwilliger, T. C, F. Dimaio, R. J. Read, D. Baker, G. Bunkóczi, P. D. Adams, R. W. Grosse-Kunstleve, P. V. Afonine, and N. Echols. 2012. "Phenix.Mr_rosetta: molecular replacement and model rebuilding with Phenix and Rosetta." *J Struct Funct Genomics* 13 (2):81–90.

Till, M., A. Robson, M. J. Byrne, A. V. Nair, S. A. Kolek, P. D. Shaw Stewart, and P. R. Race. 2013. "Improving the success rate of protein crystallization by random microseed matrix screening." *J Vis Exp* 78:50548.

14

Rosetta Structural Modeling

Phuong T. Nguyen and Vladimir Yarov-Yarovoy

CONTENTS

14.1 Introduction

Ion channels play critical roles in signal transduction in excitable cells and are primary targets of therapeutic drugs used to treat a diverse range of human diseases. Recent advances in the determination of membrane protein structures using cryo-electron microscopy (cryo-EM) and X-ray crystallography have paved the way for solving high-resolution structures of all ion channels. Computational structural modeling approaches can be useful for modeling of homologous ion channel structures, modeling of conformational changes during channel gating, modeling of ion channel modulation by small molecules and peptide toxins, and de novo design of novel ion channel modulators. This chapter briefly summarizes the underlying fundamentals of the Rosetta molecular modeling suite and its recent advances in the context of practical applications to ion channel research.

DOI: 10.1201/9781003096214-16

14.2 Rosetta Molecular Modeling

Rosetta computational modeling software was initially developed for protein structure prediction in David Baker's lab (Simons et al. 1999). The method is based on the assumption that the native state of a protein is at the global free energy minimum and a large-scale search of conformational space for protein tertiary structures is carried out to select structures that are especially low in free energy for the given amino acid sequence. Two critical components of the Rosetta method are the procedure for efficiently carrying out the conformational search and the free energy function used for evaluating possible conformations. Rosetta can model a diverse range of molecules such as protein, RNA, DNA, sugars, and ions, and perform common and advanced tasks in computational biology such as homology modeling, protein design, protein–protein docking, protein–ligand docking, loop modeling, and electron density refinement. In Rosetta, a modeling process generally uses a combination of many steps including scoring, sampling, packing, and minimization. Each Rosetta application usually has a unique implementation of these methodologies. The following sections describe the fundamental concepts underlying Rosetta scoring, sampling, packing and minimization.

14.2.1 Scoring

14.2.1.1 Rosetta Standard Full-Atom Scoring Function

The scoring function is the most fundamental aspect of molecular modeling and is accomplished by using a master equation describing molecular interactions in protein structures. One can imagine that a scoring function is like a playbook with a set of rules defining molecular interactions observed in protein structures. Molecular models of protein structure that satisfy those rules will have favorable scores. The current Rosetta standard full-atom scoring functions, called *ref2015* (Alford et al. 2017), is a weighted linear combination of multiple score terms expressed as

$$\text{score} = \sum_i w_i E_i\left(\Theta_i, \text{aa}_i\right) \tag{14.1}$$

Scaled by an associated weight w_i, each $E_i\left(\Theta_i, \text{aa}_i\right)$ has a form of a potential energy that describes changes in a degree of freedom Θ_i of a chemical property aa_i. For example, the score term dslf_fa13 describes the chemical property of disulfide bonds that have multiple degrees of freedom: the sulfur–sulfur distance (d_{SS}), the angle formed by C_β and the two sulfur atoms ($\theta_{C_\beta SS}$), the dihedral angle of the $C_\beta - S$ bond ($\phi_{C_\alpha C_\beta SS}$), and the dihedral angle of the $S - S$ bond ($\phi_{C_{\beta 1} SSC_{\beta 2}}$). Some of the Rosetta energy terms are derived from statistical analysis of high-resolution X-ray protein structures and called knowledge-based potentials, while the others are physics-based potentials. A summary of the *ref2015* energy terms is described in Table 14.1 (Alford et al. 2017).

The *ref2015* scoring function combines numerous improvements in Rosetta over the years. These include more accurate models for H-bonds, mainchain torsion angles, side chain conformations and solvation (Alford et al. 2017). In addition, further modifications have been made to allow the modeling of D-amino acids. Specifically, to accommodate the chirality of the backbone, the backbone torsions ϕ and ψ values were allowed to be inverted in the rama_prepro, omega and p_aa_p terms (Table 14.1). Other terms such as

TABLE 14.1

Definition of terms and their weight

Energy Term	Description	Weight
fa_atr	Attractive energy for atoms on different residues	1.0
fa_rep	Repulsive energy for atoms on different residues	0.55
fa_intra_rep	Repulsive energy for atoms on the same residue	0.005
fa_sol	Solvation energy for atoms on different residues	1.0
fa_intra_sol	Solvation energy for atoms on the same residues	1.0
lk_ball_wtd	Orientation-dependent solvation energy of polar atoms	1.0
hbond_lr_bb	Energy of long-range hydrogen bonds	1.0
hbond_sr_bb	Energy of short-range hydrogen bonds	1.0
hbond_bb_sc	Energy of backbone–side chain hydrogen bonds	1.0
hbond_sc	Energy of side chain–side chain hydrogen bonds	1.0
dslf_fa13	Energy of disulfide bonds	1.25
rama_prepro	Energy associated with probability of backbone ϕ, ψ angles given the amino acid identity	0.45
p_aa_pp	Energy associated with probability of amino acid identity given backbone ϕ, ψ angles	0.4
fa_dun	Energy associated with probability of finding a rotamer is native-like given backbone ϕ, ψ angles	0.7
omega	Backbone-dependent penalty for ω dihedrals that deviate from cis and trans conformation	0.6
pro_close	Penalty for an open proline ring and proline ω bonding energy	1.25
yhh_planarity	Penalty for nonplanar tyrosine $\chi 3$ dihedral angle	0.625
ref	Reference energies for amino acid identities	1.0

fa_dun, pro_close, yhh_planarity and ref were modified accordingly (Table 14.1) (Alford et al. 2017). The parameters and scaling weights of *ref2015* were also fitted to reproduce molecular data from protein structures and thermodynamic data such as density, heat of vaporization and heat capacity from a set of small molecules. As a result, one could consider the energy unit of *ref2015* is expressed in kcal/mol. However, due to the limited chemical diversity in the small molecule set and the difficulty in interpreting statistical potentials in the physical unit, the Rosetta score is often measured in the Rosetta energy unit (REU), separating itself from the physical interpretation.

14.2.1.2 Rosetta Membrane Full-Atom Scoring Function

Rosetta membrane scoring function describes molecular interactions within the membrane environment. Early work on membrane scoring function at the low-resolution centroid level and high-resolution full-atom level has enabled structure prediction of transmembrane proteins (Yarov-Yarovoy, Schonbrun, and Baker 2006). Each protein side chain is represented as a single unified atom called centroid in the centroid mode. The centroid side chain representation and a simplified centroid scoring function create a smoother energy surface for Monte Carlo sampling during the low-resolution stage. The latest Rosetta membrane full-atom scoring function named *franklin2019* allows membrane protein structure prediction and design (Alford et al. 2020). Developed based on the latest advances of the Rosetta standard scoring function *ref2015*, *franklin2019* has an additional term fa_water_to_bilayer that describes protein stability given the water-to-bilayer transfer energy $\Delta G_{\mathrm{w,l(a)}}^{\mathrm{atom}}$ of atomic groups *a*:

$$\Delta G_{\mathrm{memb}} = \sum_{r=1}^{N_{\mathrm{res}}} \sum_{a=1}^{N_{\mathrm{atom}}(r)} \left(1 - f_{\mathrm{hyd}}\right)\left(\Delta G_{\mathrm{w,l(a)}}^{\mathrm{atom}}\right) \tag{14.2}$$

The fractional hydration f_{hyd} describes an environment where an atomic group is located. The value $f_{\mathrm{hyd}} = 0$ represents when an atomic group is fully exposed to the lipid phase, whereas $f_{\mathrm{hyd}} = 1$ represents when an atomic group is fully exposed to the water phase. The transition between the water and lipid phases, f_{hyd} is modeled considering the membrane thickness f_{thk} and the geometry of a water-exposed pore f_{pore}:

$$f_{\mathrm{hyd}} = f_{\mathrm{thk}} + f_{\mathrm{pore}} - f_{\mathrm{thk}} f_{\mathrm{pore}} \tag{14.3}$$

Benchmarking of *franklin2019* has demonstrated improvements in prediction accuracy of transmembrane protein orientations in the bilayer, native structure discrimination, native sequence recovery and free energy changes upon mutations.

14.2.1.3 Other Scoring Functions

By default, Rosetta performs protein modeling in the internal coordinates space with bond lengths, angles and dihedrals fixed to ideal values. However, recent developments have shown that optimization in Cartesian space can improve Rosetta performance in the refinement of low-resolution X-ray and cryo-EM structures and the ability to distinguish native conformations (DiMaio et al. 2015). To perform optimization in Cartesian space, an additional energy term, cart_bonded, was introduced to the standard *ref2015* scoring function allowing these lengths and angles to be treated as additional degrees of freedom. The cart_bonded terms describe the deviation from the ideal values of bond lengths, angles and dihedrals with simple harmonic potentials (Alford et al. 2017):

$$E_{\mathrm{cart\_length}} = \frac{1}{2}\sum_{i=1}^{n} k_{\mathrm{length},i}\left(d_i - d_{i,0}\right)^2 \tag{14.4}$$

$$E_{\mathrm{cart\_angle}} = \frac{1}{2}\sum_{i=1}^{m} k_{\mathrm{angle},i}\left(\theta_i - \theta_{i,0}\right)^2 \tag{14.5}$$

$$E_{\mathrm{cart\_torsion}} = \frac{1}{2}\sum_{i=1}^{l} k_{\mathrm{torsion},i}\left[f_{\mathrm{wrap}}\left(\phi_i - \phi_{i,0}, \frac{2\pi}{\rho_i}\right)\right]^2 \tag{14.6}$$

Rosetta also has scoring functions for low-resolution protein conformation search and modeling of nucleic acids, carbohydrates and noncanonical amino acids. Furthermore, Rosetta allows users to construct custom scoring functions that are used to favor or disfavor molecular properties that cannot be captured by the standard scoring function or to introduce empirical knowledge directly into the modeling process.

14.2.2 Sampling

In molecular modeling, sampling can be considered as a process of changing some degrees of freedom of a molecule or a system to find its stable state. Since the stable states are unknown, a good approach is to execute conformational changes randomly. However, if the process is entirely random, chances to observe the stable states are relatively small, especially when the number of conformational changes that can be made is limited. Sampling algorithms are designed to optimize chances of observing stable states given limited random conformational changes that can be achieved by introducing a scoring function and decision-making steps to guide the sampling process. The underlying sampling algorithm in Rosetta is the Metropolis Monte Carlo algorithm, which can be described by the following master equation:

$$P_{(\mathrm{i}\to\mathrm{j})} = \begin{cases} 1 & \text{if } E_\mathrm{j} < E_\mathrm{i} \\ e^{-(E_\mathrm{j}-E_\mathrm{i})/kT} & \text{if } E_\mathrm{j} > E_\mathrm{i} \end{cases} \tag{14.7}$$

E_i, E_j are energies of a molecule or a system in states i and j, which are scores calculated by a Rosetta scoring function. $P_{(\mathrm{i}\to\mathrm{j})}$ is the transition probability from state i to state j. When a move in some degrees of freedom is made converting a system from state i to state j, E_i and E_j are calculated. If the energy is lower than the original ($E_\mathrm{j}<E_\mathrm{i}$), the move is accepted and the state j becomes current. If the energy is greater than the original ($E_\mathrm{j}>E_\mathrm{i}$), the move is accepted with a probability being equal to $e^{-(E_\mathrm{j}-E_\mathrm{i})/kT}$, which describes the relationship between state i and state j under Boltzmann distribution. In practice, a uniform random variable k is drawn from the interval [0,1]. If $k \le e^{-(E_\mathrm{j}-E_\mathrm{i})/kT}$, the move is accepted. In Rosetta, the thermal energy value kT is treated as an arbitrary number instead of having a direct relation with the system's temperature.

14.2.3 Packing

In Rosetta, packing is a process of sampling side chain conformations for a given backbone structure. The side chain conformations are called rotamers. Rosetta uses the Dunbrack backbone-dependent rotamer library, which consists of frequencies, mean dihedral angles, and variances of rotamers as a function of the backbone dihedral angles ϕ and ψ (Shapovalov and Dunbrack 2011). Packing often happens in the late stages of a modeling process and consists of three main steps:

1) A list of all possible rotamers at each position is made for given backbone coordinates.
2) Pairwise interaction energies between rotamers are calculated and stored.
3) A search for the best combination of rotamers is performed. This is a sampling process; hence the Metropolis Monte Carlo algorithm is used. A sampling at high resolution with fixed backbone and only movable side chain conformations often leads to the system being trapped within local energy minima. Rosetta allows the Monte Carlo searches to be carried out with ramping values of the thermal energy kT to help the system escape local minima. This process is called simulated annealing and is the underlying algorithm driving Rosetta side chain packing. Simulated annealing is more computationally expensive than a simple Monte

Carlo algorithm with fixed *kT*. However, since the rotamer interaction energies are precomputed, evaluating the energy of the system is extremely fast, allowing Rosetta to find optimal rotamers.

14.2.4 Minimization

A given conformation of a protein structure is not always optimal for all numbers of degrees of freedom described by a Rosetta scoring function. This is especially true when degrees of freedom have been changed in the process of sampling or packing. Minimization optimizes a given protein conformation guided by the Rosetta scoring function. Rosetta scoring functions are differentiable with energy terms defining a way to calculate the energy given a system's degrees of freedom and calculate partial derivatives for those degrees of freedom. With such property, numerical methods such as gradient descent methods can be used to search for a nearby local minimum. There are multiple flavors of gradient descent methods implemented in Rosetta, which include a single-step repetitive search algorithm (*linmin*) and quasi-Newton methods using second derivatives (*dfpmin* or *lbfgs*) (Fletcher 1987). The current default method is a light-memory flavor of the Broyden–Fletcher–Goldfarb–Shanno method (*lbfgs_armijo_nonmonotone*) that has improved performance for larger proteins.

Rosetta allows options for minimization be done in two modes: ideal mode with an internal coordinate system using *ref2015* scoring function; and non-ideal mode with a Cartesian coordinate system using *ref2015_cart* scoring function, which is the *ref2015* scoring function with the addition of the cart_bonded term and the pro_close term set to zero. Minimization can also be done with specific degrees of freedom fixed by using a MoveMap or restricted by constraints.

14.3 Homology Modeling of Ion Channels

Homology modeling is useful for structural modeling of an ion channel based on available high-resolution structures of homologous ion channels. The initial step in homology modeling is creating accurate sequence alignment between the sequence of an ion channel of interest (query sequence) and sequence(s) of known homologous structural template(s) using amino acid sequence alignment tools such as Clustal, Jalview, HHsearch and PsiBlast. RosettaCM application (Song et al. 2013) uses this sequence alignment to generate homology models of ion channels. RosettaCM homology modeling consists of the following three stages:

1) The query sequence is threaded onto each homologous template, and all the resulting structural segments are aligned in a global frame. RosettaCM then uses fragment recombination combining backbone fragments from the aligned template structures and de novo fragments to build the threaded model.
2) Further exploration of fragment recombination is combined with Cartesian space minimization to optimize the model beyond the template conformation and close the disconnected structural segments. The structures generated from the second stage have favorable backbone geometry, but side chains are only modeled at low resolution and approximated by a single "centroid" atom. The position of the centroid atom is calculated for each residue as the average position of the side chain atoms in

residues of the same amino acid identity and with φ and ψ angles in the same 10°×10° bin and taken from known protein structures in the PDB (Simons et al. 1999).

3) The Rosetta packing approach followed by a relax protocol is used to iteratively refine the full-atom representation of side chain and backbone atom conformations using the Rosetta all-atom scoring function. At least ten thousand models are generated for each RosettaCM homology run to provide sufficient conformational sampling. The top models can then be chosen using the Rosetta scoring function, available experimental data and structural quality check tools such as Molprobity. In cases when the top scoring models have significant structural differences, clustering the top models into unique conformations can be useful to select models from the most frequently sampled conformations.

The ability to use structural information from multiple homologous templates makes RosettaCM very useful for modeling ion channels. Cryo-EM and X-ray structures of ion channels often have missing loop regions and N- and C-termini. RosettaCM allows more accurate modeling of unresolved structural regions or poor sequence-identity regions using available structural fragments from multiple templates. In addition, RosettaCM can guide the conformation sampling using available experimental and co-evolutionary data-based distance constraints and modeling of ion channel complexes with nonprotein molecules such as small molecules, carbohydrates and ions.

14.4 Symmetry Modeling of Ion Channels

Symmetric assemblies are essential for the biological functions of many proteins. Structures of homomultimeric ion channels appear in cyclic point group symmetry (C group) with subunits arranged in a ring around a single rotation axis (see example symmetric ion channel structures in Figure 14.1). Rosetta can model a variety of symmetry groups such as point, helical, wallpaper and crystal symmetry using a symmetry modeling framework (DiMaio et al. 2011). The incorporation of symmetry modeling has two advantages. First, it reduces the computational resources by performing calculations mostly on a symmetric monomer and the interfaces between neighboring subunits. Second, it enforces the symmetric geometry obtained from resolved structures during modeling, which can be

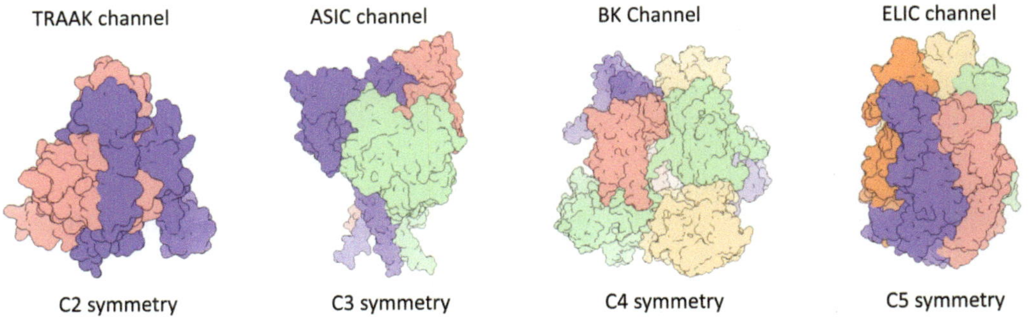

FIGURE 14.1
Types of ion channel symmetry.

essential for the channel function. For example, homotetrameric potassium channels have C4 symmetry (see BK channel in Figure 14.1). The selectivity filter has four identical loop regions that symmetrically coordinate potassium ions. Without enforcement of symmetry, modeling of these regions would result in nonidentical loop structures and loss of potassium ions coordination within the selectivity filter.

14.5 Modeling of Ion Channels with Experimental Data

One of the most valuable features of Rosetta is its ability to integrate various types of structural data into the modeling process. Unifying approaches from CS-Rosetta and PCS-Rosetta, RosettaNMR can incorporate into protein structure prediction and protein–protein docking structural data such as chemical shifts (CSs), sparse nuclear Overhauser effect (NOE), and paramagnetic NMR data such as paramagnetic relaxation enhancements (PRE), residual dipolar couplings (RDC), and pseudocontact shifts (PCS) (Leman et al. 2020). Recent advancements in cryo-EM have led to numerous ion channel structures solved at near-atomic resolution. However, accurately placing individual atoms into cryo-EM densities remains challenging and error-prone at lower resolutions. RosettaEM uses an electron density map to refine cryo-EM structures of ion channels. For example, the cryo-EM reconstruction of the TRPV1 channel using RosettaEM showed improvement in model quality while maintaining an agreement with the electron density data (Wang et al. 2016).

In addition, arbitrary experimental data can also be incorporated using the flexible constraint system in Rosetta. Rosetta constraints are semantically considered similar to the restraints in other molecular modeling approaches such as molecular dynamics simulation. The constraint sets are not fixed but instead subjected to potential energy chosen by users to reflect empirical knowledge. Applying constraints makes Rosetta versatile and valuable in modeling conformational changes in ion channels. For example, the movement of the voltage-sensing domain in voltage-gated sodium channels has been explored using disulfide cross-linking data as distance constraints (Yarov-Yarovoy et al. 2012). A preliminary template-based model was generated first to serve as a starting point for applying distance constraints between specific pairs of residues in the model. For disulfide bridge-based pairwise interaction observed experimentally, disulfide bridge interaction is enforced during modeling by specifying the cysteine pair, or, alternatively, an ~6 Å distance constraint is applied between Cβ atoms of the cysteine residues. Distance constraints between outermost carbon atoms are applied based on average distances observed between corresponding amino acids in X-ray structures to extend disulfide-locking data to modeling of interactions between native side chains. At least ten thousand models are generated for each distance constraint, and the ten lowest energy models are used as input to full-atom relax using membrane energy function without applying experimental constraints. This step allows sampling of nearby deeper energy wells that are not influenced by applying distance constraints between specific pairs of residues in a channel. After this round of modeling, the lowest energy models represent the best models of a particular conformational state of a channel. Structural models of the KCNQ1–KCNE1 channel complex have been built from experimental constraints derived from disulfide cross-linking and site-directed mutagenesis experiments using a similar approach (Kuenze et al. 2020). Distance constraints from double electron–electron resonance (DEER) experiments have been used to model structural changes in cyclic nucleotide-gated (CNG) ion channels (Evans et al. 2020).

14.6 Modeling of Ion Channels Interaction with Modulators

Small molecules and peptides are valuable molecular tools to study ion channel modulation. Studying the interaction of small molecules and peptides with ion channels at an atomic scale can provide insights into understanding the structure and function of ion channels, and for the rational design of novel therapeutics and molecular probes to study ion channel activity. Rosetta can model the interaction of ion channels with small molecules using protein–ligand docking (see an example in Figure 14.2) and interactions of ion channels with peptides using protein–protein docking.

14.6.1 Rosetta Protein–Ligand Docking

RosettaLigand (Meiler and Baker 2006) has been the most widely used application to model protein–small molecule interactions. RosettaLigand has two stages of docking: low-resolution sampling of ligand orientations within the protein receptor site and high-resolution refinement of the protein–ligand complex. The original RosettaLigand sampling protocol at the low-resolution stage used consecutive steps of translational and rotational perturbations (Meiler and Baker 2006). The recently developed Rosetta GALigandDock approach (Park et al. 2021) offers an alternative way of modeling protein–ligand interactions. The

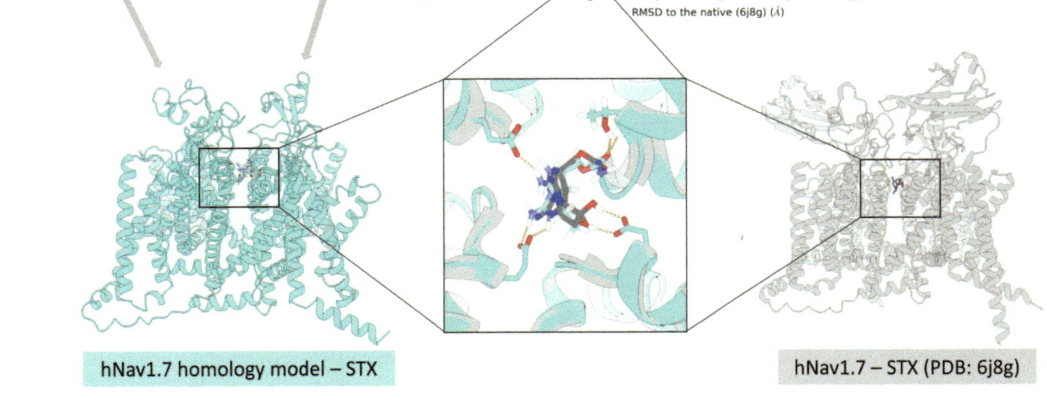

FIGURE 14.2
RosettaCM and RosettaLigand modeling of an ion channel (Nav1.7)–ligand (STX) complex.

application is built upon the new scoring function called the Rosetta generalized small molecule forcefield, RosettaGenFF, that has been parameterized using crystal structures of thousands of small molecules in the Cambridge Structural Database. Unlike protein-centric scoring functions used in RosettaLigand, RosettaGenFF parameters are optimized to identify the crystal structure conformations with the lowest energy compared to all alternatively generated ligand conformations from a single set of molecules. The GALigandDock approach is used with a newly developed genetic algorithm docking method. Instead of using Monte Carlo sampling with a single ligand conformation at the beginning (as in RosettaLigand), GALigandDock uses a starting ensemble of ligand conformations that evolve within the genetic algorithm, which allows GALigandDock to perform rapid sampling of both ligand and receptor conformations. Other utilities such as cross-docking with multiple ligands or receptors are developed to employ GALigandDock for virtual drug screening.

Both RosettaLigand and Rosetta GALigandDock are designed to be highly integrated with the Rosetta XML scripting interface, RosettaScripts, offering users a flexible way to use these applications. Thousands of docking decoys are generated for a single molecule of interest. Users can creatively utilize available filters and metrics in the RosettaScripts framework to constraint, filter and select the top ligand docking models. The following metrics are generally found to be useful:

1) total_score: Score of the whole docking protein–ligand complex. In general, only the top 5%– 10% of models based on the total score are considered.
2) interface_delta: Score of the binding interface. This metric mimics the binding free energy, calculated as:
   a. $\text{Score}_{\text{interface}} = \text{Score}_{\text{complex}} - \text{Score}_{\text{separated protein}} - \text{Score}_{\text{separated ligand}}$
   b. In general, docking poses that pass the total score filtering step can be ranked by the interface score. The top 10% of models based on interface score are often worth considering.
3) DSasa: The fractional difference in the solvent-accessible area of the ligand. The value of one means a ligand is fully buried, while the value of zero means a ligand is fully exposed. The scoring-based metrics are often accompanied by noise. This metric can be loosely used as an intermediate filter in the docking process to exclude models that are not well-engaged with the protein surface, which helps to reduce noise in the filtering steps by total score and interface score.

The docking results can be influenced by the environment, the inherent flexibility of receptors and ligands, the shape of the binding site, etc. For many cases with a well-defined binding pocket and ligand geometry, users may expect to see a funnel shape in the rmsd versus interface score plots (see an example in Figure 14.2). If this is not the case, users may need to examine all docking models that pass the filtering steps, use model clustering or apply empirical knowledge to the top model selection process. Yang et al. (2015) showed an elegant way of using RosettaLigand with Rosetta membrane scoring function and mutant cycle analysis to decipher interactions of capsaicin with TRPV1 channel binding pocket. We used RosettaLigand to identify multiple binding poses of local anesthetics in the pore lumen of a homology model of the human cardiac voltage-gated sodium channel hNav1.5 (Nguyen et al. 2019). Notably, the pore-forming domain of the voltage-gated sodium channels has a complex structure defined by pore lumen and four fenestrations. We used multiple independent starting positions of the ligands to sample binding sites within the pore thoroughly. This strategy was also employed in a study using GALigandDock to model interactions of veratridine with rat Nav1.4 channel (Craig et al. 2020).

14.6.2 Rosetta Protein–Protein Docking

Modeling of protein–protein interactions can be done using the RosettaDock approach (Gray et al. 2003). RosettaDock employs a single-trajectory Monte Carlo-based docking algorithm that explores protein–protein interactions at a low-resolution stage (centroid mode) and at high-resolution stage (all-atom mode) to refine rigid-body protein orientation and side chain conformations. Once the centroid stage is complete, the lowest energy structure of the binding complex is passed to the high-resolution stage refinement in which full side chain packing and minimization is performed using the full-atom scoring function. A newly developed low-resolution scoring function, called motif score, derived from the crystal structures of two or more interacting side chains, showed a marked improvement in accuracy over previous centroid scoring functions. Using the same low-resolution scoring scheme like motif score, RosettaSymmDock2 uses Rosetta symmetry framework to simultaneously perform docking of multiple symmetric homomers. Additionally, RosettaMPDock applies RosettaDock in conjunction with the RosettaMP framework and Rosetta membrane scoring function to explore docking two transmembrane proteins in an implicit membrane environment (Alford et al. 2015).

Protein–protein docking in Rosetta also requires generation of thousands of docking models. In general, the top 10% of docking models based on the total score are selected in the first filtering step following by another selection of the top 10% of docking models using interface score (I_sc). Docking models that pass these filtering steps can be further clustered or analyzed using empirical knowledge.

14.7 De Novo Protein Design

The most well-known functionality of Rosetta is its capability to design new proteins with novel shapes and functions or to improve existing proteins with more advanced features. Rosetta applications such as FastDesign (Bhardwaj et al. 2016) and FlexDDG (Barlow et al. 2018) can design new sequences of an existing protein binder to improve interactions with a receptor. Many other Rosetta applications can design protein binders with novel backbone topologies in various sizes. For medium- and large-sized protein binders, the MotifGraft application (Silva, Correia, and Procko 2016) can graft a functional binding motif onto a diverse set of available protein scaffolds resulting in new protein binders. FunFolDes (Silva, Correia, and Procko 2016) uses a different strategy by using an existing protein binding motif and topology constraints to guide the fragment-based folding process to generate new protein binders. Rosetta can design new mini proteins with sizes ranging from 30 to 60 amino acids stabilized by disulfide bonds, similar to cysteine knot peptide toxins that have been used as molecular tools for ion channel research (Bhardwaj et al. 2016). Relatively small cyclic peptides with mixed chirality such as 7–14 residue macrocycles can also be designed using Rosetta (Hosseinzadeh et al. 2017). Rosetta protein design approaches will be useful for designing novel ion channel modulators that can be optimized to become therapeutics or used as molecular sensors (Figure 14.3).

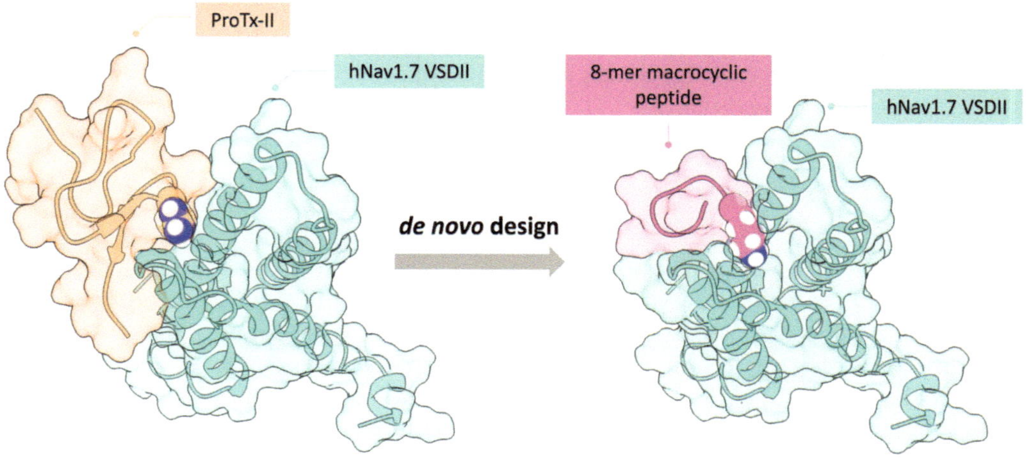

FIGURE 14.3
Rosetta design of a macrocyclic peptide based on structure of a peptide toxin (ProTx-II) in complex with an ion channel (hNav1.7).

14.8 Future Directions

Marked by the outstanding success of RosettaFold (Baek et al. 2021) and AlphaFold (Jumper et al. 2021), recent advances in machine learning and deep learning have revolutionized the field of computational protein structure prediction. Rosetta has recently employed deep-learning methods for modeling of protein complexes and protein design (Humphreys et al. 2021; Anishchenko et al. 2021). Molecular surface features derived from geometric deep learning can also be implemented in Rosetta to improve the accuracy of protein design (Gainza et al. 2020). We expect to see an exciting future with multiple new computational structural biology tools useful for ion channel research.

Resources for Learning Rosetta

Rosetta documentation, https://new.rosettacommons.org/docs/latest/Home

Rosetta tutorials, https://www.rosettacommons.org/demos/latest/Home#tutorials

PyRosetta documentation, http://www.pyrosetta.org/documentation

Suggested Readings

Alford, R. F., et al. 2015. "An integrated framework advancing membrane protein modeling and design." *PLoS Comput Biol* 11 (9):e1004398. doi:10.1371/journal.pcbi.1004398.

Alford, R. F., et al. 2017. "The Rosetta all-atom energy function for macromolecular modeling and design." *J Chem Theory Comput* 13 (6):3031–48. doi:10.1021/acs.jctc.7b00125.

Alford, R. F., et al. 2020. "Protein structure prediction and design in a biologically realistic implicit membrane." *Biophys J* 118 (8):2042–55. doi:10.1016/j.bpj.2020.03.006.
Anishchenko, I., et al. 2021. "De Novo protein design by deep network hallucination." *Nature* 600 (7889):547–52. doi:10.1038/s41586-021-04184-w.
Baek, M., et al. 2021. "Accurate prediction of protein structures and interactions using a three-track neural network." *Science* 373 (6557):871–6. doi:10.1126/science.abj8754.
Barlow, K. A., et al. 2018. "Flex ddG: Rosetta ensemble-based estimation of changes in protein-protein binding affinity upon mutation." *J Phys Chem B* 122 (21):5389–99. doi:10.1021/acs.jpcb.7b11367.
Bhardwaj, G., et al. 2016. "Accurate de novo design of hyperstable constrained peptides." *Nature* 538 (7625):329–35. doi:10.1038/nature19791.
Craig, R. A., 2nd, et al. 2020. "Veratridine: a janus-faced modulator of voltage-gated sodium ion channels." *ACS Chem Neurosci* 11 (3):418–26. doi:10.1021/acschemneuro.9b00621.
DiMaio, F., et al. 2011. "Modeling symmetric macromolecular structures in Rosetta3." *PLoS One* 6 (6):e20450. doi:10.1371/journal.pone.0020450.
DiMaio, F., et al. 2015. "Atomic-accuracy models from 4.5-A cryo-electron microscopy data with density-guided iterative local refinement." *Nat Methods* 12 (4):361–5. doi:10.1038/nmeth.3286.
Evans, E. G. B., et al. 2020. "Allosteric conformational change of a cyclic nucleotide-gated ion channel revealed by DEER spectroscopy." *Proc Natl Acad Sci U S A* 117 (20):10839–47. doi:10.1073/pnas.1916375117.
Fletcher, R. 1987. *Practical methods of optimization*. 2nd ed. New York: John Wiley & Sons.
Gainza, P., et al. 2020. "Deciphering interaction fingerprints from protein molecular surfaces using geometric deep learning." *Nat Methods* 17 (2):184–92. doi:10.1038/s41592-019-0666-6.
Gray, J. J., et al. 2003. "Protein-protein docking with simultaneous optimization of rigid-body displacement and side-chain conformations." *J Mol Biol* 331 (1):281–99.
Hosseinzadeh, P., et al. 2017. "Comprehensive computational design of ordered peptide macrocycles." *Science* 358 (6369):1461–6. doi:10.1126/science.aap7577.
Humphreys, I. R., et al. 2021. "Computed structures of core eukaryotic protein complexes." *Science* 374 (6573):eabm4805. doi:10.1126/science.abm4805.
Jumper, J., et al. 2021. "Highly accurate protein structure prediction with AlphaFold." *Nature* 596 (7873):583–9. doi:10.1038/s41586-021-03819-2.
Kuenze, G., et al. 2020. "Allosteric mechanism for KCNE1 modulation of KCNQ1 potassium channel activation." *Elife* 9. doi:10.7554/eLife.57680.
Leman, J. K., et al. 2020. "Macromolecular modeling and design in Rosetta: recent methods and frameworks." *Nat Methods* 17 (7):665–80. doi:10.1038/s41592-020-0848-2.
Meiler, J., and D. Baker. 2006. "Rosettaligand: protein-small molecule docking with full side-chain flexibility." *Proteins* 65 (3):538–48.
Nguyen, P. T., et al. 2019. "Structural basis for antiarrhythmic drug interactions with the human cardiac sodium channel." *Proc Natl Acad Sci U S A* 116 (8):2945–54. doi:10.1073/pnas.1817446116.
Park, H., et al. 2021. "Force field optimization guided by small molecule crystal lattice data enables consistent sub-angstrom protein-ligand docking." *J Chem Theory Comput* 17 (3):2000–10. doi:10.1021/acs.jctc.0c01184.
Shapovalov, M. V., and R. L. Dunbrack, Jr. 2011. "A smoothed backbone-dependent rotamer library for proteins derived from adaptive kernel density estimates and regressions." *Structure* 19 (6):844–58. doi:10.1016/j.str.2011.03.019.
Silva, D. A., et al. 2016. "Motif-driven design of protein-protein interfaces." *Methods Mol Biol* 1414:285–304. doi:10.1007/978-1-4939-3569-7_17.
Simons, K. T., et al. 1999. "Improved recognition of native-like protein structures using a combination of sequence-dependent and sequence-independent features of proteins." *Proteins* 34 (1):82–95.
Song, Y., et al. 2013. "High-resolution comparative modeling with RosettaCM." *Structure* 21 (10):1735–42. doi:10.1016/j.str.2013.08.005.
Wang, R. Y., et al. 2016. "Automated structure refinement of macromolecular assemblies from cryo-EM maps using Rosetta." *Elife* 5:e17219. doi:10.7554/eLife.17219.

Yang, F., et al. 2015. "Structural mechanism underlying capsaicin binding and activation of the TRPV1 ion channel." *Nat Chem Biol* 11 (7):518–24. doi:10.1038/nchembio.1835.

Yarov-Yarovoy, V., et al. 2006. "Multipass membrane protein structure prediction using Rosetta." *Proteins* 62 (4):1010–25. doi:10.1002/prot.20817.

Yarov-Yarovoy, V., et al. 2012. "Structural basis for gating charge movement in the voltage sensor of a sodium channel." *Proc Natl Acad Sci U S A* 109 (2):E93–102. doi:10.1073/pnas.1118434109.

15

Molecular Dynamics

Lucie Delemotte and Yun Lyna Luo

CONTENTS

15.1 Introduction

Molecular dynamics (MD) simulations of ion channels, built on high-resolution structures obtained by X-ray crystallography, cryo-electron microscopy (cryo-EM) or homology models, are frequently used to gain dynamic and mechanistic submolecular insights, and to drive drug design in the ion channel field. On one hand, various MD techniques are well suited to monitor the response of a system to an applied stimulus qualitatively, such as a change in transmembrane (TM) potential, mechanical force, pH or temperature. On the other, they enable in principle to recover complete conformational ensembles and to quantitatively compare the free energy of ensembles obtained in response to different perturbations. This chapter is not aimed to teach the reader to set up, run and analyze MD

DOI: 10.1201/9781003096214-17

simulations, but rather to introduce the most relevant and up-to-date aspects of MD simulations in ion channel biophysics.

15.1.1 Basic Introduction to MD Simulations

MD simulations hinge on a simple principle: the evolution over time of a system containing particles can be predicted by integrating the classical equations of motion, much like the trajectory of a cannonball or of a skier going down a hill (Figure 15.1A).

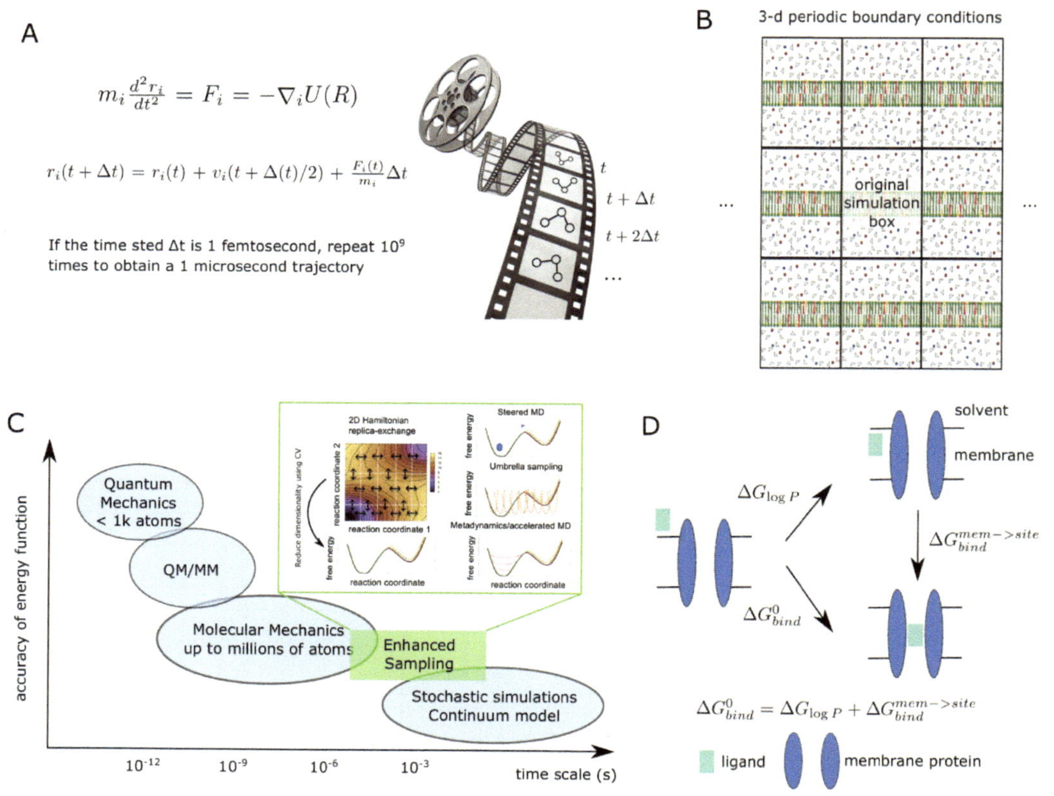

FIGURE 15.1
Basic principles of MD simulations and enhanced sampling techniques. (A) MD simulations rely on the discrete integration of the equations of motion for a set of particles representing atoms. The output of an MD simulation is a trajectory containing the position and velocities of all particles at the consecutive time steps (represented by the schematic movie roll). (B) To avoid finite size effects, a classical MD simulation considers a system replicated in three directions of space, such that the central system is in interaction with periodic images of itself on each side of the system (3D periodic boundary conditions, PBCs). The bilayer lipids are represented as green, red and yellow molecules; the water molecules as gray triangles; and the ions and red and blue spheres. (C) With each modeling technique comes a choice in terms of accuracy/time and length scale trade-off. Small systems can be modeled using a quantum description on the picosecond time scale. Large systems require a continuum description. MD simulations using molecular mechanics force fields are well suited to sample 10^5 to 10^6 particles over the nanosecond to microsecond time scale. Enhanced sampling MD simulations extend these time scales by leveraging various physical and statistical principles. (D) Ligands can bind to ion channels directly via hydrophilic pathways from solvent to channel, or partition in lipid bilayers first and then diffuse to binding sites. Drug binding can be studied using physical or alchemical-enhanced sampling simulations.

Integrating the classical equations of motion over time generates a set of atomic positions r_i and velocities v_i. The forces F_i acting on the particles with mass m_i are derived from a potential energy function $U(R)$ that depends on the particles' positions r_i. In the case of Newtonian dynamics, these forces are related to the second derivatives of the positions, their acceleration (Figure 15.1A).

A system containing an ion channel in its near-native environment (lipid bilayer and ionic solution) contains hundreds of thousands to millions of atoms of carbon, oxygen, nitrogen, phosphorus, hydrogen and ions. In classical MD simulations, these are considered as classical particles with a specified charge (often a partial charge) and mass. All these particles interact with one another, making the calculation of their evolution over time (a so-called trajectory) more complicated than that of a single cannonball. Because of the system and the interaction potential complexity, the equations of motion thus cannot be integrated analytically, and instead, time must be discretized in small increments (called time steps), and the equations of motion integrated numerically (Figure 15.1A). Several algorithms, such as the Verlet and velocity Verlet algorithms, have been derived to perform this numerical integration (Allen and Tildesley 1987). In practice, the choice of integration algorithm does not substantially affect the quality of the simulation. On the other hand, an important aspect to consider is the choice of time step: because the particles (atoms) are very light, they move rapidly, and the time steps taken to integrate the equations of motion need to be of the order of femtoseconds (10^{-15} seconds).

MD simulations generally aim at reproducing ion channel behaviors occurring on the submillisecond to second time scale. Thus, the equations of motion need to be integrated many billions of times to access these time scales. Since these calculations are extremely intensive computationally, and multiple copies (replicas) of simulations are required to obtain statistically-meaningful results, algorithms grouped under the umbrella term "enhanced sampling" simulations have been proposed to sample the Boltzmann-distributed conformational ensemble of the system in a more efficient way, without compromising the atomistic spatial resolution (Figure 15.1C; see Section 15.1.4).

When Newton's equations of motion are integrated, the conformational ensemble that is sampled by an MD simulation has a constant number of particles (N), a constant volume (V) and a constant total energy (E). This ensemble is called the microcanonical ensemble and is represented by the acronym NVE. To mimic the conditions of the biological sample, however, we are much more interested in modeling a system that is placed under constant temperature and constant volume/pressure. The canonical or NVT ensemble (constant number of particles, volume and temperature) and isothermal–isobaric or NPT ensemble (constant number of particles, pressure and temperature) are obtained through the application of various thermostatting (keeping a constant target temperature) and barostatting (keeping a constant target pressure) strategies. MD simulation software offer a range of possible algorithms, such as the Nosé–Hoover thermostat, Langevin thermostat, Berendsen barostat, Parrinello–Rahman barostat, Nosé–Hoover Langevin piston barostat and Monte Carlo barostat. Most of the barostats provide the option to impose a semi-isotropic pressure control for membrane systems, in which the normal and the lateral pressure are coupled independently. The choice of thermostat and barostat is usually not the biggest source of issues. Note, however, that the Berendsen thermostat should be avoided for long production runs due to repeated rescaling of particle velocities that can cause the flying ice cube artifact (in which a droplet of water in motion freezes artificially). In addition, strong coupling in the Langevin thermostat should be avoided for computing kinetic properties, as it causes drops in diffusion constants. For specific applications, other ensembles can also be sampled, such as the constant surface tension (NPγT) ensemble or the grand canonical (μVT) ensemble, in which membrane surface tension γ or chemical potential μ is kept constant.

The modeled system contains an ion channel surrounded by a small membrane patch (usually corresponding to two or three lipid molecule layers around the channel), and plunged in a small solution bath (extending at least 1.5 nanometers beyond the channel). To mimic the cellular environment, one needs to account for the behavior of the system at the simulation box boundaries. The usual treatment consists of applying periodic boundary conditions (PBCs) where the system is virtually replicated infinitely in all three spatial directions (Figure 15.1B). This makes it such that the system is surrounded by images of itself instead of by vacuum, and the overall system being modeled is an infinite membrane stack. Since most interactions decay fast in aqueous solution, if the copies of ion channels are separated by at least three nanometers, the environment felt by the channel is close to the native one. According to the minimum image convention, a particle interacts with the other particles closest to it, even if they are located in a periodic image. Since all periodic images are identical, interaction energy is only calculated once per pair and PBCs thus do not increase the computational cost of the calculation.

15.1.2 Force Fields and the Potential Energy Function

To integrate the equations of motion, an important ingredient is the potential energy function $U(r)$. This potential energy function, often referred to as the force field, describes the interactions between all particles in the system. Classical force fields (family names CHARMM, AMBER, OLPS) have been parametrized and refined over the years, and rely on the important principle of transferability: a carbonyl oxygen will interact in the same way with its environment wherever it is located in the system (e.g., in a protein loop or in a folded TM helix). These force fields list parameter values in large tables that parameterize a potential energy function. The functional form of the force field is generally summarized as

$$U(r) = E_{bond} + E_{angle} + E_{torsion} + E_{el} + E_{vdw} \tag{15.1}$$

where the bonded terms are expressed as

$$E_{bond} = \sum_{i,j} k_{ij}^{b}\left(r_{ij} - r_{ij}^{0}\right)^2 \tag{15.2}$$

$$E_{angle} = \sum_{i,j,k} k_{ijk}^{a}\left(\alpha_{ijk} - \alpha_{ijk}^{0}\right)^2 \tag{15.3}$$

$$E_{torsion} = \sum_{i,j,k,l} k_{ijkl}^{t}\left(1 + \cos\left(n\omega_{ijkl} - \omega_{ijkl}^{0}\right)\right) \tag{15.4}$$

where r_{ij} is the distance between atoms i and j; α_{ijk} the angle between atoms i, j and k; ω_{ijkl} the dihedral angle between i, j, k and l; and k_{ij}^{b}, k_{ijk}^{a}, and k_{ijkl}^{t} on one hand and r_{ij}^{0}, α_{ijk}^{0}, and ω_{ijkl}^{0} on the other are the force constants and equilibrium values for the bond stretch, angle bend, and dihedral torsion deformations.

A coulombic potential describes the interaction between partial charges q_i assigned to atoms:

$$E_{el} = \sum_{i,j} \frac{q_i q_j}{e_{ij} r_{ij}} \tag{15.5}$$

where q_i is the partial charge on atom i; and a van der Waals potential describes short-range repulsion and long-range attraction between pairs of atoms:

$$E_{vdw} = \sum_{i,j} \varepsilon_{ij} \left[\left(\frac{\sigma_{ij}}{r_{ij}} \right)^{12} - 2 \left(\frac{\sigma_{ij}}{r_{ij}} \right)^{6} \right] \tag{15.6}$$

where ε_{ij} and σ_{ij} are Lennard–Jones parameters for van der Waals interactions.

Polarizable force fields contain additional nonbonded terms corresponding to a multipolar expansion and/or the description of a Drude particle.

Proper treatment of long-range nonbonded interactions is essential for ion channel simulations. Van der Waals interactions (Lennard–Jones potential) decay rapidly and are typically turned off beyond a certain cutoff distance (generally 10 to 12 Å) with a smoothing function. Electrostatic interactions described by a coulomb potential decay much more slowly, which makes it necessary to consider their contribution beyond this cutoff distance. The particle mesh Ewald (PME) method is commonly used for that purpose. Membrane systems are particularly sensitive to the truncation of long-range interactions. It is advisable to use the nonbonded setup that is consistent with each force field's parameterization protocol. A set of tested protocols for several simulation programs (NAMD, AMBER, GROMACS, OpenMM, CHARMM) are available in the widely used CHARMM-GUI membrane builder (Lee et al. 2016).

These classical force fields (also called additive nonpolarizable force fields) consider particles with a fixed charge. It is now known that this approximation may fail when atomic polarization is important, such as when an ion moves across a narrow pore or a low dielectric medium (i.e., a lipid bilayer). In such a case, a polarizable force field is likely needed for accurately describing the potential energy. These consider either a multipolar expansion for each particle (and assign it a charge, a dipole, a quadruple, and possibly higher-order multipoles) or a Drude model (in which a mobile charge particle is attached to a spring). In contrast to nonpolarizable force fields, which have benefited from decades of optimization efforts, these models are much more recent. For instance, free energy profiles of cation permeation through a very simple channel (gramicidin A) have been simulated using three polarizable force fields (flucQ, AMOEBA and Drude) and have yielded encouraging results. Nevertheless, further tests are needed to evaluate their domain of applicability. Although the computational cost related to the use of such force fields is larger than for classical ones, it is much more tractable than modeling these systems at a quantum level (Figure 15.1C).

In cases where atomistic details are less important, coarse-grained models provide an alternative way to reduce the computational cost. In such models, several atoms are grouped into an interaction center (a pseudo-particle, also called a bead). The description of the system is thus coarser, but the particle number being reduced makes it possible to sample larger systems with longer time scales. In addition, the fact that pseudo-particles are heavier enables using a larger time step (tens of femtoseconds instead of 2 fs). One of these force fields, the Martini force field, has gained unprecedented success in simulating protein–lipid binding. However, the loss of atomistic interactions such as hydrogen bonds makes this type of force field only suitable to model specific processes, in which protein conformational change or a detailed description of protein–ligand interactions are not needed. Besides treating all the atoms or coarse-grained beads explicitly, ion channels have been simulated using implicit solvent and/or membrane models in which the volume occupied by water and lipids is treated as a continuum media with fixed dielectric constants. However, when the interaction between channel and environment is important, an explicit model of all the atoms is essential. There are also multi-scale approaches, in which only the water molecules beyond the simulations box are treated implicitly as reaction field, which can be useful but have not gained as much traction in the field of ion channel simulations.

15.1.3 MD Simulations Analysis and the Notion of Free Energy

The notion of free energy is crucial when attempting a quantitative understanding of the role of perturbation such as ligand binding, application of a transmembrane potential or a change in temperature. Understanding how to obtain free energies from MD simulations requires recalling some basic principles from statistical mechanics, including that of ensemble averaging, the ergodic principle, and the calculation of probability distributions. These notions are briefly defined next.

The average value of a property of interest can be computed from simulations at equilibrium by calculating the ensemble average of a function that measures the property of interest. These properties are often called *collective variables* (CVs), reaction coordinates or degrees of freedom, and can be represented by a variable $s = f(r)$ that depends on particle positions r (and more rarely particle velocities).

The ensemble average is approximated by a time average, assuming the ergodic hypothesis that states that the average of a parameter over time is equivalent to the average over the statistical ensemble.

$$s = \int s p(s) ds \approx \frac{1}{N_t} \sum_t s \tag{15.7}$$

The uncertainty of the measurement is generally estimated assuming a Gaussian distribution of the collective variable, as the standard deviation $\sigma(s) = \sqrt{\int ds\, p(s)(s-s)^2}$.

The probability distribution $p(s)$ can be estimated as

$$p(s) = \frac{1}{Z} \int \delta(s - s(r)) e^{-\frac{U(r)}{k_B T}} ds \tag{15.8}$$

where k_B is the Boltzmann constant, T the temperature of the system, $U(r)$ the internal energy of a configuration with coordinates x, and Z the partition function that lists all the possible configurations that the system can assume and serves as a normalization constant. The Dirac function $\delta(s - s(r))$ restricts the ensemble to the consideration of configurations of the system that take on the collective variable value s. Since computing Z is practically impossible, $p(s)$ is in fact defined relative to a constant, such as the distribution in bulk, for example. This is generally not a limitation, since in most applications computing relative free energies suffices, rather than absolute ones, which cancels out the need for the calculation of Z.

The calculation of probability distributions enables the direct calculation of free energies, via the Boltzmann relationship: $G(x) \propto -k_B T \log p(r)$, which relates the free energy G of a configuration with coordinates r to its probability distribution $p(r)$. The partially integrated free energy expresses the free energy profile (or surface, or landscape) along a collective variable s:

$$G(s) \propto -k_B T \log p(s) = -k_B T \log \int p(r)\delta(s - s(r))dr \tag{15.9}$$

This free energy profile is often also called a *potential of mean force* (PMF) for historical reasons. The collective variable set can be mono- or multidimensional, making the resulting surface one or multidimensional.

The free energy difference between two states A and B is related to the relative probability of the system to be found in a given state A (P_A) and in a state B (P_B), and is computed by integrating the free energy profile over the area of space that corresponds to the states:

$$\Delta G_{AB} = -k_B T \log \frac{P_B}{P_A} = -k_B T \log \frac{\int_B e^{-\frac{U(s)}{k_B T}} ds}{\int_A e^{-\frac{U(s)}{k_B T}} ds} \tag{15.10}$$

15.1.4 Enhanced Sampling Simulations Schemes

Computational techniques gathered under the general concept of "enhanced sampling" enable the sampling of larger conformational ensembles while retaining an atomistic level of description (Figure 15.1C). These algorithms act on different physical or mathematical principles known to enhance the sampling: increasing the temperature, adding external forces or bias potentials to selected collective variables, exchanging information between system replicas, restricting the sampling to specific regions of space, adaptively seeding trajectories from undersampled regions of conformational space, etc. Such schemes are particularly useful when they can be coupled to *free energy estimation methods* that make it possible to recover the conformational ensemble of interest and the probabilities associated with each configuration.

Free energies can be computed using (1) PMF-based approaches in which the conformational change is enhanced along a set of predefined collective variables (umbrella sampling, metadynamics and adaptive biasing force are variants of such schemes), and (2) alchemical approaches such as free energy perturbation or thermodynamic integration. In alchemical approaches, a bound ion or molecule can be modified into a different chemical type, and one obtains a relative binding free energy. For example, one can transform a K^+ ion into a Na^+ ion to calculate the free energy change associated with transporting a K^+ ion relative to a Na^+ ion. One could also morph a drug A into a related drug B to compute the relative binding affinity of drugs A and B in a binding site. To obtain absolute binding free energies, the ion or molecule needs to also be fully decoupled from its binding site and from the bulk environment (Figure 15.1D). Alchemical approaches do not rely on a physical pathway, but the end states need to be well defined and accurately sampled (e.g., the relative binding free energy between K^+ and Na^+ can be calculated only if the binding sites remain the same). (3) Generalized-ensemble or extended-ensemble methods rely on sampling systems in a different ensemble, sometimes by changing the temperature, and sometimes by considering a different Hamiltonian (modifying the energy function that governs the dynamics of the system).

A popular method in this category is the replica exchange molecular dynamics method (REMD), which includes temperature exchange (T-REMD) and Hamiltonian exchange (H-REMD). (4) Nonequilibrium (out-of-equilibrium) approaches hinge on steering along a collective variable. Steered and targeted MD (SMD and TMD, respectively) belong to this category. When the forces acting on the system are small enough, these calculations can in principle be "unbiased" to recover the equilibrium free energy. However, in practice, nonequilibrium approaches are better suited to explore different regions of the configurational space than to recover free energy landscapes.

Another type of strategy that promotes sampling of the conformational space is to seed trajectories from different regions of space. This can be done in an adaptive way (*adaptive seeding*) where the regions that have not been sampled much are seeded from to improve sampling specifically in those regions. The free energies (and in that case even the kinetics of processes) can be recovered using analysis methods such as *Markov state modeling (MSM)*. A specific case of adaptive seeding that has proved valuable is the *string method with swarms of trajectories*, where iterative launching of short simulations is used to converge the lowest free energy path between two available structures. This method is particularly interesting when many predefined collective variables appear important to describe the change of interest and PMF-based approaches are not well suited since they would require converging a high-dimensional CV space, which is computationally prohibited.

15.2 Applying External Stimuli in MD Simulations

Most ion channels are gated by the application of external stimuli. MD simulations enable to directly apply the stimulus of interest and to then monitor the response of the ion channel.

15.2.1 Transmembrane Potential (ΔV)

The transmembrane (TM) potential (ΔV) may be the most relevant stimulus for ion permeation or for gating of voltage-sensitive ion channels. The ΔV can be applied in two main ways, imposing an *ionic imbalance* between both sides of the membrane, using a protocol commonly referred to as computational electrophysiology (Kutzner et al. 2011), or applying an *external electric field* (E) (Figure 15.2). The former consists of creating a charge imbalance between the two sides of the membrane (Figure 15.2.B). Due to the use of PBCs, the system must be isolated from its replica in the direction normal to the membrane (otherwise the charge imbalance would effectively be canceled out). This is done by inserting a vacuum slab between images or more commonly by considering a double bilayer system in a single simulation box. In that case, the charge imbalance is imposed between the central and the outer compartment. The TM potential is related to the charge imbalance imposed Q via $\Delta V = Q / C$, where C is the membrane capacitance. When ions are conducted, the charge imbalance drops. Schemes have thus been introduced to maintain the charge imbalance by constantly monitoring the ion distribution and replenishing the ion bath. The external electric field method consists of adding a force acting along the normal to the lipid bilayer membrane on all charged particles (Figure 15.2A). Because of the presence of the low dielectric membrane, the polarization of the solvent leads to the

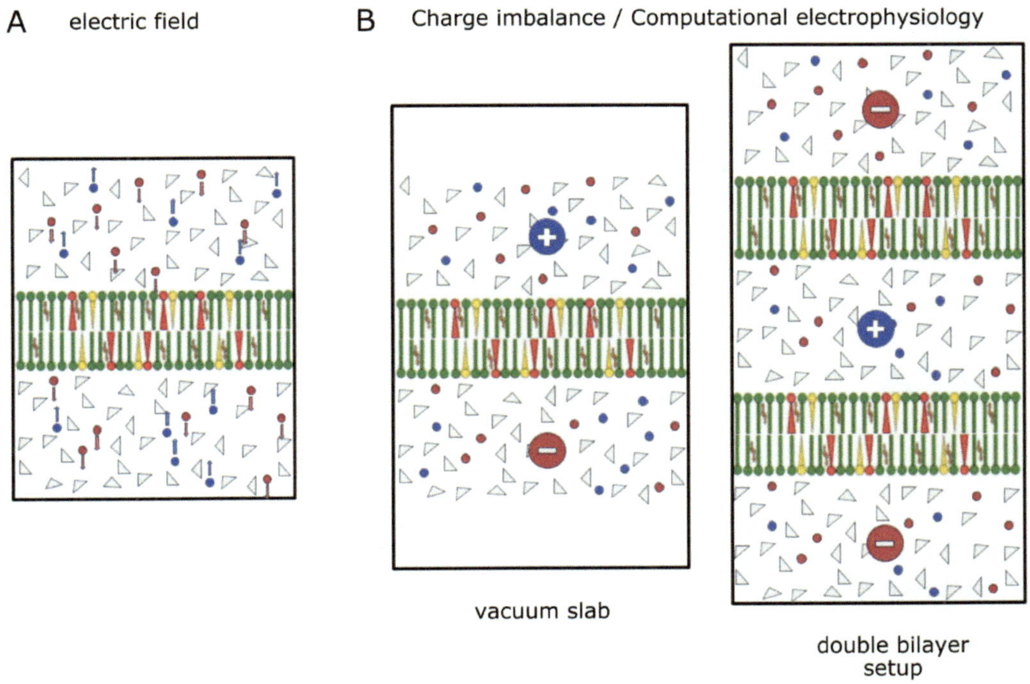

FIGURE 15.2
Two classes of methods enable the application of a transmembrane potential ΔV. (A) The electric field method, in which a force is applied to charged particles (represented here as blue arrows acting on positive charges and red arrows acting on negative charges) and (B) the charge imbalance method. Two variants of this scheme exist: one in which the periodic images are isolated from one another using a vacuum slab and the other in which a double bilayer setup is used. This class of method has also been called computational electrophysiology. The distribution of ions leads to an overall charge imbalance, represented as a blue or red symbol. Representations are the same as in Figure 15.1B.

appearance of a TM potential $\Delta V = EL_Z$, where L_Z is the size of the simulation box along the membrane normal.

Both approaches are considered valid at present, with the electric field approach being convenient to implement and the charge imbalance one intuitively appearing more realistic.

Supraphysiological potentials (~±500 to ±750 mV) are routinely applied to make it possible to observe the system's response on the MD simulation time scales. The comparison with experimental observables indicates that such high voltages do allow uncovering relevant channel conformations or model relevant phenomena. However, such high voltages significantly accelerate the response of systems such that one cannot rely on the estimation of kinetic properties, and the integrity of the membrane and protein under such high voltages should be monitored due to a risk of electroporation.

15.2.2 Mechanical Force

Mechanosensitive (MS) ion channels open upon a change in membrane tension, membrane topology, or other mechanical forces (Figure 15.3A). The free energy associated with the MS channel opening can be approximated by $\Delta G = \Delta G_{protein} - \Delta G_{memb} - \gamma \Delta A$, in which γ is the membrane tension, ΔA is the relative change in the protein projected area,

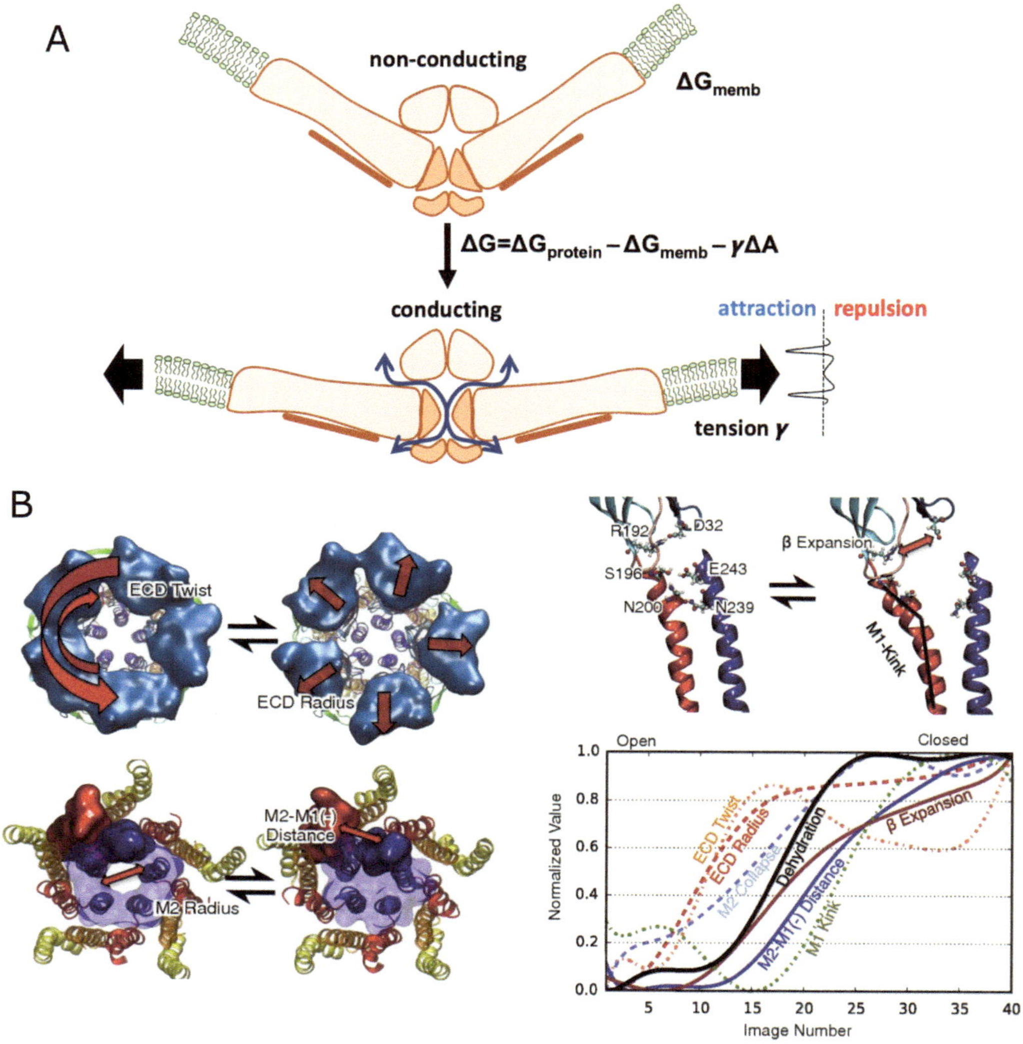

FIGURE 15.3
(A) Schematic representation of a mechanosensitive ion channel conformational change occurring upon the change of surface tension and/or membrane topology. (B) String method simulations of a pentameric ligand gated channel (left: top views; right: side views of a part of a single subunit) illustrating the various collective variables used as input to the string-of-swarms method, and the order in which they evolve, as the conformational transition between the open and closed state progresses (Lev et al. 2017). (Adapted from Lev et al. 2017. Copyright 2017 National Academy of Sciences.)

$\Delta G_{\text{protein}}$ is the free energy of protein conformational change in the absence of membrane tension, and ΔG_{memb} is the free energy cost of membrane deformations, such as bending or thinning. In MD simulations, membrane tension is a force existing at the membrane–water interface and can be controlled using the constant *surface tension ensemble* (NPγT), in which normal pressure P, surface tension γ, and temperature T are kept constant. For a flat membrane, $\gamma = \int \left(P_{zz} - \frac{P_{xx} + P_{yy}}{2} \right) dz$, in which P_{zz} is the constant normal pressure (1 atm), and P_{xx} and P_{yy} are tangential components of the pressure tensor. For a stress-free membrane, bilayer tension $\gamma \approx 0$. Note however, that although bilayer tension is guaranteed to be zero under PBCs, individual leaflet $\gamma \approx 0$ is not always satisfied in an asymmetric bilayer or curved bilayer. A positive γ increases the surface area of a lipid bilayer and decreases bilayer thickness due to volume incompressibility.

Although it is straightforward to simulate the channel under a predefined γ, care is needed to avoid membrane rupture. This is because even though the mechanical activation threshold varies between channels (e.g., ≈2 mN/m for Piezo1 versus ≈10 mN/m for MscL), large tensions are inevitably required to accelerate the opening within the simulated timescale. Gradually increasing the membrane tension can help to keep the bilayer integrity within the microsecond time scale, while instantaneously increasing the tension to 40 mN/m causes membrane rupture in less than 100 ns. Other approaches to accelerate the channel opening while maintaining the membrane integrity include directly applying forces to protein atoms, protein-bound lipids, or distributing the applied force based on the lateral distance between lipids and protein using steered MD.

15.2.3 pH, Temperature and Others

Many ion channels are sensitive to cytoplasmic or extracellular pH. Conventional MD simulations employ fixed protonation states. The effect of a change in protonation state is thus often studied by contrasting two simulations carried out with two different protonation states. To allow residues to change protonation states in response to the system's dynamics, *constant pH MD* (pHMD) simulations have been developed. For ion channels, pHMD can be conducted using an *explicit membrane* and solvent or using a *hybrid method* that samples conformational dynamics explicitly but computes the pKa of selected residues using an implicit membrane model (Huang, Henderson, and Shen 2021). *pH replica exchange* (a special case of H-REMD) simulations are frequently used to accelerate convergence of protonation and conformational space sampling. For all pHMD methods, more accurate and polarizable force fields are desirable.

Other stimuli can also be applied explicitly. For example, changes in temperature are simply modeled by changing the target temperature of the thermostat. The effect of ligand binding is modeled by inserting the ligand in its binding pocket and explicitly modeling the effect of its binding (see Section 15.6).

15.3 Sensing

Ion channels sense applied stimuli, and their sensing properties can often be detected by analyzing the MD simulations.

15.3.1 Voltage

In voltage-sensitive proteins, the conformational change occurring upon a change in ΔV is accompanied by the transfer of a *gating charge* (Q). Q is the conjugate variable of the transmembrane voltage ΔV such that multiplying these two quantities measures the excess free energy change between two states, 1 and 2, due to ΔV:

$$Q\Delta V = \Delta G_2(\Delta V) - \Delta G_1(\Delta V) \tag{15.11}$$

ΔG's are not easily estimated, but when the structures of the two activation states at both extremes of the conformational change are available, the contribution of different elements j (amino acids or other components of the systems such as bound ions) to the gating charge can be approximated by $Q^j = \sum_i q_i^j \left[f_2\left(r_i^j\right) - f_1\left(r_i^j\right) \right]$, where $f_\lambda\left(r_i^j\right)$ represents the state-specific λ coupling between the element's charge q^j and ΔV. This coupling function is estimated as the rate of change of the local electrostatic potential φ_M at the position of the element r_i^j upon application of ΔV, and the sum runs over all particles i making up the element j (Roux 2008). φ_M can be estimated from solving the Poisson equation on a grid, using as input the time-averaged charge distribution in the system samples using MD simulations.

15.3.2 Temperature

The molecular basis for temperature sensing remains at the basis of much controversy. Nevertheless, a consensus is emerging in the form of the "heat capacity hypothesis," putting forth the idea that cold and heat- sensing are two sides of the same coin, and can be related to a change in solvation interfaces. In particular, a conformational change that would promote the solvation of a hydrophobic region would lead to high temperature sensing via an increase in heat capacity. MD simulations are well placed to directly estimate changes in solvent accessibility. This is done by evaluating the water molecule occupancy along a collective variable of interest, by simply counting the time-averaged number of water molecules in a specific cavity or projecting the water density along a relevant collective variable such as the channel's pore axis.

In addition, MD simulations have been used to evaluate whether a specific conformational change (usually sampled using an enhanced sampling technique) is made easier (lowers free energy barriers) by a change in temperature (Kasimova et al. 2018). In such a case, a change in hydration can be directly related to the temperature-dependent conformational change.

15.4 Gating and Conformational Changes

Functional states associated with a channel structure are traditionally assigned by estimating the pore radius. However, it is now well known that hydrophobic gates can block ion permeation without sterically occluding the pathway. MD simulations are thus powerful to assign functional states more accurately, as they can evaluate the propensity of a pore to be hydrated directly. The channel annotation package (CHAP) tool has been developed to accomplish this in a systematic way (Klesse et al. 2019).

Gating often occurs on time scales that extend beyond the potential of classical MD simulations. The special-purpose supercomputer Anton, uniquely designed to run MD simulations by DE Shaw Research, has made it possible to run simulations reaching the multimicrosecond time scale. It has in specific cases been successful at observing ion channel gating as a result of the application of external stimuli. Access to this supercomputer is limited, however, and many studies designed to model major conformational changes, such as ion channel gating, are performed using various enhanced sampling techniques (Figure 15.1C). All these techniques come with their assumptions, and a good command of the way the sampling is enhanced is necessary to interpret the results. Nonequilibrium techniques such as steered MD usually only enable one to guess transition pathways, and the resulting structures should not be considered equivalent to experimental structures or structures obtained via more careful sampling protocols. When equilibrium methods such as PMF methods are employed, a good control of the collective variable(s) employed is necessary; for example, adding external forces to degrees of freedom assuming an iris-like model of gate opening will enhance the sampling along these specific degrees of freedom and will not necessarily evaluate if an alternative model of gating is possible.

The smallest conformational changes can be modeled precisely using PMF-based methods. For instance, to model the C-type inactivation of several potassium channels, the degrees of freedom that change along the conformational change are well identified, since high-resolution structures of the active and inactive states are resolved. In addition, the conformational changes involve small selectivity filter backbone movements that are easily sampled, and PMF-based techniques coupled to free energy estimation methods enable the accurate estimation of the free energy landscape associated with the inactivation process. For this application, the collective variables commonly used describe the conformation of the selectivity filter (measuring the diameter of the filter at its most constricted site) and the progression of the ions along the selectivity filter axis.

For larger conformational changes, reducing the dimensionality of the space to sample a one-dimensional path has proven its worth. Indeed, trying to sample a high-dimensional CV space would be computationally too expensive. For example, gating of a pentameric ligand-gated ion channel upon agonist binding has been studied using the *string method with swarms of trajectories* (Figure 15.3B). The order in which rearrangements between domains occur can be identified from the shape of the string along the different CVs. In addition, the effect of a perturbation, such as drug binding, can be looked at by comparing the free energy of the various states along the pathway in the presence and absence of different perturbations.

15.5 Ion Conduction and Selectivity

Two main categories of MD simulation protocols have been used to study conduction and selectivity: enhanced sampling free energy calculation methods and methods where conduction is modeled explicitly.

15.5.1 Enhanced Sampling Free Energy Calculation Methods

Enhanced sampling MD simulations enable the characterization of the equilibrium distribution of ions in an ion channel pore (Figure 15.4A). In general, the collective variable of interest for ion permeation is well characterized as the progression of the ion along the pore axis. Many flavors of PMF-based approaches have been applied, such as umbrella sampling, metadynamics and adaptive biasing force, using as collective variable the position of the permeating ion relative to the center of the bilayer or the center of the channel. The result of such a protocol is a free energy profile, in which valleys mark a favorable binding site for the permeating ion, and barriers mark unfavorable regions that need to be

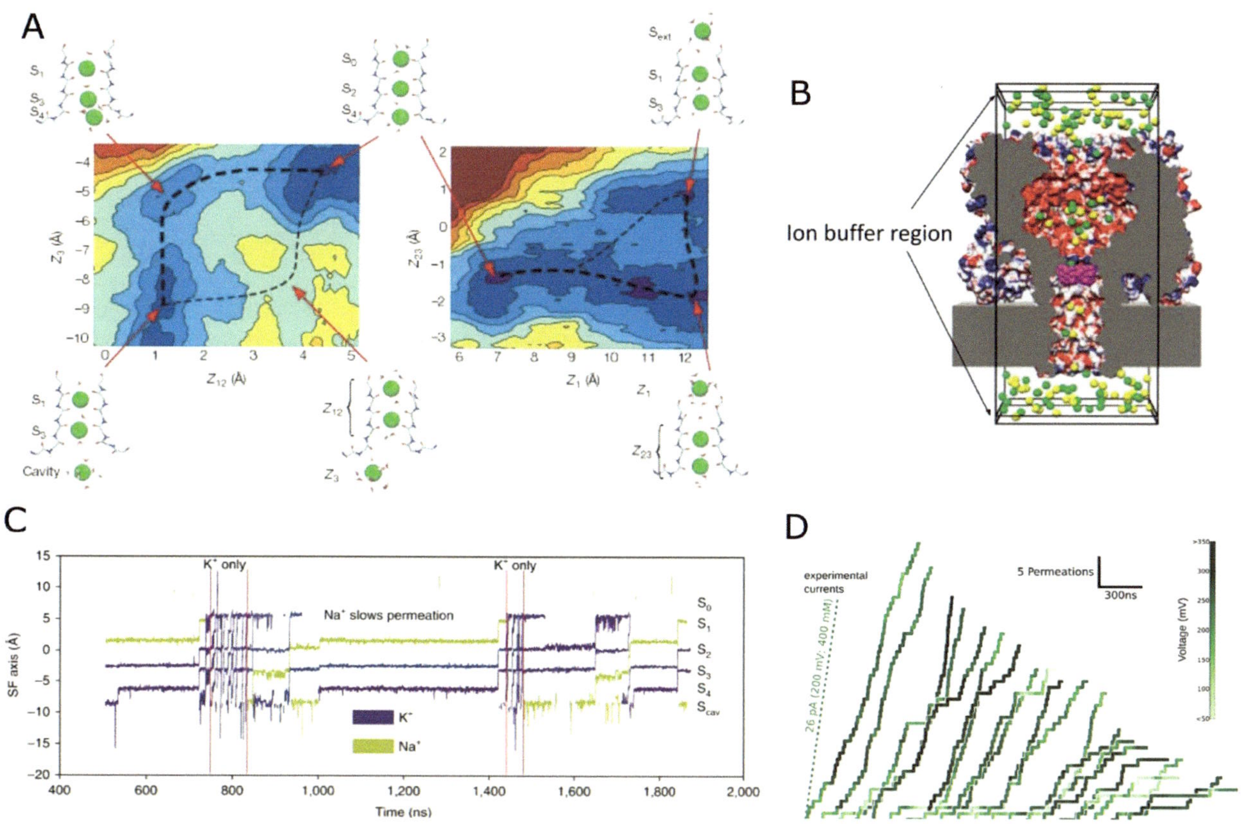

FIGURE 15.4
Approaches to study conduction and selectivity. (A) 2-D free energy landscapes to study the permeation mechanism of K^+ conduction in a potassium selective ion channel. Two landscapes are presented. In the first, the two collective variables presented represent the average position of the two outermost ions and the position of the innermost ion along the pore axis, respectively. In the other, the two collective variables presented represent the average position of the outermost ion and the position of the two innermost ions along the pore axis, respectively. Taken together one obtains a picture of the mechanism of conduction of the three ions. (Adapted with permission from Berneche and Roux 2001. Copyright 2001 Macmillan Publishers Ltd.) (B) Illustration of the system box in a GCMC/BD simulation. (Reprinted with permission from Egwolf et al. 2010. Copyright 2010 American Chemical Society.) (C) Position of the permeating ions along the selectivity filter (SF) axis in a computational electrophysiology simulation of a sodium/potassium mixture. This analysis reveals the time spent by various ions in their binding site. (Adapted with permission from Kopec et al. 2018. Copyright 2018 Macmillan Publishers Ltd.) (D) Number of permeation events along time in computational electrophysiology experiments. Repeating the simulation several times enables the evaluation of the single-channel conductance, which can be directly contrasted with the experimentally measured value. (Adapted from Kopfer et al. 2014. Copyright 2014 The American Association for the Advancement of Science.)

passed for the ion to proceed along its permeation pathway. These can be useful to attempt to rationalize the selectivity of a given pore by comparing the profiles obtained for different ions, or to pinpoint amino acids that give rise to favorable binding sites or that act as permeation barriers. Other degrees of freedom can be considered when building these free energy landscapes. Quite commonly, the ion hydration number or its radial displacement away from the central pore axis is added as a degree of freedom, enabling the characterization of the permeation and selectivity mechanism in more detail.

For ion channels in which permeation follows a single-file mechanism with a knock-on from incoming ions, it is necessary to consider the position of several ions as separate degrees of freedom as input to the enhanced sampling protocol (multi-ion collective variables; Figure 15.4A). When more than two ions permeate in a single file, the position of the two ions is often combined into one CV, enabling depiction of a 2D free energy landscape, which is more easily inspected than a 3D one.

Another type of approach that is especially valuable to study selectivity consists in directly evaluating the free energy that is gained or lost upon binding of different ions to a specific binding site. *Alchemical approaches* can be used for this purpose to slowly morph a given ion into another type, and *free energy perturbation* or *thermodynamic integration* to estimate the free energy related to this change. For example, to evaluate the difference in free energy related to a change from a K^+ to a Na^+ ion, the van der Waals radius of a K^+ ion is gradually, in a stepwise manner, decreased to match the one of a Na^+ ion. At each step, the system rearranges in response to the small radius perturbation, and the resulting free energy difference is evaluated. The overall free energy change is found by summing over the free energy differences related to each small perturbation. Conservative changes such as the preceding example are relatively easy to model, but larger changes that involve, for example, a change in charge (monovalent to divalent cation, or cation into anion) are much more difficult to converge. A negative free energy difference may indicate a preference for the ion that the system is being transformed into, and thus selectivity for said ion. However, selectivity often relies on more complicated mechanisms with multiple binding sites and permeation barriers, as well as possible multi-ion effects, such that free energy estimation methods might not be enough to paint the entire picture.

15.5.2 Methods Where Conduction Is Modeled Explicitly

Single-channel conductance can be estimated using a variety of simulation methods. At the continuum level (not a MD simulation technique per se), the *Poisson–Nernst–Planck* electrodiffusion model, in which ions are described by density distribution, is a computationally inexpensive tool to compute the ion fluxes under voltage and concentration gradient. If the flow of ions through channels can be represented by Brownian motion (chaotic and diffusive), Brownian dynamics can be used to simulate the diffusional motions of individual ions in the time scale of microseconds to milliseconds. One popular approach is *grand canonical Monte Carlo/Brownian dynamic* (GCMC/BD), in which a buffer region sufficiently remote from the channel allows the creation and annihilation of ions, thus allows simulating ion currents with rigorously controlled *concentration and transmembrane potential* (Figure 15.4B). In general, GCMC/BD treats the channel as a rigid structure with a low dielectric constant embedded in a membrane represented as a low dielectric slab. Water is represented as a continuum with a dielectric constant of 80. Thus, only the ions are moving during a GCMC/BD simulation, their dynamics being governed by the potential energy and a position-dependent diffusion constant. Note that continuum models may not be well suited to model narrow pores since modeling explicitly protein and solvent conformational change may be important.

The increased performance of MD simulation codes and the availability of powerful hardware has made it possible to study ion conduction through ion channels directly using all-atom MD simulations, subjecting the system to an electrical driving force, either by *computational electrophysiology* or by applying an *external electric field* (Figure 15.2). Since ion conduction occurs on the tens to hundreds of nanoseconds time scale, regular all-atom MD simulations appear very well suited to obtain an unbiased description of conduction mechanisms (Figure 15.4C). Nevertheless, gathering enough conduction events to obtain a reliable description of the mechanisms can correspond to hundreds of events, representing in total multiple microseconds of simulation data for each condition to be probed. This can be achieved by using, for example, the special-purpose Anton2 supercomputer or stitching together measurements from independent replicas.

Such simulations attempt to place themselves in conditions as close as possible to the experimental ones and attempt to apply TM potentials close to physiological ones. Nevertheless, to observe enough conduction events, potentials up to few hundreds of millivolts can be necessary. Often, several TM potentials of increasing magnitudes are applied resulting in G/V curves directly comparable to ones obtained by electrophysiology (Figure 15.4D). Using these approaches, one can directly count the number of ions permeating through a channel, and by dividing it by the time taken, obtain a direct estimate of the state-dependent single-channel conductance. Alternatively, conductance can be calculated from charge displacement along the z-axis, yielding cumulative currents (Figure 15.4C) (Aksimentiev and Schulten 2005). Conductance values obtained from MD simulations appear consistently lower than experimentally determined ones, possibly revealing the shortcomings of transferable force fields for such questions.

15.6 Small Molecule Modulation

The development of drugs that act as highly channel-specific modulators remains challenging. Unlike soluble proteins, ligands can reach ion channels via hydrophilic pathways (diffuse to the pore from the extracellular or intracellular sides) or lipophilic pathways (partition in lipid bilayers first before diffusing to binding sites) (Figure 15.1D). For many ion channels whose ligand binding sites or binding pathways are unknown, saturated binding simulation can be conducted in which many ligands are modeled simultaneously (flooding simulations). While care needs to be taken to avoid ligand aggregation or protein denaturing, MD simulations have identified potential binding sites and drug binding pathways that were validated using mutagenesis data. On the other hand, if the binding pocket is known, various molecular docking methods can be used to search binding poses or conduct early stage virtual screening.

With a previously known binding site, the thermodynamics (free energy differences) and kinetics (rate constants, barriers) of ligand binding at the single-channel level can be estimated from PMF-based approaches. Binding equilibrium constants K_{eq} can be, in theory, related to the free energy $w(r)$ as the function of the distance r between the ligand and binding site, $K_{eq} = \int_{site} dr\, e^{-\beta w(r)}$, wherein $\beta = 1/k_B T$. In practice, various geometric restraints of the ligand binding pathway are needed to ensure the convergence of the sampling (Woo and Roux 2005). For ion channels, many binding processes can be described by the PMF along the channel axis z, $G(z)$, then

$$K_{eq} = \pi r^2 \int_{z1}^{z2} dz\, e^{-\beta G(z)} \tag{15.12}$$

where r is the effective radius of the bulk cylindrical restraint. If the binding pathway involves permeating through lateral fenestrations, higher-dimension PMF or path sampling approaches can be used to find minimum free energy pathways between end states.

Based on PMFs, the dissociation rate constant can be estimated via Kramer's theory or transition state theory borrowed from reaction kinetics, if the PMF is dominated by a single barrier. To compute kinetic properties directly from MD simulations, a set of rare-event sampling algorithms, such as milestoning or weighted ensemble, can allow the calculation of ligand-channel dissociation rates, as well as small molecular permeation rates through ion channels. However, when the binding site and binding pose of the ligand is well defined and binding affinity is the solely desired quantity, alchemical approaches are more efficient (Figure 15.1D).

In addition, binding affinity differences between open and closed conformations of an ion channel should be computed to understand and predict the expected effect of the ligand (agonists will bind more favorably to the activated state of the protein, while silent binders will bind equally well to different functional states). Truncating the protein region that has no direct impact on binding is usually acceptable for improving the computational efficiency of such state-dependent binding investigations.

15.7 Lipid Modulation

Lipids can modulate channel function through direct binding, or through indirect, local or global mechanical property changes that alter the free energy of hydrophobic mismatch or membrane deformation. All-atom (AA) MD simulations have long been used to provide high-resolution insights into the lipid–protein interactions when bound lipid species were difficult to identify by structural methods. In recent years, Martini coarse-grained (CG) force fields have provided a way to identify potential lipid interaction sites through lipid–protein "fingerprint" simulations using realistic plasma membrane models (Corradi et al. 2018) (Figure 15.5A, B). CG models significantly accelerate lipid lateral diffusion and enable the observation of repeated binding and unbinding of lipids to channels. However, CG models require the use of protein tertiary structure constraints and thus do not enable the simulation of the dynamical response of the protein to the lipid environment. Mapping converged CG models back to all-atom models provides a possible way to model the dynamical interaction between lipids and the flexible protein with an atomistic level of resolution.

PMF-based or alchemical free energy calculations can further be used to quantify the strength and specificity of protein–lipid interactions using AA or CG models. Unlike drug-like small molecules, the slow diffusion and high flexibility of lipids often requires additional conformational/orientational restraints to improve the convergence of the calculations. Additional restraints are also needed to prevent other lipids from entering the binding site during the free energy calculation.

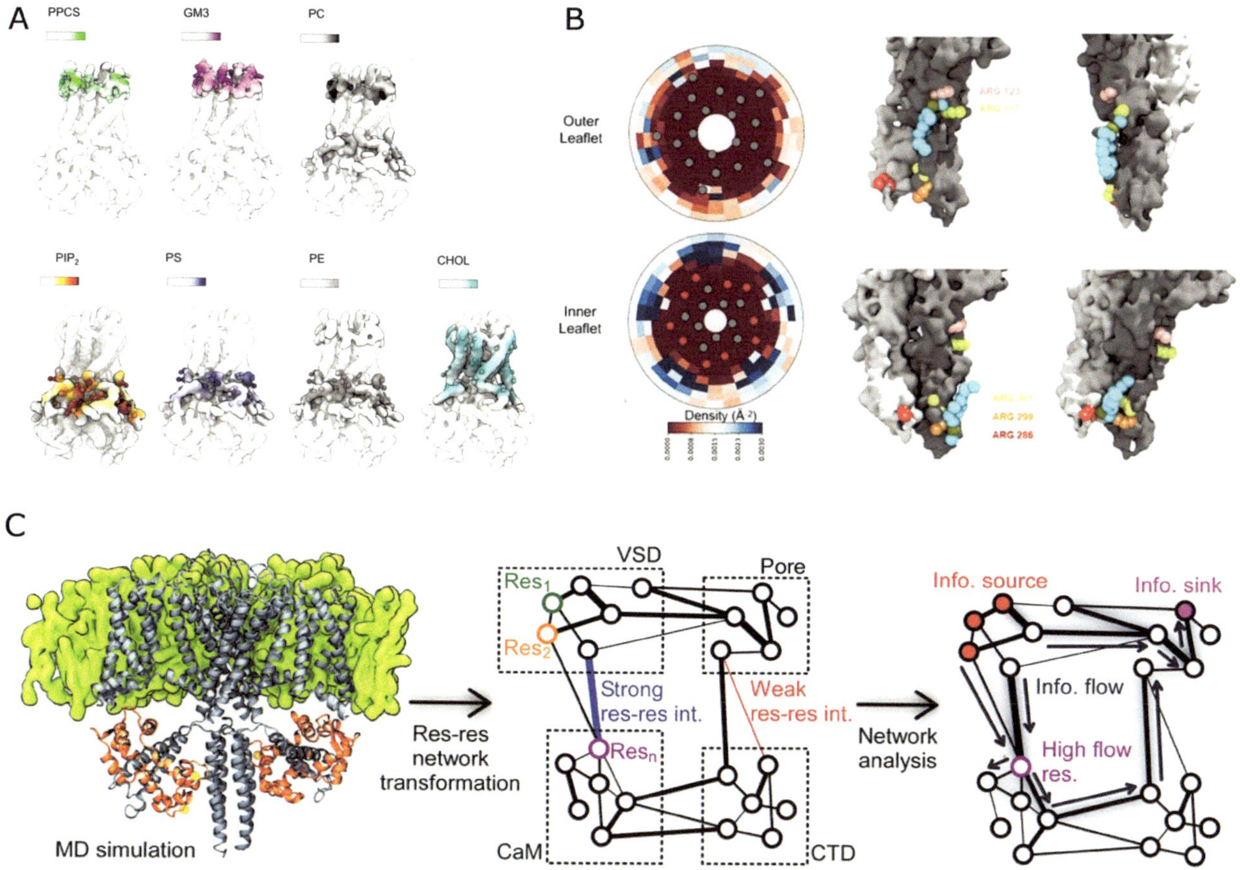

FIGURE 15.5
(A and B) Lipid binding occurs preferentially at specific sites, which can be revealed by CG MD simulations considering complex lipid membranes made of different lipid types. These can be visualized according to the residues they make extensive contacts with (A) (Duncan, Corey, and Sansom 2020) or on 2D- map projections (B) (Tong et al. 2019). (Panel A adapted from Duncan, Corey, and Sansom 2020. Licensed under a Creative Commons Attribution NonCommercial License 4.0 (CC BY-NC). Panel B adapted from Tong et al. 2019. Licensed under Creative Commons BY 4.0.) (C) Network analysis can be used to map allosteric pathways linking an allosteric site (source region) and the gate (sink region). In this representation, residues represent nodes of the network and the edges map if residues interact with one another and whether their motions are correlated. (Adapted from Kang et al. 2020. Licensed under a Creative Commons Attribution NonCommercial License 4.0 (CC BY-NC).)

15.8 Allostery

To quantitatively characterize allosteric signal transmission, an estimation of the free energy of the change occurring at the gate in the presence and absence of an allosteric signal would theoretically be necessary. However, converging such free energy estimates is very difficult to achieve due to the limited time scales MD simulations can cover.

When allosteric signaling is mediated by a folded structure, which is often the case in ion channels, representing the protein as a residue interaction network can be valuable. The network then represents residues as nodes and interactions between adjacent residues as edges (Figure 15.5C). The edges can be weighted according to the strength of the correlated motion of neighboring residues, via the estimation of Pearson's correlation coefficient of deviations away from the equilibrium position, or of mutual information. This network can be analyzed using network analysis tools. A particularly useful analysis type involves estimating information transfer from the allosteric site to the gate, using either random walk or Laplacian formulations. The residues that carry large information flow are deemed important for allosteric signaling and their importance can be verified via mutagenesis experiments.

Comparing the network and its analysis in different protein states, or isoforms, can be useful to reveal state-dependent allosteric signal transmission. This is because even though the time required for a gating conformational change may span milliseconds, the time scale required to transmit motion between adjacent amino acid pairs is much faster. While these types of analyses do not measure free energy changes and cannot thus be considered quantitative, they can provide useful insights in the absence of free energy landscapes.

15.9 In Silico Mutagenesis

Genetic mutations in ion channels can be at the origin of channelopathies and excitability disorders. In addition, mutagenesis is an extremely useful biophysical technique to probe the role of selected residues.

Mutations can be easily inserted in the models used for MD simulations, and the effect of a mutation is often rationalized by comparing simulations of the wild-type and the mutant channels.

Alchemical approaches are often used to characterize the free energy change associated with a specific mutation. A prominent example is in silico alanine scanning, which consists in systematically mutating residues in a specific protein segment by alanine, mimicking what is done in the experimental technique with the same name. By carrying out such calculations in two ion channel states, one can understand if a mutation stabilizes one more than the other, which can be directly verified by mutagenesis experiments.

15.10 Conclusion

It can be said that we have entered an era where advances in both computational and membrane protein structure determination technologies have made MD simulations an indispensable tool in ion channel research. For decades, great efforts have been made in solving the limitations of force fields and sampling time scales. In the next ten years, more mature polarizable force fields, multiscale simulations (QM/MM, all-atom/coarse-grained or continuum models), and machine learning methods to improve sampling and analysis are likely to make MD simulations even more significant for ion channel research.

Suggested Readings

Aksimentiev, A., and K. Schulten. 2005. "Imaging alpha-hemolysin with molecular dynamics: ionic conductance, osmotic permeability, and the electrostatic potential map." *Biophys J* 88 (6):3745–61. doi:10.1529/biophysj.104.058727.

Allen, M. P., and D. J. Tildesley. 1987. *Computer simulation of liquids*. Oxford: Clarendon Press.

Berneche, S., and B. Roux. 2001. "Energetics of ion conduction through the K+ channel." *Nature* 414 (6859):73–7. doi:10.1038/35102067.

Corradi, V., E. Mendez-Villuendas, H. I. Ingolfsson, R. X. Gu, I. Siuda, M. N. Melo, A. Moussatova, L. J. DeGagne, B. I. Sejdiu, G. Singh, T. A. Wassenaar, K. Delgado Magnero, S. J. Marrink, and D. P. Tieleman. 2018. "Lipid-protein interactions are unique fingerprints for membrane proteins." *ACS Cent Sci* 4 (6):709–17. doi:10.1021/acscentsci.8b00143.

Duncan, A. L., R. A. Corey, and M. S. P. Sansom. 2020. "Defining how multiple lipid species interact with inward rectifier potassium (Kir2) channels." *Proc Natl Acad Sci U S A* 117 (14):7803–13. doi:10.1073/pnas.1918387117.

Egwolf, B., Y. Luo, D. E. Walters, and B. Roux. 2010. "Ion selectivity of alpha-hemolysin with beta-cyclodextrin adapter. II. Multi-ion effects studied with grand canonical Monte Carlo/Brownian dynamics simulations." *J Phys Chem B* 114 (8):2901–9. doi:10.1021/jp906791b.

Huang, Y., J. A. Henderson, and J. Shen. 2021. "Continuous constant pH molecular dynamics simulations of transmembrane proteins." *Methods Mol Biol* 2302:275–87. doi:10.1007/978-1-0716-1394-8_15.

Kasimova, M. A., A. Yazici, Y. Yudin, D. Granata, M. L. Klein, T. Rohacs, and V. Carnevale. 2018. "Ion channel sensing: are fluctuations the crux of the matter?" *J Phys Chem Lett* 9 (6):1260–4. doi: 10.1021/acs.jpclett.7b03396.

Klesse, G., S. Rao, M. S. P. Sansom, and S. J. Tucker. 2019. "CHAP: a versatile tool for the structural and functional annotation of ion channel pores." *J Mol Biol* 431 (17):3353–65. doi:10.1016/j.jmb.2019.06.003.

Kopec, W., D. A. Kopfer, O. N. Vickery, A. S. Bondarenko, T. L. C. Jansen, B. L. de Groot, and U. Zachariae. 2018. "Direct knock-on of desolvated ions governs strict ion selectivity in K(+) channels." *Nat Chem* 10 (8):813–20. doi:10.1038/s41557-018-0105-9.

Kopfer, D. A., C. Song, T. Gruene, G. M. Sheldrick, U. Zachariae, and B. L. de Groot. 2014. "Ion permeation in K(+) channels occurs by direct Coulomb knock-on." *Science* 346 (6207):352–5. doi:10.1126/science.1254840.

Kutzner, C., H. Grubmuller, B. L. de Groot, and U. Zachariae. 2011. "Computational electrophysiology: the molecular dynamics of ion channel permeation and selectivity in atomistic detail." *Biophys J* 101 (4):809–17. doi:10.1016/j.bpj.2011.06.010.

Lee, J., X. Cheng, J. M. Swails, M. S. Yeom, P. K. Eastman, J. A. Lemkul, S. Wei, J. Buckner, J. C. Jeong, Y. Qi, S. Jo, V. S. Pande, D. A. Case, C. L. Brooks, 3rd, A. D. MacKerell, Jr., J. B. Klauda, and W. Im. 2016. "CHARMM-GUI input generator for NAMD, GROMACS, AMBER, OpenMM, and CHARMM/OpenMM simulations using the CHARMM36 additive force field." *J Chem Theory Comput* 12 (1):405–13. doi:10.1021/acs.jctc.5b00935.

Lev, B., S. Murail, F. Poitevin, B. A. Cromer, M. Baaden, M. Delarue, and T. W. Allen. 2017. "String method solution of the gating pathways for a pentameric ligand-gated ion channel." *Proc Natl Acad Sci U S A* 114 (21):E4158–67. doi:10.1073/pnas.1617567114.

Roux, B. 2008. "The membrane potential and its representation by a constant electric field in computer simulations." *Biophys J* 95 (9):4205–16. doi:10.1529/biophysj.108.136499.

Tong, A., J. T. Petroff, 2nd, F. F. Hsu, P. A. Schmidpeter, C. M. Nimigean, L. Sharp, G. Brannigan, and W. W. Cheng. 2019. "Direct binding of phosphatidylglycerol at specific sites modulates desensitization of a ligand-gated ion channel." *Elife* 8. doi:10.7554/eLife.50766.

Woo, H. J., and B. Roux. 2005. "Calculation of absolute protein-ligand binding free energy from computer simulations." *Proc Natl Acad Sci U S A* 102 (19):6825–30. doi:10.1073/pnas.0409005102.

16

Genetic Models and Transgenics

Andrea L. Meredith

CONTENTS

16.1 Introduction

Genetics provides powerful tools for dissecting the function of ion channels in physiological contexts. The function of ion channels can be decoded by phenotype, by genetic engineering in animal models and through human channel variant–phenotype linkages. The evolution of our understanding of ion channel function has been stimulated by the development of genetically tractable model organisms (1970s), recombinant DNA technology and genetically modified mice (1980s), genome sequencing and annotation (late 1990s–early 2000s), human single-nucleotide polymorphism (SNP) mapping (2000s), and gene editing (2010s). The resulting discoveries complemented biophysical studies of ion channel properties and revealed multifaceted interactions in vivo, including associations with >100 human disorders. This chapter covers widely used model organisms and transgenic methodologies and tools.

DOI: 10.1201/9781003096214-18

16.2 Linking Ion Channels to Phenotypes: Forward Genetics

16.2.1 Model Organisms

Beginning with Hodgkin and Huxley's biophysical studies of the giant squid axon, simple model organisms have revealed fundamental properties of the major ion channel classes. Moreover, given the ubiquity of ion channels in prokaryotes and eukaryotes, even the simplest single-celled organisms have facilitated formation of the primary connections between physiology and ion channel function. This function features in the rich diversity of behaviors exhibited across phyla, exemplified by modulation of membrane potential by light in the "giant unicell" *Acetabularia,* hyphal tip growth in the "nerve spore" *Neurospora,* escape response in the "swimming neuron" *Paramecium*, aggregation in the "social amoebae" *Dictyostelium* and motility in the "plant-animal chimera" *Clamydomonas* (Martinac et al., 2008; Kamata, 1934). Mediating growth, reproduction and survival behaviors, these single-celled organisms possess specific channel-based ion transport and osmoregulation systems, intracellular Ca^{2+} signaling pathways, and membrane excitation networks (Hille, 2001), making them valuable models for understanding ion channel-related physiology.

Further up the phylogenetic tree, multicellular organisms from plants to mammals use ion channels with evolutionarily conserved functional mechanisms on a backdrop of increasing cellular and system specialization. Understanding the basis for specialized electrical signaling requires knowledge of the molecular constituents of ion channels. Despite the rich diversity of electrical behaviors and ion transport mechanisms studied in simple model organisms, the confluence of experimental tractability and development of molecular genetic tools contributed to the rise of particular animal models, notably fly and mouse, for the study of ion channel function in vivo (Table 16.1).

16.2.2 Cloning

Early electrophysiological preparations such as squid axon and electric fish preceded the molecular biology and genetic revolutions. Biochemical purification enabled identification of the first ion channel proteins, isolated from the electric organs of the *Torpedo* ray (nicotinic acetylcholine receptor; Noda et al., 1982) and the eel *Electrophorus electricus* (voltage-gated Na^+ channel; Noda et al., 1984). Subsequent amino acid sequencing was used to generate DNA probes and screen cDNA libraries. The resulting coding sequences could be used to express channels in *Xenopus* oocytes or to identify other similar channel sequences by homology screening. While these methods identified channel genes, additional steps were required to discover their roles in native physiological contexts.

Parallel to biochemical purification, other researchers, notably Seymour Benzer at Caltech, were pioneering a different concept. Their approach centered on identifying genes linked to a functional role in animals, such as a behavioral alteration. In the early 1970s, Benzer's theory that mutations in single genes would produce a measurable effect on animal behavior was controversial, but quickly vetted using *Drosophila melanogaster* as a model organism. This approach for identifying genes, dubbed *forward genetics,* denotes methods for identifying a genotype based on a phenotype. The genetic makeup of an organism or, more colloquially, the particular sequence of a gene is the *genotype. Phenotype* is the constellation of observable traits.

With the rise of *Drosophila* as a model organism for the genetic dissection of behavior, gene products encoding ion channels not amenable to biochemical purification could be

TABLE 16.1

Comparison of Standard Model Organisms

Organism	Genome	Genes Similar to Human	Generation Time	Experimental Advantages	Experimental Disadvantages
Yeast *Saccharomyces cerevisiae*	6352 genes	46%	2 hours	Inexpensive; stocks can be frozen; easy to clone genes; powerful modifier screening; possess all basic eukaryotic cellular components; *Saccharomyces* Genome Database	No distinct tissues; electrophysiology possible but not widely performed
Worm *Caenorhabditis elegans*	21,187 genes	43%	3 days	Inexpensive and tiny; can be stored frozen; only 959 cells; transparent development; hermaphrodite self-fertilization and sexual reproduction; complete morphological characterization by serial EM and lineage mapping; laser ablation of single identified cells; easy gene cloning (transposon tagging, SNP mapping, gene rescue, genome-wide deletion collection); comprehensive proteome (ORFeome) and promoter libraries; modifier screening; large collections of mutant, expression (GFP-tagged) and RNAi lines; reverse genetics by RNAi (can be fed to worms); WormBase	Limited genetic, anatomical and behavioral similarities to human; electrophysiology difficult but possible
Fly *Drosophila melanogaster*	15,431 genes	61%	10 days	Inexpensive; oldest animal model with elegant forward genetic tools and easy gene cloning; modifier screening; precise temporal and spatial control of gene expression (Gal4-UAS); can make chimeric animals; large collections of mutant, expression, P-element lines; several reverse genetics techniques; more complex behaviors and similarities to human disease; amenable to drug discovery; electrophysiological access in larvae and adult muscle; FlyBase	Freezing protocols still under development; RNAi must be injected; electrophysiology limited in the adult brain
Zebrafish *Danio rerio*	26,206 genes	76%	10–12 weeks	Simplest vertebrate model with good forward and reverse genetics; external fertilization and development; large number of offspring per female; organ similarity to mammals (except lungs, mammary glands); transparent embryo optimal for imaging organogenesis; good transgenics for protein visualization; can make chimeric animals; reverse genetics by morpholino (antisense oligonucleotide knockdown) or genome editing with engineered nucleases (TALEN); in vivo and ex vivo electrophysiology recording capabilities; Zebrafish Model Organism Database	Many duplicate genes; small but requires specialized housing and Institutional Animal Care and Use Committee (IACUC) protocols; homologous recombination limited by availability of embryonic stem (ES) cells; few cell lines available for cellular or biochemical studies
Mouse *Mus musculus*	20,210 genes	99%	6–8 weeks	Mammalian model possessing essentially all human tissues and cell types; inbred and outbred strains; cryopreservation; in vitro fertilization; reverse genetics by targeted homologous recombination (extensive resources available); can make chimeric animals; extensive molecular genetic tools (BAC libraries, gene trap libraries; expression arrays); well-characterized primary, ES and stem cells for culture; superior in vivo and ex vivo electrophysiology recording capabilities; public and commercial mutant repositories; Mouse Genome Informatics	Expensive to house; longer generation time; incomplete genome coverage for conditional expression drivers (Cre lines); forward genetics laborious; requires IACUC protocols
Human *Homo sapiens*	19,040 genes		12+ years	>5000 genetically based diseases; familial pedigree analysis to link gene mutations and disease; detailed self-reporting of traits; genome-wide association studies (GWAS) linking gene haplotypes, traits and disease risk; population studies (genome, transcriptome, proteome, epigenome and microbiome); Online Mendelian Inheritance in Man	Limited experimental access; requires extensive Institutional Review Board (IRB) protocols

Sources: Barbazuk et al. (2000); Church et al. (2009); Howe et al. (2013); Jorgensen and Mango (2002); Kile and Hilton (2005); St Johnston (2002); Woods et al. (2000). Adapted from Bier and McGinnis (2004).

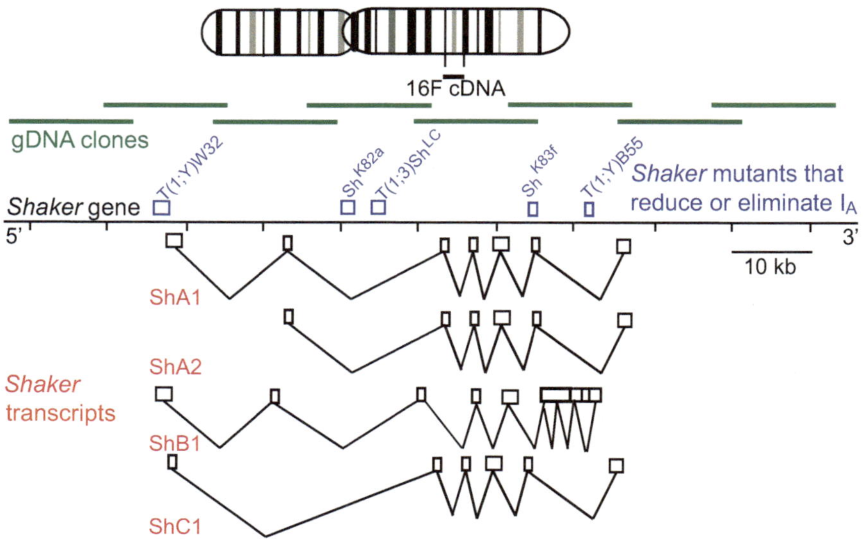

FIGURE 16.1
Positional cloning of *Shaker.* Abnormal banding patterns in *Drosophila* polytene chromosomes resulting from *Shaker* chromosomal aberrations were mapped to band 16F. Genomic DNA (gDNA) sequence encoding the *Shaker* locus was obtained by chromosomal walking, using a cDNA clone previously mapped to 16F to isolate a series of overlapping genomic clones from a library. Physical mapping by Southern blot revealed that independent *Shaker* mutations covered >65 kb. Restriction fragments were used to probe a cDNA library and identify the *Shaker* coding region. Boxed regions indicate areas of hybridization between genomic and cDNA sequences. Four clones were isolated (ShA1, ShA2, ShB1 and ShC1) and shown to encode three distinct splice variants. (Adapted from Papazian et al., 1987.)

cloned, and, importantly, linked to phenotypes. *Shaker* was the first ion channel identified using a forward genetics approach (Baumann et al., 1987; Papazian et al., 1987). Like the nomenclature of other mutants according to their phenotypic presentation, *Shaker* flies exhibited leg shaking under ether. Electrophysiological analysis of independent *Shaker* lines revealed alterations in I_A, a fast transient K^+ current in muscle, suggesting the underlying mutation would be in a gene-encoding structural component of I_A. A *positional cloning* strategy (using chromosomal location to identify genomic DNA sequence) was employed to locate the *Shaker* gene, which was shown to encode a K^+ channel with structural features similar to the previously cloned Na^+ channel (Figure 16.1). Other *Drosophila* ion channel genes were identified by forward genetics, including K^+ channels (*slowpoke, ether a go-go* and *seizure*) (Atkinson et al., 1991; Wang et al., 1997; Warmke et al., 1991) and the *transient receptor potential* cation channel (Montell and Rubin, 1989). Later, with the development of additional positional cloning resources in mice, strains harboring spontaneous mutations such as *lurcher, tottering* and *weaver,* were used to independently identify the mammalian ion channel genes GluRδ2, CACNA1a and GIRK2, respectively (Fletcher et al., 1996; Patil et al., 1995; Zuo et al., 1997).

Once a sequence was identified, cDNA or genomic libraries could be screened at low stringency to identify related channels (*homologs*), *paralogs* (gene duplications within families) or *orthologs* in other species. After the cloning of *Shaker,* three additional K^+ channel subtypes in flies (*Shab, Shaw* and *Shal*) and the mammalian voltage-gated K^+ channels were identified by homology screening (Tempel et al., 1987; Butler et al., 1989; Wei et al., 1990).

16.2.3 Phenotypic Screens

Phenotype-based screens identify genes, gene mutations and physiological roles for gene products. In the case of *Shaker,* a specific ion channel subunit could now be linked to a deficit in I_A and prolonged transmitter release at fly neuromuscular junctions. One advantage to forward genetic strategies was the feasibility of large-scale screening, which did not necessarily require labor-intensive electrophysiology. High-throughput behavioral or morphological screens could be performed to isolate mutants based on predictions for the physiology involving ion channel functions. Early *Drosophila* screens evaluated neural development, locomotor and flight behaviors, sensory responses, learning and memory, and sensitivity to toxins.

Large-scale mutagenesis screening also made it possible to perform unbiased interrogations of the entire genome, without a priori knowledge of the genomic location or molecular composition of ion channels. Small organisms, inexpensive to house, with relatively short generation times were particularly amenable to high-throughput screens. Positional or molecular cloning techniques applied to these model organisms (Table 16.1) were designed to clone ion channels and their modulatory subunits, identify the genetic and physiological pathways in which they act, and reveal their relative roles or redundancy.

In forward genetic screens, mutations are introduced into the genome by chemical exposure or irradiation (Figure 16.2). The recovery of mutants in subsequent *felial* (F) generations, progeny resulting from a genetically defined parental cross, is governed by Mendelian inheritance. F_1 progeny can be screened for dominant alleles, or recessive alleles in F_2. Dominant alleles are most often gain-of-function alterations in gene function, since they produce an effect in the presence of the *wild-type* (normal, unmutagenized) gene product.

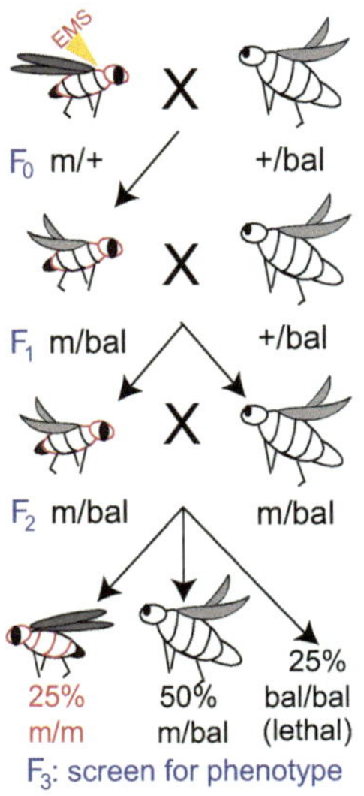

FIGURE 16.2
Mutagenesis screens. The chemical mutagen ethyl methane sulphonate (EMS) introduces random mutations into the *Drosophila* genome. In a typical screen for recessive alleles (variant forms of the gene), mutagenized male flies (m/+; F_0 generation) are crossed to females carrying balancers (specialized chromosomes harboring visible markers that cannot recombine). F_1 progeny will have a random assortment of mutant alleles in trans with the balancer. Flies that carry the marker trait (eye color, wing shape, etc.) and the mutant allele are selected and backcrossed to the balancer stock to generate F_2 progeny carrying the mutation over the balancer. Intercrossing produces a Mendelian ratio of homozygous mutant (m/m) progeny, manifesting recessive phenotypes. bal/bal progeny die due to a recessive lethal mutation.

Hypermorphs increase normal gene function (such as caused by a duplication), *antimorphs* antagonize normal gene function (dominant negative) and *neomorphs* confer a new function to the gene (permitting novel interactions or expression domains). More commonly, recessive phenotypes are desired, expressed in animals with a *homozygous* genotype (two copies of the allele). Recessive screens allow for recovery of deleterious loss-of-function mutations such as *hypomorphs* (partial loss of function) or *null* (complete loss of function). Several mutations mapping to the same gene may form an *allelic series*, a set of mutations with progressive severity. An allelic series is illustrated by the mouse CACNA1a (P/Q Ca^{2+} channel) gene with the *rocker*, *tottering* and *rolling* being milder, progressing to the more severe *leaner* allele (Zwingman et al., 2001).

Once a gene is identified, additional screens can be designed to further reveal interacting gene products. In *sensitized* screens, a second site mutation is recovered that modifies the primary mutation within the same strain. *Suppressor* screens identify compensatory mutations, while *enhancer* screens identify mutations that act to amplify or worsen the primary mutation. Sensitized screens can also be used to isolate second-site mutations within the same gene that interact with the primary mutated residues (double-mutant cycle analysis). Sensitized screens have been used to clone plant K^+ channels, investigate the structure and gating of mammalian K_{ir} channels, and identify pharmacological channel modulators (Minor, 2009).

16.3 Targeted Alteration of Ion Channel Function: Reverse Genetics

Forward genetics isolated many genes encoding ion channel subunits and relied on random mutagenesis in model organisms suitable for high-throughput phenotypic screening. In contrast, *reverse genetics* allows the introduction of a *targeted* alteration into a known gene. Engineering mutations into precise genomic loci was based on a series of fundamental molecular biological advances. The discovery of restriction enzymes allowed DNA to be cut at specific sequences.* Ligases could then be used to join together fragments of DNA with compatible ends.† Recombinant DNA could be created and amplified using polymerase chain reaction (PCR).‡ This series of techniques facilitated the creation of ion channel cDNAs for introduction into plasmid-based expression vectors and also the ability to manipulate DNA sequences for targeted homologous recombination.§ Together with CRISPR/Cas DNA editing,¶ these transgenic methods are the mainstay for altering ion channel function via reverse genetics.

Mouse (*Mus musculus*) is a widely used laboratory animal model based on amenability to reverse genetic methods, in vivo and ex vivo electrophysiological recordings, and shared genomic and physiological organization with humans. Reverse genetic alterations are divided into targeted (*homologous*) and nontargeted (*random*) mutations. The following sections focus on transgenic manipulations in mouse, although many of these approaches are routinely used in other organisms (Table 16.1).

* Nobel Prize in 1978 to Werner Arber, Dan Nathans and Hamilton Smith.
† Nobel Prize in 1980 to Paul Berg, Walter Gilbert and Fred Sanger.
‡ Nobel Prize in 1993 to Kary B. Mullis and Michael Smith.
§ Nobel Prize in 2007 to Martin Evans, Oliver Smithies and Mario Capecchi.
¶ Nobel Prize in 2020 to Emmanuelle Charpentier and Jennifer Doudna.

16.3.1 Nontargeted Transgenics

Nontargeted mutagenesis generates transgenic animals with a randomly inserted copies of a gene fragment into the genome (Palmiter et al., 1982). A minimal *transgene* is comprised of a promoter (sequences that direct transcription), protein coding sequence (cDNA) and polyadenylation signal (sequences that terminate transcription and regulate mRNA stability). Promoter and polyadenylation sequences may be derived from constitutively expressed housekeeping genes (β-actin, β-globin or human growth hormone), strong viral promoters (CMV or SV40) and/or tissue-specific transcriptional regulatory elements. The protein coding sequence is usually comprised of an intron-less cDNA containing a translation start (ATG) and stop codon. The linearized construct is introduced into fertilized eggs by pronuclear injection, creating stable transgenic integrations in every cell that can be transmitted through the germline. This method is fast and reasonably efficient, with about 10%–40% transgenic progeny. Nontargeted strategies are commonly used to express ion channels with protein tags, point mutations or within expression domains. Transgenic lines can be crossed onto knockout backgrounds to remove interactions with the wild-type gene product. Alternately, dominant effects can be characterized in the presence of the wild-type allele.

Drawbacks of this method include imprecise transgene expression due to both positional effects and incomplete regulatory elements. Transgenes incorporate at a single site, usually as a concatemer of several to hundreds of copies. *Positional effects* produced by copy number variation and the influence of surrounding sequence necessitates evaluating several lines for the appropriate transgene expression. In practice, the number of synthetic regulatory elements with defined expression patterns is small. Transgenes containing more comprehensive regulatory sequences are generated from *bacterial artificial chromosomes* (BACs), stable plasmids containing large inserts >100 kb. BAC libraries are available for most model organisms. Due to their size, BAC-based transgenes are less susceptible to positional effects, integrate in fewer copies, and allow delivery of long regions of genomic regulatory sequence, even if the sequence has not yet been characterized. However, compared to conventional transgenic constructs, *BAC recombineering* requires additional time and expertise (Testa et al., 2003). BAC-based strategies can also be used for targeted homologous recombination (Valenzuela et al., 2003).

16.3.2 Targeted Homologous Recombination

Mutation of ion channel genes is traditionally achieved by homologous recombination. This method is used to generate loss-of-function alleles through deletion of essential coding or noncoding genomic sequence (*knockout*), or gain-of-function alleles through insertion of sequence that alters expression or function (*knock-in*). A basic transgenic construct consists of two regions with homology to the target gene ("arms") and selection cassettes (Figure 16.3). The targeting construct is electroporated into *embryonic stem (ES) cells,* a recombination-competent dividing cell line that can integrate into blastocyst stage embryos. Transgenes are recombined into specific genomic sites defined by the homologous sequence arms on the targeting construct. The size and degree of homology matched with the target locus determines the frequency of recombination. Usually quite rare, the efficiency of intact targeting events can be enhanced by asymmetric homology arms (long and short arms), use of isogenic DNA between the targeting construct and ES cell line, and avoiding repetitive DNA. Correct integration events are determined by PCR and Southern blot analysis. Injection of transgenic ES cells into host blastocysts creates targeted germline-transmissible modifications in animals (Figure 16.4).

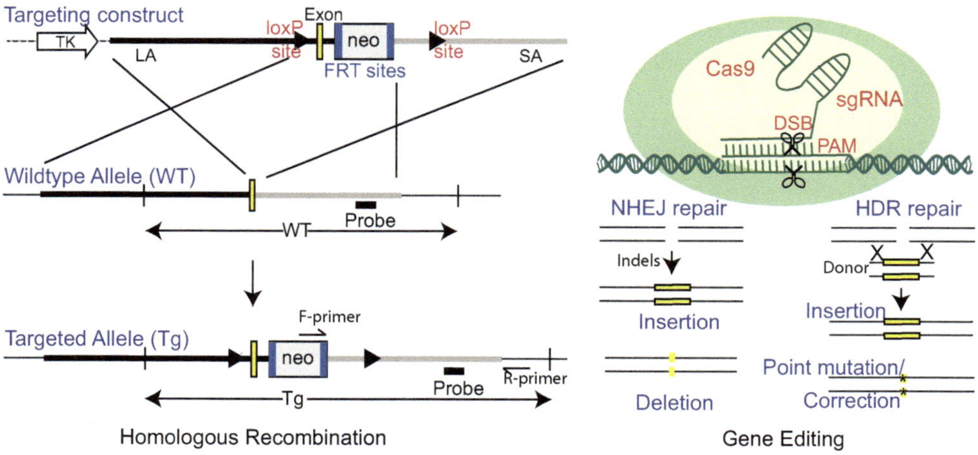

FIGURE 16.3
Homologous recombination and CRISPR. (Left) Homologous recombination strategy introducing *loxP* recombinase sites flanking a critical exon (Ex). Homology regions: long arm (LA) and short arm (SA). Thymidine kinase (TK) is lost with correct recombination events and eliminates clones with random integrations under negative selection (gancyclovir). *Neomycin* (neo) confers survival under positive selection (G418) for clones that undergo a double cross-over event resulting in intact integration of both homology arms. A second recombinase, flippase recognition target (FRT), can be used to remove *neo.* Several hundred ES clones typically survive positive–negative selection, usually with fewer than five producing correct integrations. (Right) CRISPR gene editing strategies to create insertions, deletions or point mutations. The Cas9 endonuclease binds the single-guide RNA (sgRNA), an RNA duplex unit containing sequences that base pair with the DNA target. Cas9 catalyzes a double-strand break (DSB) three to four nucleotides upstream of the protospacer adjacent motif (PAM). Cells repair DSBs by non-homologous end-joining (NHEJ) or homology-directed repair (HDR). NHEJ-mediated mutations are produced by insertions/deletions (indels) of variable length at the site of the break, often yielding frameshifts that knock out gene function. In HDR, nucleotide substitutions or additions on a donor template are inserted at the DSB. Targeting and repair are efficient, generally not requiring positive–negative selection. Several hundred pronuclear injections are typically performed (one to two days' work), often resulting in up to ten founders carrying the desired mutation.

A typical timeline for generating a transgenic mouse line is about a year with expertise and luck. This estimate is prolonged by the number of steps required to make the targeting construct, efficiency of recombination, and quality of the ES cells, which affects their ability to contribute the germline. Targeted ES cells generated by high-throughput screens can provide a shortcut. Repositories containing annotated gene trap lines are consolidated at the Mutant Mouse Resource & Research Center (MMRRC; https://www.mmrrc.org/about/resources.php), although most ES lines have not been preverified as functional nulls. The International Knockout Mouse Consortium (IKMC; https://www.mousephenotype.org/about-impc/about-komp/) has developed a high-throughput gene trap strategy to create a knockout of every gene in mice. Many knockout mouse lines have been successfully made using repositories, at lower effort and cost.

CFTR was among the first gene loci to be targeted by homologous recombination in mice (Snouwaert et al., 1992). This fundamental first step was quickly followed by the targeted knock-in of ΔF508, a mutation that accounts for ~70% of human cystic fibrosis cases (Grubb and Boucher, 1999). Numerous knockouts in voltage and ligand-gated ion channels followed. To date, virtually all the ion channel pore-forming subunits have been knocked out in mouse (>100), many providing models for human channelopathies.

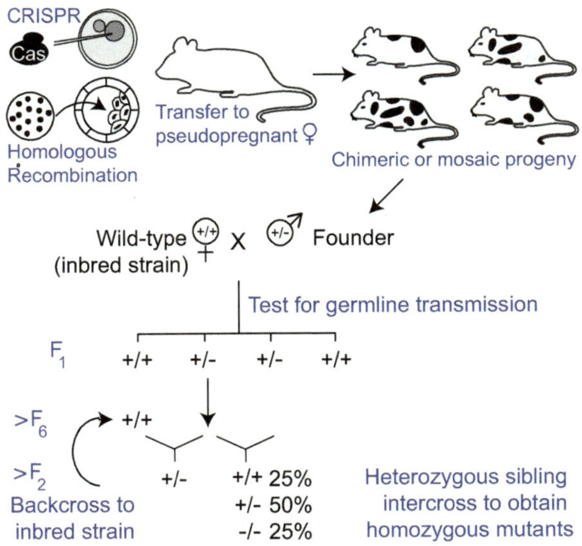

FIGURE 16.4
Generation of a transgenic mouse line. Correctly targeted ES cell clones are injected into blastocysts. CRISPR components are injected into fertilized one-cell zygotes. Embryos are transferred to primed females. Chimeric or mosaic founders are mated to wild-type mice. Progeny are genotyped to screen for germline transmission of the transgene. These steps have higher success rates with CRISPR compared to homologous recombination. To establish a transgenic mouse line, heterozygous progeny are backcrossed >6 generations to the desired inbred strain for homologous recombination methods or >2 generations for CRISPR. To generate recessive phenotypes, homozygous transgenic (–/–) animals are generated from heterozygous intercrosses with Mendelian ratios.

16.3.3 Gene Editing

Targeted genetic manipulation in mice has become significantly faster and less labor-intensive by moving beyond first-generation homologous recombination strategies. *Gene editing* can also be used in models not amenable to homologous recombination, such as human cell lines. ZFN (zinc-finger nuclease), TALEN (transcription activator-like effector nuclease) and CRISPR/Cas9 (clustered regulatory inter-spaced short palindromic repeat) incorporate these strategies into modular nuclease and DNA binding activities to recruit cleavage and repair events to specific sites. The targeting specificity is rapidly modifiable through alteration of the DNA binding template, eliminating the laborious screening and cloning required to isolate kilobase-length DNA stretches required for homologous recombination. Due to the ease of obtaining components and wide access to facilities that perform the technique, gene editing is considered "off-the-shelf" technology.

In mice, CRISPR/Cas9 is routinely used to produce deletion, replacement and insertion of DNA (Doudna and Charpentier, 2014). The basic system consists of the Cas9 endonuclease and a guide RNA that directs Cas9 to the target sequence in the genome (Figure 16.3). The components are typically microinjected into one-cell embryos, where Cas9 binds to the sgRNA and cleaves the target site (Figure 16.4). DNA modification occurs during the repair process, either by an error-prone mechanism that results in insertion or deletion (*indels*) at the cleavage site or by a template-based repair mechanism that results in incorporation of new sequences. Gene knockouts can be generated from indel repair events, while knock-ins (point mutations, protein tags and reporters) can be introduced by template-based repair. The repair template (*donor*) can be either a single-stranded oligonucleotide (<50 bp alterations) or double-stranded plasmid DNA (larger alterations).

Single-guide (*sgRNA*) sequences contain short sequences (17–24 nucleotides) complementary to the target DNA and an RNA scaffold sequence for Cas9 binding. Guide design centers on identification of a protospacer adjacent motif (PAM) near the target, which is essential for cleavage but not part of the sgRNA. Donor sequences introducing point mutations or new sequences vary in length but contain 5′ and 3′ homology arms flanking the target. Guides and donor oligonucleotides can be reliably designed using online tools and are typically synthesized by companies. Minimization of off-target effects is achieved through coordinated guide design and target site optimization. For difficult targets, efficiency can be further improved by using multiple guides, titrating the guides to the desired level of activity at the target and establishing an in vitro cleavage assay to test guide effectiveness prior to embryo injections (or using commercially available prevalidated guides).

Typical timelines can be approximately four months to obtain founder mice, mostly depending on the queue for the transgenic facility performing the microinjections. Correct editing events are determined by PCR from founder tissue samples. As with homologous recombination, undesired mutations and deletions need to be identified. Most strategies opt for breeding out extraneous indels rather than whole-genome sequencing. At a minimum, the regions surrounding the target site are sequenced, where off-target effects are the most problematic because they cannot be bred out. Because there is no chimerism between mouse strains with CRISPR/Cas9 systems, experiments can be initiated at earlier generations (F2 or F3) compared to homologous recombination.

Many modifications of ion channel genes have been generated in less than a decade of using CRISPR/Cas9. With the genomic sequence of most channel subunits known, the flexibility of this technique will rapidly increase the number of targeted genes, types of modifications, and afford novel animal models, expanding the repertoire of physiological questions that can be addressed. Gene editing also creates new possibilities to correct disease-associated mutations in animal models and human patients.

16.3.4 Conditional Site-Specific Recombination

Conventional targeting strategies generate a permanently modified genomic locus in every cell. Complications from this approach result from lethality, developmental defects, pleiotropy or phenotypic compensation. Conditional gene modification based on site-specific recombination of DNA mitigates such issues through more precise control over the tissue and temporal aspects of expression from the transgenic locus. The first widely used conditional modification system in mice was *Cre/loxP* (Gu et al., 1993; Orban et al., 1992). *Cre* (*c*auses *re*combination) is an integrase-type enzyme from bacteriophage that catalyzes a recombination event between two DNA sequences, termed *loxP* (*lo*cus of *x*-over in *P1*) sites. Depending on the orientation of the *loxP* sites, the intervening DNA may be deleted, translocated or inverted (Figure 16.5). Cre/*loxP* recombination is highly specific, with little endogenous recombination in mice, and efficient, nearing 100% recombination in some cases.

A transgene containing an entire gene sequence or an essential functional domain, such as the pore of an ion channel, is *floxed* (*fl*anked with *loxP* sites) for deletion. The sites are introduced into the genomic locus by homologous recombination, and a floxed mouse line is generated using standard methods (Figure 16.6). A second line expressing Cre recombinase must also be obtained. Many Cre lines have been generated as randomly integrated transgenes, containing regulatory sequences driving Cre expression in a specific tissue or temporal pattern. The two mouse lines are crossed, yielding double transgenic Cre^+, floxed progeny. Double heterozygotes are intercrossed to generate F_2 animals with two copies of

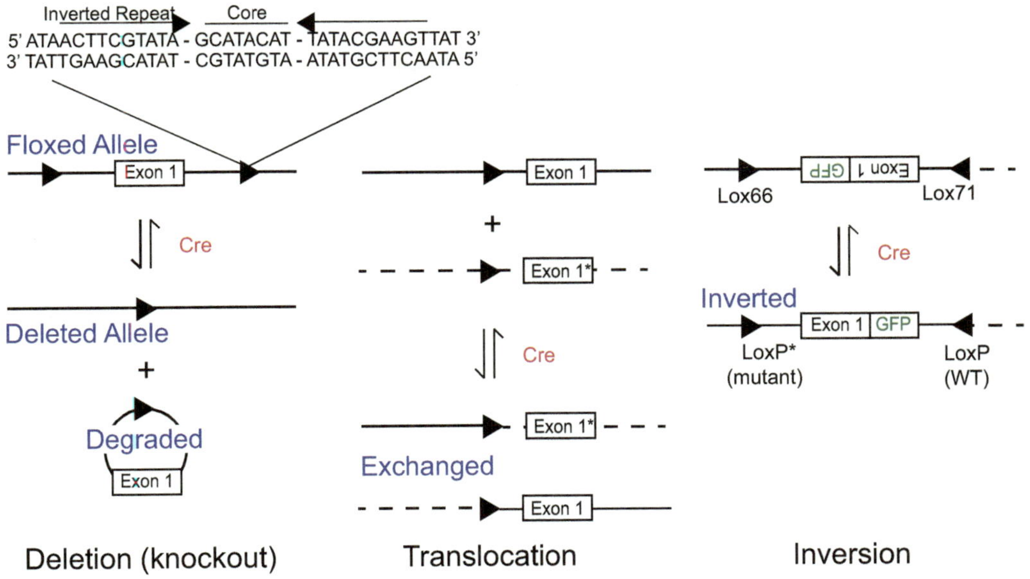

FIGURE 16.5
Cre-mediated recombination. Cre catalyzes several types of recombination events at *loxP* sites, determined by the orientation and location of *loxP* sites within the DNA. A 34 bp *loxP* site consists of two 13 bp inverted repeats surrounding an 8 bp nonpalindromic core sequence. Placement of two direct repeat *loxP* sites in cis results in intramolecular recombination that deletes the intervening DNA. This recombination event is limited to the forward reaction due to degradation of the excised sequence. Trans arrangement results in intermolecular recombination and reciprocal exchange of DNA, a reaction that may proceed in both directions. Inverted, cis-orientation *loxP* sites result in inversion of the intervening sequence. This can be biased toward a single reaction product by using distinct *loxP* sites. Recombination between lox66 and lox71 regenerates one wild-type and one mutant *loxP* site, which inefficiently binds Cre and prevents the reverse reaction.

the floxed allele (homozygous) and at least one copy of the Cre transgene. Homozygous Cre is generally not required for efficient recombination. Following the tissue and temporal expression pattern of Cre, these animals will be chimeric for the conditionally modified floxed allele.

Conditional recombination has a variety of applications besides gene inactivation. Expression can be turned on via conditional recombination, such as by introducing a flox–stop–flox cassette. The presence of the stop sequence suppresses gene expression until Cre-mediated excision of the stop sequence. Similarly, expression from a cDNA can be basally silenced by inverting the sequence in an intron. Activation occurs by Cre, inverting the cDNA into frame for transcription (Figure 16.5). This technique can be used to conditionally express markers, such as fluorescent proteins, or exchange wild-type for mutant sequence, such as introducing a disease-linked point mutation. To avoid positional effects, selectable markers used for screening recombinant ES cells are typically deleted. This can involve a complex strategy using a triple *loxP* allele. ES clones that underwent a single recombination event deleting the selectable marker, but retaining two *loxP* sites flanking the target, are then used to generate the mouse line. The correct recombination event is very rare. More commonly, a second recombinase system operating at distinct sequences, Flp/FRT (flippase/flippase recognition target), is used to flank the selection cassette (FRT sites; Liu et al., 2002). Concurrent incorporation of Cre/*loxP* and Flp/FRT is compatible

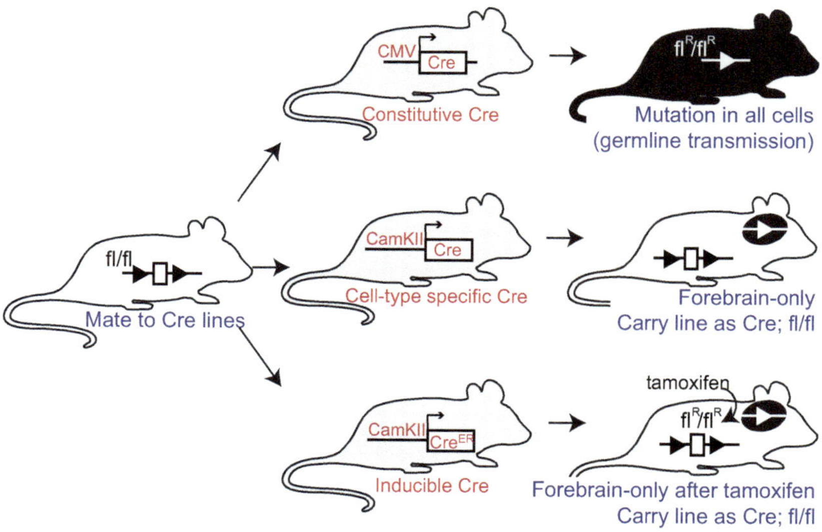

FIGURE 16.6
Cre-mediated tissue-specific and inducible gene modifications. A mouse line carrying a floxed (fl/fl) allele is mated to three representative Cre lines. (Top) The CMV viral promoter drives Cre expression everywhere, including the germline. The floxed allele undergoes recombination in every tissue (fl^R/ fl^R) and is stably heritable. (Middle) Tissue-specific expression of an ~8 kb fragment of the calcium/calmodulin-dependent protein kinase II (CamKII) gene drives forebrain-specific Cre expression (Tsien et al., 1996). Because the germline does not undergo recombination in a brain-specific Cre line, the line is maintained as a double transgenic. (Bottom) Fusion of the Cre protein to a mutated estrogen receptor (Cre^{ER}) renders Cre translocation to the nucleus tamoxifen-dependent (Feil et al., 1996), combining tissue specificity with temporal control of Cre activity. This line is also maintained as a double transgenic.

with most targeting cassettes, providing formulaic removal of the selectable marker and conditional gene modification (Figure 16.3).

A major consideration for conditional strategies is availability of mouse lines with the desired recombinase expression pattern. In 1996, there were about 10 published Cre mouse lines. By 2021, there were over 3720 Cre or Flp mouse lines, which along with floxed lines, have been cataloged in the Jackson Laboratory Cre Portal (https://www.jax.org/research-and-faculty/resources/cre-repository), supported by the National Institutes of Health. Besides lines where Cre is driven by tissue-specific promoters, there are other types of inducible lines. Ligand activation of Cre can be conferred by fusion to a mutated human estrogen receptor (Cre^{ER}; Feil et al., 1996). Cre^{ER} has low affinity for endogenous estrogens and is activated by the antagonist tamoxifen. In the presence of tamoxifen, usually delivered to mice through their water bottle, Cre^{ER} is translocated to the nucleus where it can catalyze recombination at *loxP* sites. Another type of inducible system involves driving Cre expression via tet-ON or tet-OFF transgenes (Utomo et al., 1999). The strategy requires generation of double or triple transgenic mice. In addition to the floxed allele, a second tetracycline-responsive promoter element (TRE; tetO) Cre transgene is needed. This line is crossed with either a reverse tetracycline-controlled transactivator protein (rtTA) or tetracycline-controlled transactivator protein (tTA) under the control of tissue-specific promoters. Alternately, the transactivators and Cre can be expressed from the same transgene, reducing the number of transgenes to just two. In the final transgenic line, Cre expression and *loxP* recombination in the appropriate tissues can be regulated with the tetracycline analog, doxycycline. Although inducible strategies provide an additional layer of specificity and control to conditional recombination, time and cost are also significant factors.

16.3.5 Genetically Encoded Tools for Studying Ion Channel Function

Beyond altering genes, characterization of ion channel physiology in cells is further accomplished through expression of fluorescent biomarkers or optical biosensors. The use of genetically encoded proteins is less invasive for living tissues. Imaging-based physiological measurements can also be combined with standard transgenic animal models. *Aequorea*-based GFP derivatives (EGFP, EYFP, EBFP2, Cerulean3, sfGFP, Citrine and Venus) and red fluorescent proteins derived from *Discosoma* (DsRed, mCherry and tdTomato) are routinely used in transgenes to mark cells that undergo DNA recombination, target specific cells for electrophysiological recordings, track protein expression and localization, and visualize cellular signaling (Livet et al., 2007).

Genetically encoded biosensors are increasingly capable of tracking the spatial and temporal dynamics of cellular activity. Early GFP-based Ca^{2+} sensors Cameleon and GCamP undergo changes in fluorescence upon Ca^{2+} binding (Miyawaki et al., 1997; Nakai et al., 2001). "Optical highlighters" have applications in channel trafficking and localization (Patterson, 2002). These proteins increase, decrease, or change their fluorescence after activation by a specific wavelength and come in three types: photoactivable, photoswitchable and photoconvertible. Channel and rhodopsin-based voltage sensors undergo changes in fluorescence with membrane potential, making action potentials and non-Ca^{2+} electrical signaling accessible (Bando et al., 2019; Siegel and Isacoff, 1997; Jin et al., 2012). Novel windows into glutamatergic neurotransmission became possible with iGluSnFR (Marvin et al., 2013), as did a whole host of other signaling pathways revealed by new biosensors. The challenges of low signal-to-noise ratios, kinetic limitations on temporal tracking and short lifetimes compared to biological timescales have been mitigated enough to open a new field of optical electrophysiology (Fan et al., 2020).

Beyond passive monitoring, direct manipulation of membrane potential is achieved with targeted expression of optically gated ion channels (*optogenetics*). Light-driven control has the advantage of tight spatial and temporal resolution over channel gating. The most well known are Channelrhodopsin-1 and 2, both intrinsically light-sensitive ion channels. Identified from algae, neither requires additional cofactors for activity. ChR2 photostimulation effectively depolarizes mammalian neurons, driving action potential activity (Boyden et al., 2005). Other light-sensitive channels inhibit activity, such as the halorhodopsin Cl^- pump (NpHR) with an excitation wavelength distinct from ChRs (Han and Boyden, 2007; Zhang et al., 2007). The precise control of membrane potential via optogenetics is subject to many factors related to the biophysical properties of the light-modulated channel, as well as expression levels. Since the illumination field affords a high-level of spatial control, viral delivery has been successful for targeting optogenetic proteins to the desired location. This approach opened primates to neurophysiological manipulation (Han et al., 2009).

16.3.6 Genetic Integrity and Maintenance of Mouse Lines

Once transgenic founders have been obtained, the genetic background is cleared of extraneous mutations by backcrossing to the desired inbred strain for >6 successive generations (*isogenic* line generation). A fully backcrossed line is *congenic* to the parent inbred strain, differing at only one locus (the transgenic allele). Congenic lines maximize the attribution of a phenotypic consequence to the engineered mutation. A mixed genetic background can exhibit variable *penetrance* (severity of a phenotype among transgenic animals) due to the influence of modifier alleles and extraneous mutations. Periodic backcrossing (every two to three years) to the inbred strain is essential, necessitating transgenic breeding colony that tracks filial generations (F_1, F_2, etc.). In the case of randomly integrated transgenics, keeping track of animals with one versus two copies of the transgenic (*hemizygous* versus homozygous) may also be important.

Recessive mutations account for a major segment of standard genetic manipulations. After a transgenic line is established, a typical mating scheme involves intercrossing heterozygote animals to allow the expression of recessive phenotypes in homozygous progeny (Figure 16.4). If the recessive phenotype does not affect reproduction, homozygous offspring can be intercrossed to carry the line. However, due to segregation of random alleles over successive generations (*genetic drift*), maintaining separate crosses to generate controls and transgenic experimental animals is not considered ideal. More commonly, colonies are maintained by heterozygous crosses that produce the homozygous and wild-type controls from the same litter.

16.3.7 Phenotypic Characterization

Transgenic mouse models are typically generated to identify physiological or pathophysiological roles for ion channels. An initial broad phenotypic characterization is essential to interpreting the experimental data obtained. Beyond the primary phenotype of interest, transgenic lines are routinely assessed for basic developmental defects, overall health, neurological reflexes, motor and sensory function, learning and memory, emotionality, and social behaviors. Considerations for the optimal phenotypic tests, experimental design and caveats to data interpretation are covered by an excellent comprehensive reference (Crawley, 2007). Phenotypic characterization is increasingly quantitative, with commercially available automated screening and data collection developed for a wide variety of behaviors. Many institutions and commercial vendors support "phenotyping cores." A principal consideration for phenotypic characterization is the selection of appropriate controls. Minimally, transgenic and control cohorts should be age- and sex-matched littermates, which share a greater genetic and environmental similarity than inbred wild-type mice maintained separately.

Because mice exhibit strain-dependent traits and disease susceptibility, transgenic alleles can be crossed onto sensitized backgrounds to enhance the phenotypic presentation or examine interactions with susceptibility loci. There are inbred strain-based differences in seizure threshold, hearing loss, learning and memory tasks, and drug sensitivity, to name a few. The Mouse Phenome Database (https://phenome.jax.org/) is a valuable resource for selecting strains based on specific traits. Strain and transgenic phenotype information has been consolidated into several widely used databases across model organisms (Table 16.1), including the Knockout Mouse Phenotyping project (https://www.mousephenotype.org/) and Mouse Genome Informatics (MGI) Database (https://www.informatics.jax.org/) (Eppig et al., 2012).

16.4 Ion Channels in Human Physiology and Disease

16.4.1 Linkage and Single-Nucleotide Polymorphisms

Human allelic variation accounts for individual differences in physical traits, personality and cognitive function, and disease risk. The human genome contains three billion base pairs of DNA and >20,000 genes, creating a challenge in associating disease phenotypes with the specific sequences of a gene (*allele*). Identifying alleles associated with traits or disease in humans was previously limited by the low resolution of genetic markers used for linkage analysis. A *genetic marker* is a stretch of DNA of known location identifying a specific variation in size or sequence (*polymorphism*). Common genetic markers include restriction fragment length polymorphisms (RFLPs), microsatellite or variable number tandem repeats, and *single-nucleotide polymorphisms* (SNPs). The Human Genome Project, completed in 2003, vastly expanded the regions and density of the genome covered by various types of genetic markers. These advancements in the technology for making the connections between genotype and phenotype have resulted in an increased understanding of the mechanisms of human disease. Combinations of these markers are also used to identify individuals, such as for forensic analysis.

Associations are made through cosegregation of a set of *linked* markers with the disease in a family pedigree or across a population (genome-wide association studies). A minimal set of linked markers identifies a candidate genomic region that may contain the disease gene, indicated by a statistically high LOD (logarithm of odds) score. Genes between the markers are identified. For mutations changing the protein sequence (*nonsynonymous*), subsequent mechanistic investigation is performed to investigate the causal nature of the mutation in the disease phenotype.

Not all associations between a specific ion channel sequence and phenotype are clear. Detailed familial pedigrees are not widely available for linkage analysis, and many complex traits and diseases are *polygenic*, attributed to the function of multiple genes. *Genome-wide association studies (GWAS)* harness the power of a systems approach for identifying genetic markers present in a population. These studies rely on SNPs, genetic markers defined by sequence variation occurring at only one base. SNPs are found on average every 1200 base pairs, creating variation across populations. A mutation is a deleterious SNP with a low population prevalence. Several linked SNPs within a gene (*haplotype*) may also be associated with disease or disease risk. The correlation of haplotypes and disease requires a comprehensive human SNP map, a central aim of the HapMap and 1000 Genomes projects, which genotyped hundreds of individuals and identified ten million SNPs (International HapMap Consortium et al., 2007).

Because much of the basic function of ion channels is well understood, there is the prediction that a functional alteration imparted by an SNP would manifest in the disease etiology. However, using idiopathic epilepsy as a model, it is not yet feasible to correlate an individual's SNP profile and the theoretical variation in ion channel function with epilepsy phenotypes (Klassen et al., 2011). This lack of correlation may be influenced by several factors, including an incomplete understanding of the gene interactions within an individual's genome. In addition, a large number of SNPs are located in noncoding regions and have unknown function. Another distinct factor is *epigenetic* modification of DNA, which produces functional changes in a gene product by mechanisms distinct from sequence variation (genomic imprinting is an example). As a result of these unknowns, attribution of traits and diseases with particular SNPs continues to be a significant biological frontier. Data sets generated by these investigations are searchable using the Phenotype-Genotype Integrator (PheGenI), an integrated repository combining the National Human Genome Research Institute's catalog of GWAS data with several National Center for Biotechnology Information (NCBI) databases, including the database of Genotypes and Phenotypes (dbGaP), Online Mendelian Inheritance in Man (OMIM), Genotype-Tissue Expression (GTEx), ClinVar and the SNP database (dbSNP). Most can be searched by phenotype, SNP, gene and chromosomal interval.

16.5 Summary

Ion channels are central to the physiology of cells and animals. Genetic methods in a diverse set of model organisms enabled each fundamental advance in our understanding of ion channel function in vivo, from cloning the genomic sequences, identifying the phenotypes stemming from ion channel dysfunction, to the linkage with human traits and disease. With centralized repositories for existing transgenic models, widespread access to transgenic techniques and further optimization of targeted genomic manipulation, animal models can be routinely engaged in lab research programs.

Targeted mutations have been made in almost every ion channel gene, shaping decades of research through the analysis of single-gene alterations on inbred backgrounds. This approach led to the identification of the principal physiological roles for each ion channel family and their mechanistic causality in human disease (*channelopathies*). Going forward, multifactorial genetic strategies made possible by gene editing technology will transform approaches to increasingly complex physiological questions as well as gene-based therapeutic strategies. For ion channels, which exist not as singular entities but as components of membrane excitability and ion transport systems, this next frontier in physiology will integrate, rather than exclude, genetic variability. Recombinant inbred strains, where each individual animal possesses a unique and random combination of inherited alleles, can be consistently produced by a defined cross between two inbred strains (Complex Trait Consortium, 2004). Animal models incorporating recombinant inbred strains are being used to address the relative contributions of genetic, developmental, and environmental factors in the expression of traits and disease, mimicking the genetic heterogeneity of human populations.

Suggested Readings

This chapter includes additional bibliographical references hosted only online as indicated by citations in blue color font in the text. Please visit https://www.routledge.com/9780367538163 to access the additional references for this chapter, found under "Support Material" at the bottom of the page.

Atkinson, N. S., G. A. Robertson, and B. Ganetzky. 1991. "A component of calcium-activated potassium channels encoded by the Drosophila *slo* locus." *Science* 253:551–5.

Bier, E., and W. McGinnis. 2004. "Model organisms in development and disease." In *Molecular basis of inborn errors of development*, edited by C. J. Epstein, E. P. Erickson, and A. Wynshaw-Boris, 23–40. New York: Oxford University Press.

Boyden, E. S., F. Zhang, E. Bamberg, et al. 2005. "Millisecond-timescale, genetically targeted optical control of neural activity." *Nat Neurosci* 8:1263–8.

Butler, A., A. G. Wei, K. Baker, et al. 1989. "A family of putative potassium channel genes in *Drosophila*." *Science* 243:943–7.

Complex Trait Consortium, G. A. Churchill, D. C. Airey, H. Allayee, et al. 2004. "The collaborative cross, a community resource for the genetic analysis of complex traits." *Nature Genet* 36:1133–7.

Crawley, J. N. 2007. *What's wrong with my mouse? Behavioral phenotyping of transgenic and knockout mice*. 2nd ed. Hoboken, NJ: Wiley-Interscience.

Eppig, J. T., J. A. Blake, C. J. Bult, J. A. Kadin, J. E. Richardson, and the Mouse Genome Database Group. 2012. "The Mouse Genome Database (MGD): comprehensive resource for genetics and genomics of the laboratory mouse." *Nucleic Acids Res* 40:D881–6.

Feil, R., J. Brocard, B. Mascrez, et al. 1996. "Ligand-activated site-specific recombination in mice." *Proc Natl Acad Sci USA* 93:10887–90.
Grubb, B. R., and R. C. Boucher. 1999. "Pathophysiology of gene-targeted mouse models for cystic fibrosis." *Physiol Rev* 79 (Supplement):S193–214.
Gu, H., Y. R. Zou, and K. Rajewsky. 1993. "Independent control of immunoglobulin switch recombination at individual switch regions evidenced through Cre-loxP-mediated gene targeting." *Cell* 73:1155–64.
Han, X., and E. S. Boyden. 2007. "Multiple-color optical activation, silencing, and desynchronization of neural activity, with single-spike temporal resolution." *PLoS One* 2:e299.
Han, X., X. Qian, J. G. Bernstein, et al. 2009. "Millisecond-timescale optical control of neural dynamics in the nonhuman primate brain." *Neuron* 62:191–8.
International HapMap Consortium, K. A. Frazer, D. G. Ballinger, et al. 2007. "A second generation human haplotype map of over 3.1 million SNPs." *Nature* 449:851–61.
Jin, L., Z. Han, J. Platisa, et al. 2012. "Single action potentials and subthreshold electrical events imaged in neurons with a fluorescent protein voltage probe." *Neuron* 75:779–85.
Kamata, T. 1934. "Some observations on potential differences across the ectoplasm membrane of paramecium." *J Exp Biol* 11:94–102.
Klassen, T., C. Davis, A. Goldman, et al. 2011. "Exome sequencing of ion channel genes reveals complex profiles confounding personal risk assessment in epilepsy." *Cell* 145:1036–48.
Liu, P., N. A. Jenkins, and N. G. Copeland. 2002. "Efficient Cre-loxP-induced mitotic recombination in mouse embryonic stem cells." *Nat Genet* 30:66–72.
Livet, J., T. A. Weissman, H. Kang, et al. 2007. "Transgenic strategies for combinatorial expression of fluorescent proteins in the nervous system." *Nature* 450:56–62.
Martinac, B., Y. Saimi, and C. Kung. 2008. "Ion channels in microbes." *Physiol Rev* 88:1449–90.
Minor, D. L., Jr. 2009. "Searching for interesting channels: pairing selection and molecular evolution methods to study ion channel structure and function." *Mol Biosyst* 5:802–10.
Miyawaki, A., J. Llopis, R. Heim, et al. 1997. "Fluorescent indicators for Ca2+ based on green fluorescent proteins and calmodulin." *Nature* 388:882–7.
Montell, C., and G. M. Rubin. 1989. "Molecular characterization of the Drosophila trp locus: a putative integral membrane protein required for phototransduction." *Neuron* 2:1313–23.
Nakai, J., M. Ohkura, and K. Imoto. 2001. "A high signal-to-noise Ca^{2+} probe composed of a single green fluorescent protein." *Nat Biotechnol* 19:137–41.
Noda, M., S. Shimizu, T. Tanabe, et al. 1984. "Primary structure of Electrophorus electricus sodium channel deduced from cDNA sequence." *Nature* 312:121–7.
Noda, M., H. Takahashi, T. Tanabe, et al. 1982. "Primary structure of alpha-subunit precursor of Torpedo californica acetylcholine receptor deduced from cDNA sequence." *Nature* 299:793–7.
Orban, P. C., D. Chui, and J. D. Marth. 1992. "Tissue- and site-specific DNA recombination in transgenic mice." *Proc Natl Acad Sci USA* 89:6861–5.
Palmiter, R. D., R. L. Brinster, R. E. Hammer, et al. 1982. "Dramatic growth of mice that develop from eggs microinjected with metallothionein-growth hormone fusion genes." *Nature* 300:611–5.
Papazian, D. M., T. L. Schwarz, B. L. Tempel, et al. 1987. "Cloning of genomic and complementary DNA from *Shaker*, a putative potassium channel gene from *Drosophila*." *Science* 237:749–53.
Patterson, G. H. 2002. "Highlights of the optical highlighter fluorescent proteins." *J Microsc* 243:1–7.
Siegel, M. S., and E. Y. Isacoff. 1997. "A genetically encoded optical probe of membrane voltage." *Neuron* 19:735–41.
Snouwaert, J., K. K. Brigman, A. M. Latour, N. N. Malouf, R. C. Boucher, O. Smithies, B. H. Koller. 1992. "An animal model for cystic fibrosis made by gene targeting." *Science* 257, 1083–8.
Tempel, B. L., D. M. Papazian, T. L. Schwarz, et al. 1987. "Sequence of a probable potassium channel component encoded at *Shaker* locus of *Drosophila*." *Science* 237:770–5.
Testa, G., Y. Zhang, K. Vintersten, et al. 2003. "Engineering the mouse genome with bacterial artificial chromosomes to create multipurpose alleles." *Nat Biotechnol* 21:443–7.

Tsien, J. Z., D. F. Chen, D. Gerber, et al. 1996. "Subregion- and cell type-restricted gene knockout in mouse brain." *Cell* 87:1317–26.

Utomo, A. R., A. Y. Nikitin, and W. H. Lee. 1999. "Temporal, spatial, and cell type-specific control of Cre-mediated DNA recombination in transgenic mice." *Nat Biotechnol* 17:1091–6.

Wang, X. J., E. R. Reynolds, P. Deak, et al. 1997. "The *seizure* locus encodes the *Drosophila* homolog of the HERG potassium channel." *J Neurosci* 17:882–90.

Warmke, J., R. Drysdale, and B. Ganetzky. 1991. "A distinct potassium channel polypeptide encoded by the Drosophila eag locus." *Science* 252:1560–2.

Wei, A., M. Covarrubias, A. Butler, et al. 1990. "K^+ current diversity is produced by an extended gene family conserved in Drosophila and mouse." *Science* 248:599–603.

Zhang, F., L. P. Wang, M. Brauner, et al. 2007. "Multimodal fast optical interrogation of neural circuitry." *Nature* 446:633–9.

Zwingman, T. A., P. E. Neumann, J. L. Noebels, et al. 2001. "Rocker is a new variant of the voltage-dependent calcium channel gene *Cacna1a*." *J Neurosci* 21:1169–78.

17

EPR and DEER Spectroscopy

Eric G. B. Evans and Stefan Stoll

CONTENTS

17.1 Introduction

This chapter aims to provide a brief introductory overview of the use of electron paramagnetic resonance (EPR) spectroscopy for studying the structure and dynamics of proteins such as ion channels. We will focus on two particular EPR methods: continuous-wave (CW) EPR and double electron–electron resonance (DEER) spectroscopy, a pulse EPR method.

EPR is a magnetic-resonance technique that selectively probes paramagnetic centers, i.e., centers with unpaired electrons, and their nanoscale environment. Most proteins do not have an intrinsic EPR or DEER response (except metalloproteins with paramagnetic metal ions such as copper, iron or iron–sulfur clusters). Therefore, EPR usually requires the introduction of paramagnetic spin labels at specific sites in the protein sequence. This approach is known as site-directed spin labeling (SDSL).

With CW EPR, it is possible to measure the local mobility at a spin-labeled site, providing fingerprint information about local structure and conformational dynamics as well as local solvent accessibility. DEER, on the other hand, utilizes doubly spin-labeled proteins and measures the full distribution of absolute distances between the two spin labels. From this distribution, detailed insight into the structure and energetics of the protein conformational ensemble can be gained.

While this chapter is only able to give a brief overview, there are extensive – and more detailed – resources available in the literature. These include reviews (Jeschke 2012, 2013; Mchaourab, Steed, and Kazmier 2011; Fanucci and Cafiso 2006), protocols (Bordignon,

DOI: 10.1201/9781003096214-19

Kucher, and Polyhach 2019; Pliotas 2017; Tilegenova et al. 2018), a recent benchmark and guideline paper (Schiemann et al. 2021), a multiauthor book (Timmel and Harmer 2014), and several relevant chapters in a more general EPR textbook (Goldfarb and Stoll 2018).

17.2 Site-Directed Spin Labeling (SDSL)

In order to perform CW EPR or DEER spectroscopy on a protein, it needs to be site-specifically labeled with a spin label. By far the most commonly employed spin label goes by the not-very-specific acronyms MTSL or MTSSL (methane thiosulfonate spin label). It is a nitroxide radical with a five-membered 3-pyrroline ring as the head group, an N–O group that hosts the EPR-active unpaired electron with its spin, four adjacent methyl groups that sterically protect the reactive N–O group and a tether for disulfide ligation to cysteine (Cys). When attached to a Cys, the resulting residue is called R1. It is shown on the left side of Figure 17.1.

Compared to many common fluorescent labels, R1 has a shorter tether and a much smaller head group. Nevertheless, its side chain has five dihedral degrees of freedom, providing substantial side chain flexibility and a large range of rotameric states. The two dihedral angles closest to the backbone have preferential values of around +60°, 180° and −60°, and a hydrogen bond between one of the sulfurs and the backbone might further restrict them. The S–S bond dihedral angle is constrained to near +90° or −90°, and the fourth and fifth angles are the most flexible.

R1 is introduced into proteins as follows. All wild-type cysteines in a protein of interest are tested for reactivity with MTSL, and the reactive ones are mutated out. Then, sites are selected for spin labeling. Selection criteria include proximity to the surface, location on protein domains or secondary structure elements of interest, and noninterference with protein function. These sites are mutated to Cys, and reacted with MTSL to give R1. For CW EPR, one site is labeled, whereas for DEER, a pair of sites is labeled. This pair may be two residues within a single chain, one site in each of two proteins that form a complex or a single residue in a protein homooligomer. The resulting spin-labeled protein must be tested for integrity and functional competence, for example, with electrophysiological recordings or binding assays. Labeling efficiency is determined by comparing protein concentration with spin label concentration as determined by CW EPR. Labeling efficiency should be high, but does not need to approach 100%.

Besides MTSL/R1, there exists a great variety of other SDSL approaches: different attachment chemistries (iodoacetamide, maleimide), other nitroxide spin labels (shorter

FIGURE 17.1
The R1 side chain (left), obtained by reacting Cys with the spin label reagent MTSL. The five side-chain dihedral angles χ_1 through χ_5 are indicated. The unpaired electron (red dot) is delocalized over the N–O group (red). DEER measures the distance distribution between two spin labels (right).

or longer side chain, different chemical structure) and other spin label types (trityl radicals, copper(II), gadolinium(III)). Also, a range of methods exist to introduce spin labels in a bioorthogonal fashion, without the need of eliminating wild-type cysteines. However, compared to MTSL/R1, the use of these methods is much less common.

17.3 CW EPR Spectroscopy

The spectroscopically active part of the spin label is the unpaired electron that is delocalized over the N–O group. The unpaired electron possesses a magnetic dipole moment. In both CW EPR and DEER, the sample is placed in an external magnetic field, B_0. The magnetic moment of the unpaired electron aligns with B_0, akin to a compass needle. Microwave radiation is used to apply a torque to the magnetic moment and change its orientation. If not aligned with B_0, the magnetic moment precesses around it. The frequency of this precession is proportional to B_0. In a molecular context, the precession frequency carries detailed information about the local environment and is the quantity that provides structural information. Microwave radiation interacts with the unpaired electron only when the microwave frequency matches the precession frequency in the given B_0. This condition is called resonance.

For technical reasons, CW EPR spectroscopy is not performed by scanning the frequency of the incident microwave radiation while keeping B_0 fixed, but rather by keeping the microwave frequency fixed and sweeping the magnetic field (Figure 17.2). Absorption is observed at frequency and field values characteristic for the spin label. The most commonly employed frequencies are around 9.5 GHz (called X-band) and

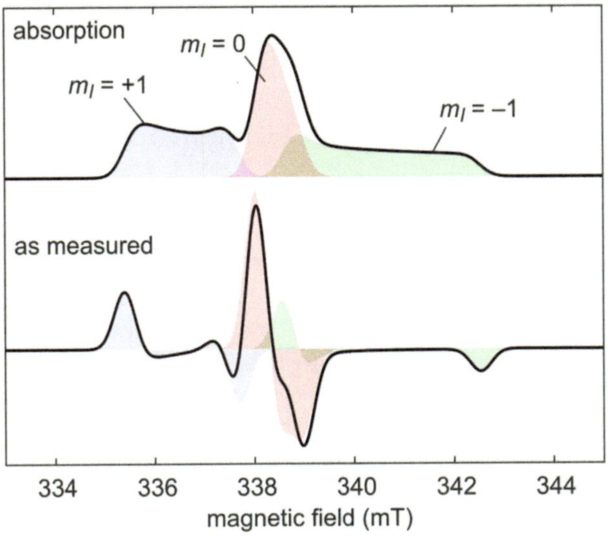

FIGURE 17.2
9.5 GHz CW EPR spectrum of a sample with immobilized randomly oriented nitroxide spin labels. (Top) Absorption spectrum, with ^{14}N component spectra indicated in color and labelled by the m_I state of the nitrogen nucleus. (Bottom) Spectrum as measured with field modulation and lock-in detection, appearing as first derivative of the absorption spectrum.

around 34 GHz (called Q-band). Additionally, the absorption spectrum is not recorded directly: For improved sensitivity, magnetic-field modulation and lock-in detection are used, providing what is essentially the first derivative of the absorption spectrum. For a more thorough introduction to CW EPR instrumentation, see chapter 1 in Goldfarb and Stoll (2018).

Figure 17.2 (bottom) shows the CW EPR spectrum of a sample with immobilized nitroxide spin labels as measured, and (top) the corresponding absorption spectrum, obtained by integration. The spectrum extends over about 8 mT, and consists of three component spectra, indicated by color. These three components are due to the interaction between the unpaired electron and the nearby magnetic nitrogen nucleus (isotope ^{14}N). The ^{14}N nucleus can be equally likely in one of three states (denoted by m_I of –1, 0 and +1), and the unpaired electron senses this state. Therefore, in a sample, a third of the proteins have spin labels in each of the three m_I states. The EPR spectrum of the unpaired electron depends on m_I, and therefore three different partially overlapping subspectra are observed (indicated by color in Figure 17.2). The resonance frequency of the unpaired electron also depends on the orientation of the nitroxide with respect to the applied magnetic field. Therefore, the three subspectra have significant widths if the nitroxides assume random orientations within the sample.

17.3.1 Mobility

The CW EPR spectrum in Figure 17.2 is representative of immobilized nitroxides, such as on a protein in a frozen solution. In liquid environments, such as in cells or buffered solutions, the CW EPR spectrum changes as a function of the reorientational mobility of the spin label. This is shown in Figure 17.3. Each of the three subspectra narrows as the mobility of the spin label increases. The time scale of the reorientational motion is characterized by the rotational correlation time, τ_c. The EPR spectrum is sensitive to motions in the 0.1 to 100 ns range, with faster and slower motions becoming indiscernible. In the fast-motion regime (τ_c shorter than about 1 ns), the spectrum consists of three nonoverlapping lines, one from each m_I state, with different widths and amplitudes. In the slow-motion regime (τ_c between about 1 and 10 ns), the three lines are broadened and overlap. As the motion slows further (τ_c longer than 50–100 ns), the EPR spectrum approaches the rigid-limit spectrum shown in Figure 17.2.

The overall reorientational mobility as manifested in the CW EPR spectrum is a convolution of three microscopically distinct types of stochastic motions: First, the overall tumbling of the protein. For large proteins, and especially those embedded in detergent, liposomes or nanodiscs, the slow time scale of this motion is outside the EPR sensitivity window. Second, the local backbone motion. This can fall into the EPR sensitivity window and is faster for spin labels in loops than in helices. Third, the reorientational motion of the spin label side chain itself. This is also typically within the EPR sensitivity window, unless the side chain is buried and locked in a specific rotameric state. The side chain mobility depends on the local environment, and, if surface exposed, also on the viscosity of the solvent.

To extract mobility metrics from the measured CW EPR spectrum, two approaches are commonly applied. In one approach, a motional model is fitted to the data. A range of such models are possible. The simplest one approximates the spin label as a spherical particle tumbling in an isotropic viscous environment and depends only on one parameter, τ_c. More sophisticated models take the restricted nature of the reorientational motion into account and provide insights into the degree of local ordering.

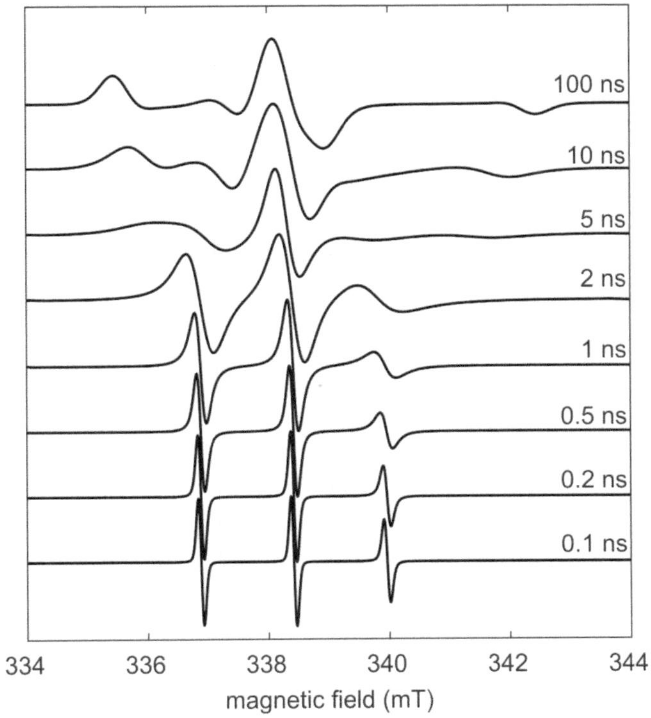

FIGURE 17.3
The effect of spin label mobility on the room-temperature 9.5 GHz CW EPR spectrum of a nitroxide spin label, with increasing mobility from top to bottom. The associated rotational correlation time, τ_c, is indicated.

A second approach foregoes the fitting of motional models and instead describes mobility in semiquantitative terms by calculating empirical parameters directly from the experimental CW EPR spectrum. One of these parameters is the inverse of the peak-to-trough width of the central line (at 338–339 mT in Figure 17.3). It is a measure of mobility – larger values (narrower central lines) correspond to higher mobility. A second empirical parameter is the inverse of the overall spectral spread, quantified by the variance of the magnetic field over the spectrum. Again, larger values (narrower overall spectrum) correspond to higher mobility. Together, these two parameters can be used to identify the type of local environment of the spin label (helix vs. loop, buried vs. surface). By scanning a spin-label site over all (feasible) residues of a protein, these mobility parameters provide detailed insight into the topology of the protein in the absence of a high-resolution structure.

17.3.2 Solvent Accessibility

In addition to mobility, CW EPR can be used to determine the location of a protein-attached spin label relative to protein, membrane and solvent. These measurements utilize additional paramagnetic probes that partition selectively into either the aqueous solvent or the lipid membrane. Nickel(II)–EDDA (ethylenediaminetetraacetate) and chromium(III) oxalate, for example, partition predominantly into the aqueous solvent, whereas O_2 gas enriches mostly in the lipid phase. Collisions of these probes with the nitroxide decrease the excitation lifetime T_{1e} of the unpaired electron, enhancing relaxation toward thermal equilibrium. T_{1e} is referred to as the longitudinal relaxation time.

The CW EPR spectrum of the spin label is not directly sensitive to T_{1e}, but its overall intensity is. At low power, P_{mw}, of incident microwave radiation, the spectral intensity is proportional to $\sqrt{P_{mw}}$. As power is increased, the spectral intensity begins to deviate from this proportionality, eventually becoming "saturated," where further increases of P_{mw} no longer increase the intensity of the CW EPR spectrum and can even decrease it. The characteristic power for this saturation process is called the half-saturation power, $P_{1/2}$, and it is a function of T_{1e}. The $P_{1/2}$ can be determined by recording the spectral intensity I (as measured by the peak-to-trough amplitude of the central peak) as a function of microwave power and fitting it with

$$I\left(P_{mw}\right)=c\frac{\sqrt{P_{mw}}}{\left[1+\left(2^{1/\epsilon}-1\right)P_{mw}/P_{1/2}\right]^{\epsilon}} \quad (17.1)$$

where the three adjustable parameters are an overall scaling factor c, an exponent ε that ranges between 1/2 and 3/2, and the half-saturation power $P_{1/2}$. By analyzing changes in $P_{1/2}$ for a spin-labeled protein upon alternate exposure to O_2 and Ni(II)–EDDA or Cr(III)–oxalate, insight can be gained into whether the spin label, and the associated residue on the protein, is solvent exposed or buried within the membrane. By performing a scan over a series of sites, the localization and orientation of protein residues relative to the membrane can be determined (Altenbach et al. 1990).

17.4 DEER Spectroscopy

DEER, occasionally also referred to as pulsed electron double resonance (PELDOR), is a pulse EPR experiment that measures the full distance distribution between the electron spins of two spin labels, which might be located on the same protein or on different proteins (see Figure 17.1, right). This provides direct structural insight, including information about structural heterogeneity and conformational equilibria.

17.4.1 Principles

DEER signal. DEER is a pump–probe experiment that uses two slightly different microwave frequencies that selectively excite the two electron spins. One spin (called the observer spin) is excited by one of these frequencies. The other spin (called the pump spin) is then excited by the other frequency, and the effect of this pump spin excitation on the observer spin is measured. The resulting DEER signal has an oscillatory shape, as shown in Figure 17.4 (left). The pump and observer spins are coupled via the magnetic dipole–dipole interaction, which depends on the inverse cube of their distance, r^{-3}. Longer distances lead to weaker interactions, which lead to slower modulations in the DEER signal. An approximate relationship between the modulation period T and the dominant distance r is

$$T/\mu s \approx \frac{(r/\mathrm{nm})^3}{52} \qquad r/\mathrm{nm}/\approx\sqrt[3]{52(T/\mu s)} \quad (17.2)$$

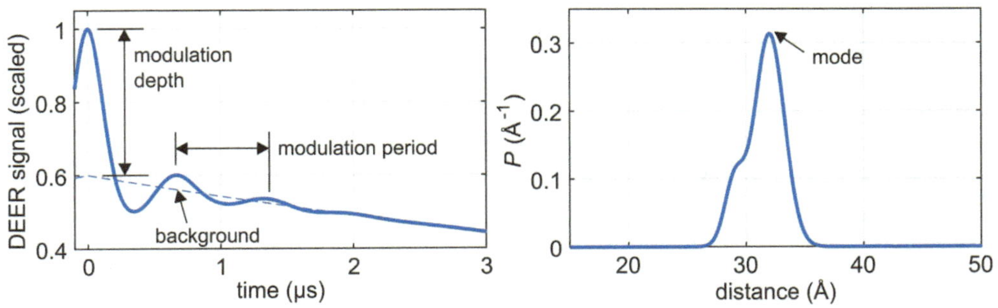

FIGURE 17.4
DEER time-domain trace (left) and associated distance distribution P (right).

So, a dominant distance of 3.2 nm (or 32 Å) will result in a modulation period of 0.63 μs. Due to the cubic dependence, DEER is extremely distance-sensitive and can resolve distance changes in the subangstrom range. For example, the modulation periods for 32 and 32.5 Å differ by 0.030 μs, a difference that is easily resolved. The details of the modulations in the DEER signal encode the full distance distribution between the two spin labels, which is plotted as the probability density, $P(r)$, as a function of interspin distance, r, where r is very often in angstroms and $P(r)$ in Å^{-1} (Figure 17.4, right). It is important to note that the distances obtained by DEER are absolute and do not require any calibration.

Interprotein background. DEER measures spin–spin distances irrespective of whether the two spin labels are on the same protein (or protein complex) or whether they are on different proteins in the bulk solution. For a protein sample in a frozen solution, the latter distances are generally not of direct interest as they only contain information about the stochastic interprotein distances, rather than the intraprotein distances that reveal the conformational distribution. In the DEER signal, interprotein spin label pairs contribute an additional monotonic, approximately exponential decay that is superimposed on the more resolved modulations from intraprotein spin label pairs (see Figure 17.4, left). The decay rate of this so-called background signal is proportional to the overall spin label concentration in the sample.

Spectral selectivity. Even though DEER is a pump–probe experiment that selectively excites one of the two spin labels at a time, it uses two chemically identical labels. This is possible since it is not the chemical structure alone that determines the absorption spectrum of a nitroxide: as shown in Figure 17.2, the state of the magnetic nitrogen nucleus affects the spectrum significantly. In each of the three possible ^{14}N states (m_I of –1, 0 and +1), the absorption spectrum is shifted. In any given sample, nitroxides stochastically assume one of the three states with equal likelihood. Pump and observer frequency are chosen to excite separate m_I subspectra, and therefore a significant fraction of proteins will contribute to the DEER signal.

Data analysis. To extract the underlying distance distribution, a physical model is fitted to the DEER time trace. The simplest model is

$$V(t) = V_0\left[(1-\Delta)+\Delta\int_0^\infty K(t,r)P(r)\,\mathrm{d}r\right]\mathrm{e}^{-kt} \tag{17.3}$$

Its parameters are the overall amplitude V_0, the modulation depth Δ (see Figure 17.4, left), the background decay rate k and the distance distribution $P(r)$. The function $K(t,r)$, called the dipolar kernel function, describes the time-domain oscillation as a function of distance r.

There are two approaches for representing the distance distribution, either as a nonparametric histogram (a range of distance values with associated probabilities) or as a parametric model represented in terms of a linear combination of simple basis functions such as Gaussians (see Figure 17.5). Note that these basis functions are for mathematical simplicity and do not necessarily represent distinct backbone or spin label side chain subpopulations. For the nonparametric model, an additional smoothness requirement, most commonly achieved by Tikhonov regularization, is added to make the fitting problem mathematically tractable. With either approach, the quality of the extracted distance distribution, in particular the distance range and the uncertainties, strongly depends on the quality of the time-domain data: the longer the time trace, and the lower the noise, the better.

Distance range and deuteration. The distance range that is most comfortably accessed by DEER is 20 to 60 Å. Longer distances are more challenging because protons in the environment of the spin labels limit the phase memory time (decoherence time), T_M, of the unpaired electrons, and thereby the length of the measurable DEER time-domain trace. Using deuterons instead of protons prolongs T_M and therefore the accessible distance range. Using perdeuterated solvents is standard practice, and by using perdeuterated proteins, the range has been extended to 160 Å for soluble proteins. For membrane proteins, the hydrogens in the enclosing detergent or lipid molecules limit these possibilities.

Types of distance distributions. The type of distance distributions can vary greatly from protein to protein, and even from site pair to site pair within the same protein. The most easily interpreted are narrow distributions with a single mode (monomodal), which manifest in DEER as clearly resolved modulations with several observable periods. For broader distributions, the DEER modulations are rapidly damped, such that often not even one full modulation is apparent. For multimodal distributions, several dominant modulation frequencies appear in the DEER trace.

Equilibria. DEER reveals not an average spin–spin distance distribution, but the full distance distribution in the measured ensemble of proteins in the sample. This provides access to thermodynamic information, such as equilibrium constants and free energies of conformational change. By measuring DEER distance distributions as a function of an

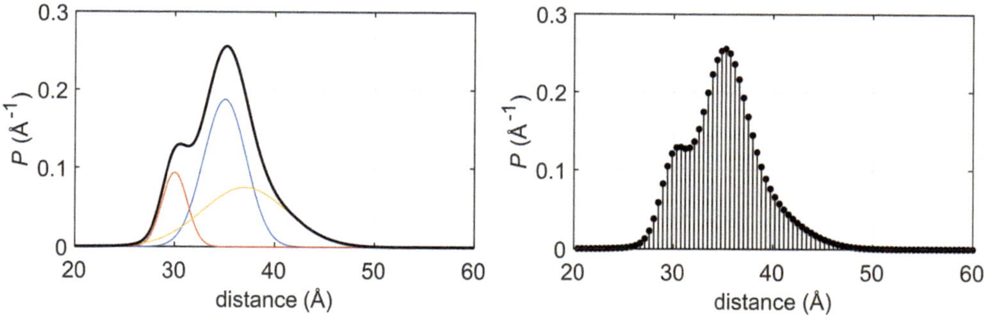

FIGURE 17.5
Two approaches to represent distance distributions. (Left) Parametric model using a linear combination of basis functions. In this example are three Gaussians, giving a total of eight free parameters (three positions, three widths, three intensities, minus the normalization condition $\int P(r)\mathrm{d}r = 1$). (Right) Nonparametric model using a discretized histogram over evenly spaced distances. In this example are 100 free parameters (101 distances, minus the normalization condition).

effector such as a ligand, the effector-induced shifting of the conformational equilibrium can be directly observed. From the measured distributions, fractional populations of a two-state model can be extracted, which then give access to the equilibrium constant for the conformational change. Using a global-fitting approach for a titration series for this analysis, it is not necessary to obtain distributions of the two pure states.

Figure 17.6 shows an illustration of the application of DEER to determine the dissociation constant K_D and the state-specific distance distributions for a protein–ligand binding equilibrium. A series of samples with varying ligand concentration is prepared, and DEER is measured. The progression of distance distributions (Figure 17.6, left) reflects the shifting equilibrium. The appearance of an isosbestic point at about 37 Å is diagnostic of a two-state equilibrium. For the analysis, a simple two-state model is fitted globally to the experimental distributions. The model represents the total P as a linear combination, $P = (1 - x_b)P_u + x_b P_b$, where x_b is the fraction of bound protein, and P_u and P_b are the pure-state distributions for the unbound and bound state, respectively. These can be represented as either Gaussian mixtures or as nonparametric histograms. The fit provides P_u and P_b and x_b, the latter being a function of ligand concentration (Figure 17.6, right). The extracted x_b dependence is then fitted with the two-state equilibrium model

$$x_b = \frac{1}{2}\left(s - \sqrt{s^2 - 4c_{P0}c_{L0}}\right) \tag{17.4}$$

where $s = K_D + c_{P0} + c_{L0}$, c_{P0} is the total protein concentration (bound and unbound), and c_{L0} is the total ligand concentration. This yields the dissociation constant K_D.

By analyzing the dissociation constant in terms of standard free energy ΔG° via $\Delta G^{\circ} = -RT \ln K_D$, the energetics of the binding equilibrium can be determined. Interpretation of these numbers needs to be done carefully, however, as DEER is measured on rapidly frozen samples and therefore does not directly provide energetics at physiological temperatures. Nevertheless, it provides a means to compare energetics of different ligands or different protein variants in a series of similar systems, via differences in standard free energy, indicated as $\Delta\Delta G^{\circ}$.

Learning about models more complex than a two-state equilibrium from DEER data is challenging for two reasons. First, more complex models often involve states that do not

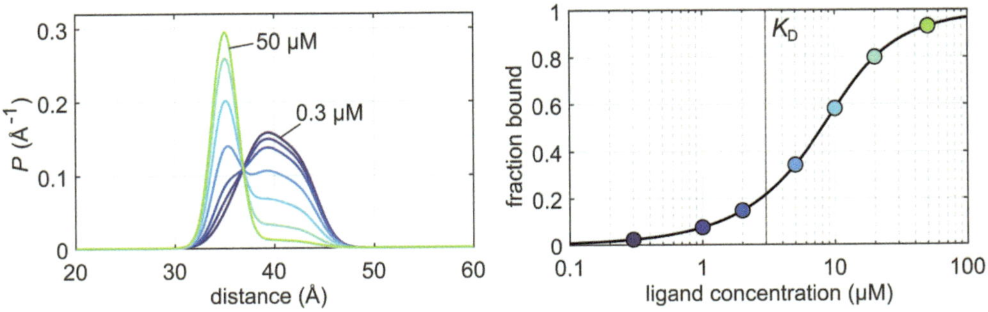

FIGURE 17.6
Titration of a binding equilibrium using DEER. (Left) Progression of distance distributions as a function of ligand concentration. (Right) Fraction bound protein as a function of ligand concentration. Parameters: total protein concentration 10 μM; dissociation constant 3 μM.

differ in conformation (for example, a ligand-free and ligand-bound open state). Second, the distance resolution of DEER distributions is generally insufficient to support the extraction of more than two states, unless these differ significantly in their distance distributions.

Kinetics. DEER can be combined with rapid freeze-quenching (RFQ) techniques to provide insight into the kinetics of conformational changes on the submillisecond to millisecond time scale (Collauto et al. 2017; Hett et al. 2021). RFQ setups are used to produce a series of samples where a spin-labeled protein and a stimulus are rapidly mixed and then freeze-quenched after a controlled short time in the 1–100 ms time scale. Analyzing the series of DEER distributions with kinetic models provides access to rate constants. These experiments are challenging and often require relatively large sample volumes.

17.5 Practical Aspects

A recent white paper gives extensive guidelines for performing DEER experiments using nitroxide spin labels (Schiemann et al. 2021). It also shows a multilaboratory benchmark study that demonstrates the reproducibility of DEER. This white paper, and the references cited therein, are an excellent resource regarding the practical aspects of DEER. Here, only the essentials are mentioned.

Samples. Typical CW EPR and DEER experiments with nitroxide spin labels require on the order of 10 μL of sample containing spin-labeled protein with a spin label concentration of approximately 1–20 μM. Membrane proteins can be in detergent, in liposomes or in nanodiscs. Note that proteins reconstituted in liposomes have been shown to crowd together in the membrane in frozen solutions, speeding electron decoherence and lowering sensitivity. Dilution with unlabeled channel proteins during reconstitution reduces this effect, but can also limit absolute sensitivity (Endeward et al. 2009). Detergent micelles and nanodiscs do not suffer this problem. The sample should be free of unreacted spin labeling reagent. Labeling efficiencies below 100% (i.e., when some Cys are unlabeled) do not present a fundamental problem, as unlabeled proteins are not visible in EPR and DEER, but sensitivity is reduced. Samples for DEER have additional requirements: (a) the buffer should contain at least 10%–20% of glycerol as a cryoprotectant to prevent protein aggregation during freezing; (b) the buffer should be perdeuterated (D_2O and d_8-glycerol) to provide a long phase memory time for adequate sensitivity; (c) samples are plunge-frozen in liquid nitrogen or liquefied MAPP gas.

Experimental conditions. EPR spectroscopy can be performed at various microwave frequencies and magnetic fields, as well as sample temperatures. For CW EPR mobility and solvent accessibility studies, experiments are commonly run at room temperature at about 9.5 GHz (X-band) and magnetic fields around 0.33 mT. For DEER, current practice is to run experiments at about 50 K at Q-band frequencies (around 34 GHz) using observer and pump frequencies separated by about 70–100 MHz. The magnetic field employed is around 1.2 T. DEER experiments at X-band are also feasible but have lower sensitivity.

17.6 Applications to Ion Channels

Spin-labeling EPR has played a prominent role in elucidating the structures and dynamics of membrane proteins, including G protein-coupled receptors, transporters and ion channels (for review, see Mchaourab, Steed, and Kazmier 2011). Spin-labeling experiments gave early insight on the structural topologies, local solvent environments, and dynamics of these proteins within lipid membranes, often long before high-resolution structures were available. CW EPR methods like spin-label mobility analysis and power saturation remain useful tools for studying local dynamics and microenvironments, and as reporters of conformational change. In the past 15 years, DEER has also emerged as a powerful technique for probing the structures and, importantly, the conformational energetics of proteins. EPR and DEER can therefore provide insight that goes beyond static structure in studies of ion channel mechanism.

17.6.1 Local Dynamics and Solvent Environment

As described in Section 17.3, the CW EPR spectrum of a spin-labeled protein reports on the local mobility of the spin label as well as on the solvent microenvironment. Mobility is obtained directly from the EPR line shape, whereas the solvent environment can be inferred from power saturation measurements in the presence of relaxing agents. An additional perturbation of the CW EPR spectrum is observed under conditions where two or more spin labels are very close together (<15 Å). Magnetic interactions between spin labels in such close proximity give rise to spectral broadening large enough to observe by CW EPR.

CW EPR played an important role in early knowledge of the structure and gating mechanisms of potassium (K^+) channels. The bacterial K^+ channel, KcsA, was exhaustively spin-labeled by Perozo and colleagues, where mobility and accessibility data from EPR revealed the channel architecture, including the location of transmembrane helices relative to the pore and the identity of the pore lining residues (Perozo, Cortes, and Cuello 1998). Residues at or near the gate were identified from spectral broadening caused by strong magnetic interactions, and these residues were found to move away from each other during pH-dependent gating. Subsequent EPR studies revealed the coupling between the KcsA activation gate and the closing of an inactivation gate located at the selectivity filter, providing insight into a hallmark of voltage-gated K^+ channels known as C-type inactivation (Cordero-Morales et al. 2007; Cuello et al. 2010). More recently, CW EPR has been used to map the topology of KCNE1 subunits in lipid bilayers (Sahu et al. 2015) and to characterize ligand-induced rearrangements of bacterial Mg^{2+} channels (Dalmas et al. 2014) and pentameric ligand-gated ion channels (pGLICs) (Schmandt et al. 2015).

17.6.2 Oligomerization

DEER distance distributions can provide direct insight into channel symmetry and subunit oligomerization. Because the DEER experiment reports distance distributions from all spin-label pairs in the sample, proteins containing a single spin label per polypeptide chain will exhibit peaks in the DEER distribution with characteristic distance ratios and relative intensities for specific symmetries. The effects of oligomerization on the resulting DEER distributions are illustrated in Figure 17.7. For tetramers and higher-order oligomers, multiple peaks are expected in the distance distribution even when the channel adopts a single conformation. These distances – and their relative populations – are diagnostic of the oligomerization state. Analysis of this symmetry under varying conditions (ligands, lipid environment, pH, etc.) can reveal potential changes in channel quaternary structure during function. An important caveat is that reliable resolution of longer distances requires that high-quality DEER data be collected out to long dipolar evolution

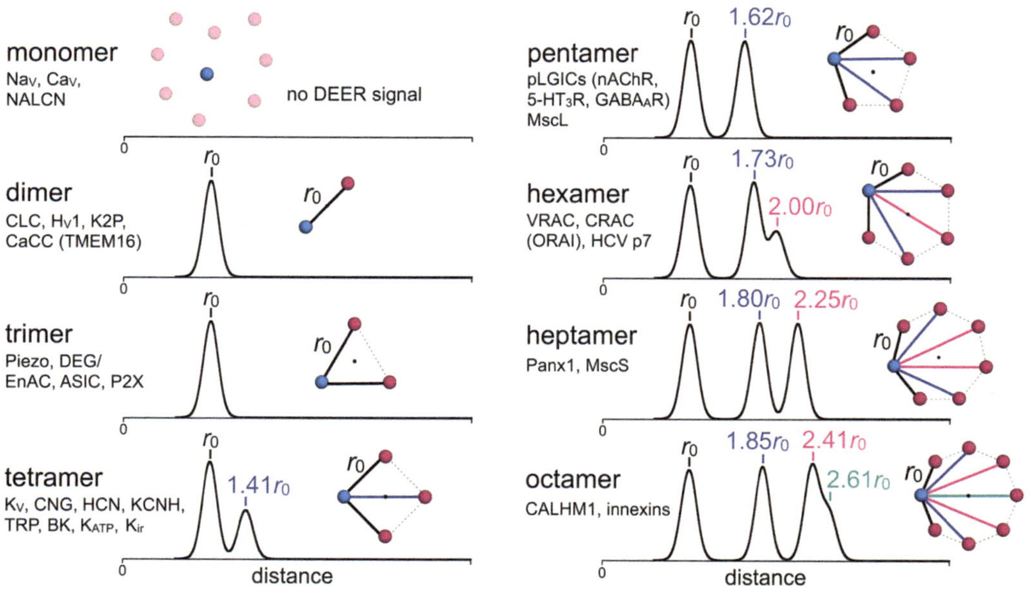

FIGURE 17.7
Effects of oligomerization state on DEER distance distributions. Simulated distance distributions for singly spin-labeled symmetric homooligomers are shown along with some representative channels from each class. DEER distributions are normalized to the maximum of each distribution. Distance ratios relative to the nearest-neighbor distance, r_0, are indicated in color.

times. This is often not possible due to spin decoherence. Therefore, longer distance peaks, especially for higher-order oligomers, may be difficult or impossible to distinguish from background.

It is also possible to determine the oligomerization state of a singly spin-labeled protein from the modulation depth, Δ. Δ is determined from the DEER time trace (see Figure 17.4) and depends on the number of spins in the oligomer, n, via $\Delta = 1-(1-\lambda)^{n-1}$, where λ is the pump pulse inversion efficiency. For example, DEER performed with a pump pulse efficiency of 0.3 would give a 30% modulation depth for a singly spin-labeled homodimer; however, the same experiment on a homotrimer would produce a modulation depth of 51%. The oligomeric state can therefore be inferred even though both samples give only a single peak in the distance distribution (see Figure 17.7). Care must be taken with this approach, as it requires that Δ is known accurately and that λ is determined separately with precise calibration experiments.

Oligomeric channels with more than two spin labels per channel present additional challenges in calculation of the distance distribution. With trimers and higher-order assemblies, it is possible for two or more spins in the same channel to be inverted by the pump pulse. This can result in "ghost peaks" in the distance distribution. These multispin artifacts can be reduced by decreasing the microwave power of the pump pulse (effectively lowering λ) and by power-scaling the time-domain data prior to calculating the distance distribution (Valera et al. 2016).

In rare cases, angular correlations between spin labels on the same channel can distort the distance distributions. This can occur if the spin label is sufficiently hindered such that it can effectively occupy only one rotameric state. This is rarely encountered for surface-exposed sites, but was observed for R1 attached at the outer pore helix of KcsA (Endeward et al. 2009). Angular correlations can be checked for by performing DEER at several different pump–probe frequency offsets.

17.6.3 Structure and Conformational Changes

The DEER experiment provides a full distance distribution between all pairs of spin labels present in the sample, often at subangstrom resolution. However, in the context of an ion channel structure, this is still very sparse information. A very large number of distance distributions from site-pairs placed strategically throughout a protein sequence could potentially constrain a protein structure for de novo modeling; however, this is generally not a practical approach. A primary strength of DEER is, instead, the ability to report conformational changes. Additional information on the inherent heterogeneities within and between states, many of which may not be amenable to high-resolution structural study, may also be gained. A typical situation one might encounter is when a structure is available in one state (closed or inactivated, perhaps) but not in another (open/activated). In such cases, a relatively small number of DEER experiments may be enough to confirm or reject a hypothesis. DEER distances can be compared to the existing high-resolution structure under similar conditions, and data obtained for the state lacking a solved structure can be used as restraints in modeling programs such as Rosetta to make predictions regarding molecular mechanism.

As an example, Figure 17.8 (right) shows DEER distributions from the bacterial cyclic nucleotide-gated ion channel, SthK, a homotetramer (Evans et al. 2020). The channel is spin-labeled in the C-linker domain, a structural element that allosterically couples ligand binding to pore opening. Addition of cAMP shifts the distance distribution to longer distances while retaining the characteristic distance ratios expected for a symmetric tetramer ($r_1 \approx 1.4r_0$), allowing for structural interpretation. The DEER distribution shown in Figure 17.8 also reveals that cGMP, which binds with similar affinity to the ligand-binding

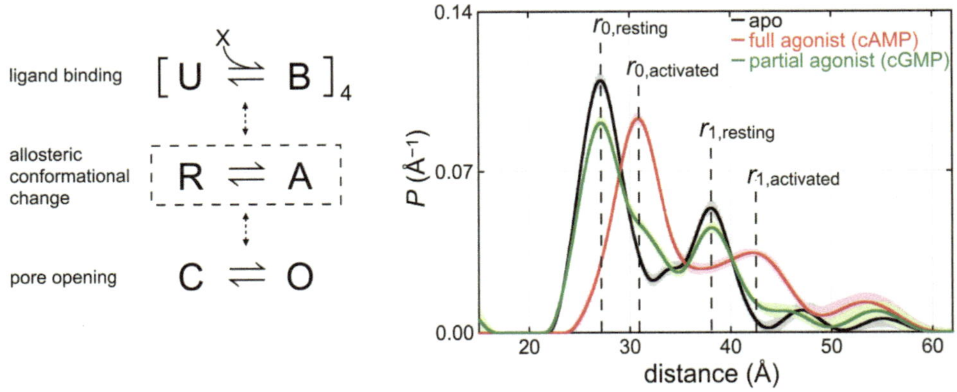

FIGURE 17.8
Conformational equilibrium measurements with DEER. (Left) Modular gating scheme for a purely ligand-gated channel with four identical binding sites allosterically coupled to the channel pore (see Chapter 10 in this volume). Ligands can be either bound (B) or unbound (U) with ligand (X). The binding equilibrium is allosterically coupled to a conformational change characterized by resting (R) and activated (A) conformations, which affects the state of the channel pore, which can be closed (C) or open (O). (Right) DEER distance distribution of the C-linker residue D261R1 in the homotetrameric bacterial CNG channel SthK. Pairs of symmetry-related peak distances, r_0 and r_1, are indicated for both the resting and activated conformations. Partial agonism by cGMP (green) compared to cAMP (red) is apparent in the distribution. (Data from Evans et al. 2020.)

domain as cAMP, is much less efficient at promoting the allosteric conformational change, providing insight into the energetic coupling that underlies channel function (see Figure 17.8, left). DEER has been employed to determine ligand-induced motions of a number of ion channels, including HCN channels (Puljung et al. 2014; DeBerg et al. 2016), pLGICs (Dellisanti et al. 2013; Schmandt et al. 2015) and bacterial Mg^{2+} channels (Dalmas et al. 2014).

The example shown in Figure 17.8 highlights the type of independent energetic information that can be gained from DEER. Thermodynamic free energies of conformational change, obtained from DEER titration curves or estimated from relative affinities, can be integrated with information from electrophysiology and direct measurements of ligand binding (e.g., fluorescence anisotropy or radiolabeling assays) to gain a more complete picture of channel energetics. The thermodynamic and structural information attained with DEER can therefore play an important, complementary role to high-resolution X-ray and cryo-EM structures in understanding the allosteric mechanisms that govern ion channel function.

Suggested Readings

Altenbach, C., T. Marti, H. Gobind Khorana, and W. L. Hubbell. 1990. "Transmembrane protein structure: spin labeling of bacteriorhodopsin mutants." *Science* 248 (4959):1088–92.

Bordignon, E., S. Kucher, and Y. Polyhach. 2019. "EPR techniques to probe insertion and conformation of spin-labeled proteins in lipid bilayers." In *Lipid-protein interactions: methods and protocols*, edited by J. H. Kleinschmidt, 493–528. New York: Springer New York.

Collauto, A., H. A. DeBerg, R. Kaufmann, W. N. Zagotta, S. Stoll, and D. Goldfarb. 2017. "Rates and equilibrium constants of the ligand-induced conformational transition of an HCN ion channel protein domain determined by DEER spectroscopy." *Phys Chem Chem Phys* 19 (23):15324–34.

Cordero-Morales, J. F., V. Jogini, A. Lewis, V. Vásquez, D. Marien Cortes, B. Roux, and E. Perozo. 2007. "Molecular driving forces determining potassium channel slow inactivation." *Nat Struct Mol Biol* 14 (11):1062–9.

Cuello, L. G., V. Jogini, D. Marien Cortes, A. C. Pan, D. G. Gagnon, O. Dalmas, J. F. Cordero-Morales, S. Chakrapani, B. Roux, and E. Perozo. 2010. "Structural basis for the coupling between activation and inactivation gates in K+ channels." *Nature* 466 (7303):272–5.

Dalmas, O., P. Sompornpisut, F. Bezanilla, and E. Perozo. 2014. "Molecular mechanism of Mg2+-dependent gating in CorA." *Nat Commun* 5 (1):3590.

DeBerg, H. A., P. S. Brzovic, G. E. Flynn, W. N. Zagotta, and S. Stoll. 2016. "Structure and energetics of allosteric regulation of HCN2 ion channels by cyclic nucleotides*." *J Biol Chem* 291 (1):371–81.

Dellisanti, C. D., B. Ghosh, S. M. Hanson, J. M. Raspanti, V. A. Grant, G. M. Diarra, A. M. Schuh, K. Satyshur, C. S. Klug, and C. Czajkowski. 2013. "Site-directed spin labeling reveals pentameric ligand-gated ion channel gating motions." *PLoS Biol* 11 (11):e1001714.

Endeward, B., J. A. Butterwick, R. MacKinnon, and T. F. Prisner. 2009. "Pulsed electron–electron double-resonance determination of spin-label distances and orientations on the tetrameric potassium ion channel KcsA." *J Am Chem Soc* 131 (42):15246–50.

Evans, E. G. B., J. L. W. Morgan, F. DiMaio, W. N. Zagotta, and S. Stoll. 2020. "Allosteric conformational change of a cyclic nucleotide-gated ion channel revealed by DEER spectroscopy." *Proc Natl Acad Sci* 117 (20):10839–47.

Fanucci, G. E., and D. S. Cafiso. 2006. "Recent advances and applications of site-directed spin labeling." *Curr Opin Struct Biol* 16 (5):644–53.

Goldfarb, D., and S. Stoll, eds. 2018. *EPR spectroscopy - fundamentals and applications*. Chichester, UK: Wiley.

Hett, T., T. Zbik, S. Mukherjee, H. Matsuoka, W. Bönigk, D. Klose, C. Rouillon, N. Brenner, S. Peuker, R. Klement, H.-J. Steinhoff, H. Grubmüller, R. Seifert, O. Schiemann, and U. B. Kaupp. 2021. "Spatiotemporal resolution of conformational changes in biomolecules by combining pulsed electron–electron double resonance spectroscopy with microsecond freeze-hyperquenching." *J Am Chem Soc* 143 (18):6981–9.

Jeschke, G. 2012. "DEER distance measurements on proteins." *Annu Rev Phys Chem* 63:419–46.

Jeschke, G. 2013. "Conformational dynamics and distribution of nitroxide spin labels." *Prog Nucl Magn Reson Spectrosc* 72:42–60.

Mchaourab, H. S., P. R. Steed, and K. Kazmier. 2011. "Toward the fourth dimension of membrane protein structure: insight into dynamics from spin-labeling EPR spectroscopy." *Structure* 19 (11):1549–61.

Perozo, E., D. Marien Cortes, and L. G. Cuello. 1998. "Three-dimensional architecture and gating mechanism of a K^+ channel studied by EPR spectroscopy." *Nat Struct Biol* 5 (6):459–69.

Pliotas, C. 2017. "Chapter eight - ion channel conformation and oligomerization assessment by site-directed spin labeling and pulsed-EPR." In *Methods in enzymology*, edited by Christine Ziegler, 203–42. Cambridge: Academic Press.

Puljung, M. C., H. A. DeBerg, W. N. Zagotta, and S. Stoll. 2014. "Double electron–electron resonance reveals cAMP-induced conformational change in HCN channels." *Proc Natl Acad Sci* 111 (27):9816–21.

Sahu, I. D., A. F. Craig, M. M. Dunagan, K. R. Troxel, R. Zhang, A. G. Meiberg, C. N. Harmon, R. M. McCarrick, B. M. Kroncke, C. R. Sanders, and G. A. Lorigan. 2015. "Probing structural dynamics and topology of the KCNE1 membrane protein in lipid bilayers via site-directed spin labeling and electron paramagnetic resonance spectroscopy." *Biochemistry* 54 (41):6402–12.

Schiemann, O., C. A. Heubach, D. Abdullin, K. Ackermann, M. Azarkh, E. G. Bagryanskaya, M. Drescher, B. Endeward, J. H. Freed, L. Galazzo, D. Goldfarb, T. Hett, L. E. Hofer, L. F. Ibáñez, E. J. Hustedt, S. Kucher, I. Kuprov, J. E. Lovett, A. Meyer, S. Ruthstein, S. Saxena, S. Stoll, C. R. Timmel, M. Di Valentin, H. S. McHaourab, T. F. Prisner, B. E. Bode, E. Bordignon, M. Bennati, and G. Jeschke. 2021. "Benchmark test and guidelines for DEER/PELDOR experiments on nitroxide-labeled biomolecules." *J Am Chem Soc* 143 (43):17875–90.

Schmandt, N., P. Velisetty, S. V. Chalamalasetti, R. A. Stein, R. Bonner, L. Talley, M. D. Parker, H. S. Mchaourab, V. C. Yee, D. T. Lodowski, and S. Chakrapani. 2015. "A chimeric prokaryotic pentameric ligand–gated channel reveals distinct pathways of activation." *J Gen Physiol* 146 (4):323–40.

Tilegenova, C., B. W. Elberson, D. Marien Cortes, and L. G. Cuello. 2018. "CW-EPR spectroscopy and site-directed spin labeling to study the structural dynamics of ion channels." In *Potassium channels: methods and protocols*, edited by S.-L. Shyng, F. I. Valiyaveetil, and M. Whorton, 279–88. New York: Springer New York.

Timmel, C. R., and J. R. Harmer. 2014. *Structural information from spin-labels and intrinsic paramagnetic centres in the biosciences, structure and bonding*. Berlin, Heidelberg: Springer.

Valera, S., K. Ackermann, C. Pliotas, H. Huang, J. H. Naismith, and B. E. Bode. 2016. "Accurate extraction of nanometer distances in multimers by pulse EPR." *Chemistry – Eur J* 22 (14):4700–3.

Index

Note: Locators in *italics* represent figures in the text.